PHL
54060000103894

THE
PAUL HAMLYN
LIBRARY
DONATED BY
THE PAUL HAMLYN
FOUNDATION
TO THE
BRITISH MUSEUM
opened December 2000
WITHDRAWN

ANTIQUARIAN HOROLOGICAL SOCIETY MONOGRAPH NO. I

HEAVENLY CLOCKWORK

Fig. 1. A pictorial reconstruction of the astronomical clock-tower built by Su Sung and his collaborators at K'ai-fêng in Honan province, then the capital of the empire, in A.D. 1090. The clock-work, driven by a water-wheel, and fully enclosed within the tower, rotated an armillary sphere on the top platform and a celestial globe in the upper storey; puppet figures giving notice meanwhile of the passing hours and quarters by signals of sight and sound. (Original drawing by John Christiansen.)

HEAVENLY CLOCKWORK

THE GREAT ASTRONOMICAL CLOCKS OF MEDIEVAL CHINA

JOSEPH NEEDHAM,
WANG LING,
AND
DEREK J. DE SOLLA PRICE

SECOND EDITION

WITH SUPPLEMENT BY
JOHN H. COMBRIDGE

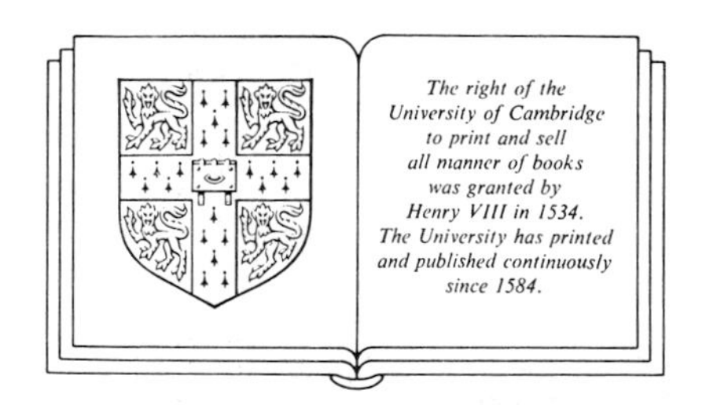

CAMBRIDGE UNIVERSITY PRESS
CAMBRIDGE
LONDON NEW YORK NEW ROCHELLE
MELBOURNE SYDNEY

Published by the Press Syndicate of the University of Cambridge
The Pitt Building, Trumpington Street, Cambridge CB2 1RP
32 East 57th Street, New York, NY 10022, USA
10 Stamford Road, Oakleigh, Melbourne 3166, Australia

Monograph No. 1
Antiquarian Horological Society,
New House, High Street, Ticehurst,
Wadhurst, Sussex TN5 7AL

First published 1960
Second edition 1986

Printed in Great Britain at the University Press, Cambridge

British Library cataloguing in publication data

Needham, Joseph
Heavenly clockwork: the great astronomical clocks of medieval China. – 2nd ed. rev. –
(Antiquarian Horological Society monograph; no. 1)
1. Astronomical clocks – China – History
I. Title II. Wang Ling III. De Solla Price, Derek J. IV. Series
681.1'13 QB107

Library of Congress cataloguing in publication data

Needham, Joseph, 1900–
Heavenly clockwork.
(Antiquarian Horological Society monograph; no. 1)
Bibliography: p.
Includes index.
1. Astronomical clocks – China – History.
I. Wang Ling.
II. Price, Derek J. de Solla (Derek John de Solla), 1922–1983.
III. Title. IV. Series.
QB107.N32 1986 529'.78 85-31363

ISBN 0 521 32276 6

CONTENTS

LIST OF ILLUSTRATIONS

CHRONOLOGICAL TABLE

SHANG (YIN) kingdom	c. 1500–c. 1030 B.C.	
CHOU (feudal age) early	1030–722	
Springs and Autumns period	722–480	
Warring States period	480–221	
CH'IN dynasty	221–207	
HAN dynasty Former (Western)	B.C. 202–9 A.D.	
HSIN interregnum	9–23	
Later (Eastern)	25–220	
	103	Ecliptic ring added to armillaries
	120–40	Work of Chang Hêng (first use of water-power for rotating astronomical instruments)
SAN KUO (Three Kingdoms)	221–265	
	220–48	Work of Lu Chi
	219–57	Work of Wang Fan
	250	Ko Hêng
	247	Liu Chih
CHIN dynasty	265–420	
CH'IEN CHAO dynasty	304–29	
	323	K'ung T'ing
(LIU) SUNG dynasty	420–79	
	436	Ch'ien Lo-Chih
(NORTHERN) WEI dynasty	386–556	
	415	Hsieh Lan
Northern and Southern splinter States	479–581	
SUI dynasty	581–618	
	595–610	Work of Kêng Hsün
T'ANG dynasty	618–906	
	633	Li Shun-Fêng
	725	I-Hsing and Liang Ling-Tsan
Five Dynasties period	907–60	
LIAO dynasty	907–1125	
SUNG dynasty (Northern)	960–1126	
	975–9	Work of Chang Ssu-Hsün
	1000	Han Hsien-Fu
	1050	Shu I-Chien
	1074	Shen Kua
	1075 c.	Su Sung's mission to the LIAO

SUNG dynasty (Northern) (*cont.*)	1086–94	Work of Su Sung
	1124	Wang Fu
(Southern)	1127–1279	
	1170	Efforts of Chu Hsi
	1172	First printing of Su Sung's book
CHIN (Tartar) dynasty	1115–1234	
YUAN (Mongol) dynasty	1260–1368	
	1275	Kuo Shou-Ching
	1354	Toghan Timur
MING dynasty	1368–1644	
	1370 *c.*	Chan Hsi-Yuan
	1583	Coming of the Jesuits
CH'ING (Manchu) dynasty	1644–1911	
KUOMINTANG Republic	1911–48	
PEOPLE'S Republic	1949–	

NOTES AND CONVENTIONS

1. We distinguish between Glosses and Commentaries in the main text and occasionally elsewhere. The Glosses appear in the original text as small characters in double rows within the main rows of standard-size characters; they are usually single sentences, and we have incorporated them in the translated text inside single angle-brackets ⟨ ⟩.

Commentaries, on the other hand, are paragraphs which follow the sections of the main text, in characters of standard size but differentiated from it by the fact that they begin a few lines lower on the page. We have incorporated these as separate paragraphs, enclosed in double angle-brackets ⟨⟨ ⟩⟩.

Both types of addition occur only in sections A–D and F–N, and we believe them to be coeval with, and perhaps by the same hand as, the sections E and P–S (cf. Fig. 2).

Minor glosses, if purely repetition, have been omitted.

2. It must be remembered that the text is not intended as one continuous narrative. About half of it consists of captions describing at length the successive illustrations.

3. Round brackets ( ) are used to indicate editorial additions whether by way of explanation, amplification, or adaptation to the grammar of the English language. Square brackets [ ] have been reserved for the reference numbers which direct the reader to the lists of Chinese characters (pp. 229 ff.).

4. An effort has been made to render consistently the technical engineering terms involved in these texts. This has sometimes necessitated overlooking certain minor variations in the Chinese technical terms. Phrases which we believe to be synonyms will therefore be seen in the first list of Chinese characters (pp. 229 ff.).

5. Attention may be directed to the fact that in medieval Chinese engineering descriptions back and front, left and right, are often referred to as 'north' and 'south', 'east' and 'west', respectively. The upper parts of a machine are called 'heavenly' and the lower parts 'earthly'. The text is sometimes inconsistent with the orientation of the illustrations. This is probably due to the fact that Chinese wood blocks were made by pasting on the drawing (or page of writing) to the piece of pear or other wood which the carver would then carve. As the paper was almost transparent, lateral inversion of a drawing could arise. It is also very likely that the draftsmen were not consistent in the aspect from which they regarded the apparatus.

6. The Chinese foot (*ch'ih*) varied in length, during the period under discussion, from 9 in. to a little over 12 in. of our measure, being almost identical with our foot during the Sung dynasty in which Su Sung worked. The Chinese foot was always subdivided into 10 'inches', but to facilitate translation we have silently ignored the duodecimal implications of the word 'inch', and the relatively unimportant variations in its absolute length.

Similar conventions relating to the numbers of hours and quarters in the day and night will be found discussed in the Appendix.

7. Braces { } are used in the Foreword and Supplement to the second edition to indicate references to pages or Figures in the original text, general bibliography, and tables of Chinese characters.

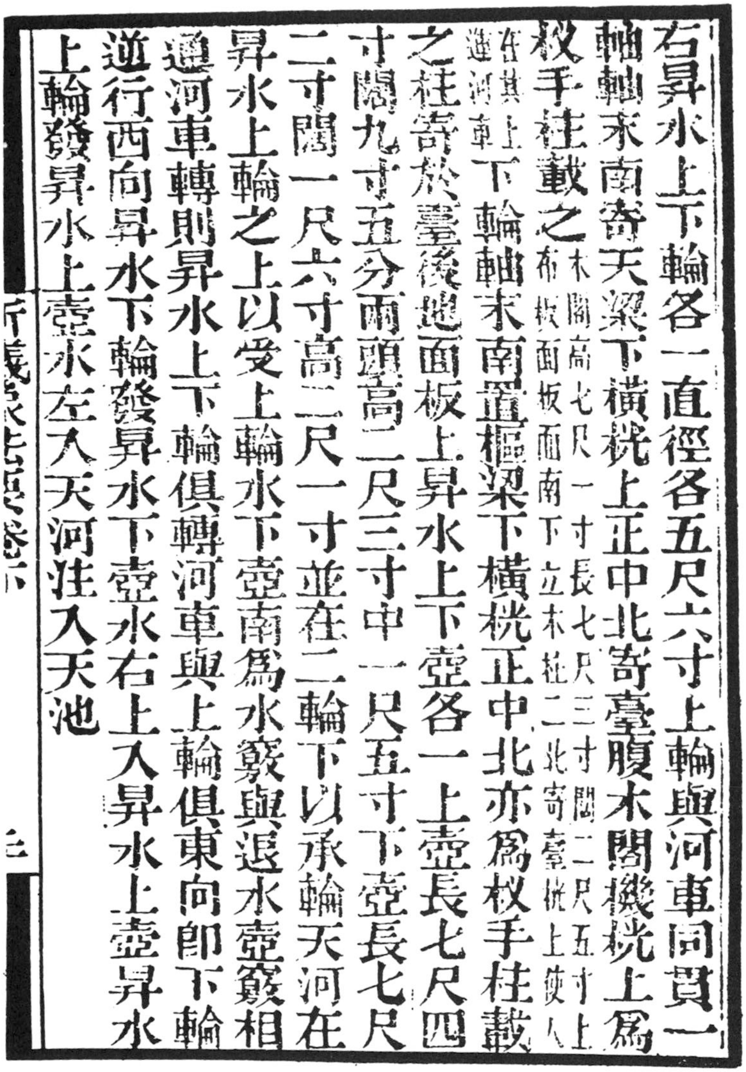

右昇水上下輪各一直徑各五尺六寸上輪與河車同貫一
軸軸末南寄天梁下橫桄上正中北寄臺腹木闑機桄上爲
杈手柱載之（木閣高七尺一寸長七尺三寸闊二尺五寸上布板面板面南下立木柱二北寄臺桄上使人
在其上運河車）下輪軸末南置樞梁下橫桄正中北亦爲杈手柱載
之柱寄於臺後趨面板上昇水上下壺各一上壺長七尺四
寸闊九寸五分兩頭高一尺三寸中一尺五寸下壺長七尺
二寸闊一尺六寸高二尺一寸並在二輪下以承輪天河在
昇水上輪之上以受上輪水下壺南爲水竅與退水壺竅相
通河車轉則昇水上下輪俱轉河車與上輪俱東向即下輪
逆行西向昇水下輪發昇水下壺水右上入昇水上壺昇水
上輪發昇水上壺水左入天河注入天池

新儀象法要卷下　三

Fig. 2. A page of text (ch. 3, p. 20*a*) of the *Hsin I Hsiang Fa Yao* (section L, p. 42 below), showing a gloss in small characters. This section describes the noria wheels which replenished the water-supply tanks for the clock's driving-wheel.

FOREWORD TO THE SECOND EDITION

JOSEPH NEEDHAM

SINCE the original publication of *Heavenly Clockwork* in 1960, textual studies and practical work have substantially increased our knowledge, and correspondingly modified some of the views which we then held about the text of Su Sung's *Hsin I Hsiang Fa Yao*, about the astronomical clock-tower which he sponsored, and about some of the other time-keeping devices mentioned by us.

Distance has for many years separated us from Wang Ling, now retired from the Institute of Advanced Studies at Canberra, Australia; and a recently lamented death has sadly finalised our already long separation by distance from Derek J. de Solla Price, lately Professor of the History of Science at Yale University, New Haven, Connecticut, U.S.A. Without their help, it was impossible to undertake the rewriting of this work for a projected new edition. Instead, pp. 1–205 have been reprinted without amendment, and I have pleasure in commending to readers the Supplement, on pp. 206–15, which has been prepared by John H. Combridge, whose collaboration in our studies began as a spare-time hobby in 1961.

The attention of readers is especially drawn to the following matters, referred to both on the reprinted pages and in the Supplement:

NOTES AND CONVENTIONS {pp. xii–xiii}

The 'Glosses' and 'Commentaries' have been shown to be the results of Shih Yuan-Chih's editorial conflation, in A.D. 1172, of two or more interdependent texts describing and illustrating features of astronomical clock-towers built in 967–9, 1078–85, and 1086–9, with Su Sung's memorial of 1092 included as an introduction. The latter led to the attribution of the whole work to Su Sung.

INTRODUCTION {p. 3}

The synopsis of the operation and timing of the clockwork mechanism should be read in conjunction with Section 1 of the Supplement, on pp. 206–9.

CHAPTER II {p. 14}

An 1844 quarto reprint of the 1172 edition of the *Hsin I Hsiang Fa Yao* is held by the Cambridge University Library [class-mark FC. 55·84/19 (Wade B 1258)], and contains woodblock illustrations substantially clearer in some parts than those of the 1922 octavo reprint used for Figs. 2, 3, 5, 7–22, 28–9 and 70. Further octavo reprints were published in 1935–7 and 1969 (see Shih Yuan-Chih (ed.) (1), in the Supplementary Bibliography, p. 218).

CHAPTER III {*p.* 20}

It has been shown (see Supplement, p. 208) that in both their practical work and its subsequent recording, Su Sung and his collaborators made extensive use of the then-surviving illustrated descriptions of the 976–9 and 1078–85 astronomical clock-towers.

CHAPTERS IV AND V

These chapters should be read, throughout, in the light of the information set out in Section 1 of the Supplement, on pp. 206–9.

CHAPTERS VI AND VII

Revised ideas about steelyard clepsydras {pp. 86ff}, monumental Striking Clepsydras {pp. 135–40}, Ming sand-clocks {pp. 154–61}, Wang Chêng's verge-and-foliot clock {pp. 146–7}, and the Korean armillary clock {p. 162 and Fig. 59}, are referred to in Section 2 of the Supplement, on pp. 209–14.

CHAPTER X

Revised ideas about compartmented cylindrical clepsydras {pp. 191–5}, and early European mechanical clocks {pp. 195–6}, are referred to in Section 3 of the Supplement, on pp. 214–15.

East Asian History of Science Library, Cambridge

1986

INTRODUCTION

It is generally allowed that the invention of the mechanical clock was one of the most important turning-points in the history of science and technology. Not only was it the earliest complex device, heralding a whole age of machine-making, but also its regular imitation of the natural motion of the sun and the heavens fascinated men and exerted no small influence on their philosophy and theology. If such a major innovation had come as a single stroke of genius we should expect to find some record of the inventor and his work; if on the other hand there had been gradual evolution over a long period of time we should be able to trace some stages in its development. Strangely enough the origin of the mechanical clock has long been shrouded in mystery and neither of these expectations has been fulfilled—we have known of no inventor and no stages along the way.

According to the view accepted until recently,[1] the problem of slowing down the rotation of a wheel so as to make it keep a constant speed continuously in time with the apparent daily turning of the heavens was first solved in Europe in the early fourteenth century A.D. by the use of the verge-and-foliot escapement fitted to a weight-driven mechanism. Before that time, it was thought, there had been only sundials and elementary water-clocks—after that date there were fully fledged mechanical clocks all over Europe. Refinements such as the spring drive (*c.* 1475), pendulum control (*c.* 1650) and the anchor escapement (1680) constituted a more or less continuous process of subsequent development.

Recent research has shown, however, that the first mechanical time-keepers were not so much an innovation as had been supposed.[2] They descended, in fact, from a long series of automatically rotated star-maps, planetary models and other devices designed primarily for exhibition and demonstration rather than for accurate time-keeping. Although such mechanisms are of the greatest interest as the earliest complex scientific machines, it has not hitherto been possible to adduce more than a few specimens, fragmentary remains, and literary descriptions tantalisingly incomplete. They give no clue to the crucial problem of the origin of the

[1] For a statement of the customary views on the origin of the mechanical clock see Sarton (1), vol. 3, pp. 1540 ff. Similar accounts are given by Beckmann (1), vol. 1, pp. 340 ff.; Usher (1), 2nd edition, pp. 191 ff., 304 ff.; Frémont (1); Baillie (1, 2); Howgrave-Graham (1); Saunier (1).

[2] A fuller account of such 'clockwork before the clock' has been given in Price (1).

mechanical escapement—a problem which has resisted all scholarly penetration for more than a century.

The examination of certain medieval Chinese texts, the relevance of which had not previously been realised, now permits us to establish the existence of a long tradition of astronomical clock-making in China between the seventh and fourteenth centuries A.D., and perhaps even having its origins as early as the second century A.D. These texts are remarkable for the wealth of historical and technical detail they provide, portraying the events, the people and the machines so clearly and so vividly that they cannot yet be matched by any comparable corpus from the history of science or technology in the West before very recent times. It happens, moreover, that these great Chinese clocks constitute an unsuspected missing link between the early water-clocks and later mechanical clocks found in the West. They are powered by a water-wheel but governed by an escapement device which checks the motion of the wheel and intermittently releases it to work astronomical devices and jacks, bells and gongs, and other indicators of the time of day and night.

From the available modern printed editions of these Chinese texts—treatises on clocks and on astronomy, dynastic histories, encyclopaedias and other sources, we have been able to trace the Chinese tradition. In the West, early history must rely largely on manuscript sources, but thanks to the early invention of printing in China (eighth or ninth century A.D.) and the attention given there to the proper keeping of records, the modern editions faithfully reproduce earlier texts, and these earlier texts, in their turn, may be facsimile editions of still more ancient books which had become scarce and were thought worthy of preservation. Often we have not only the printed word, but the original diagrams as well, faithfully copied from edition to edition and thus handed down to us in unbroken lineage.

The key text for our study is the *Hsin I Hsiang Fa Yao* (New Design for a (Mechanised) Armillary (Sphere) and (Celestial) Globe), written by Su Sung in A.D. 1090 (Fig. 3), the appropriate sections of which we have fully translated. Like so much evidence for the history of science in all parts of the world, its importance has probably been concealed because of its very technical appearance, and perhaps also in this case because Chinese specialists have not in general been familiar with the history of horology as a whole. So complete is the description in this text that it has been possible to prepare detailed working drawings of the mechanism, and to identify more than 150 technical terms of eleventh-century Chinese mechanics. It must be observed that such elucidation of technical terms is especially vital because of the peculiarities of Chinese philology. The help given by this one text has now made it possible to understand several other texts and fundamental problems which had previously been concealed or misinterpreted.

Su Sung's 'clock' was, in fact, a great astronomical clock-tower more than thirty feet high, surmounted by a huge bronze power-driven armillary sphere for observation, and containing, in a chamber within, an automatically rotated celestial globe with which the observed places of the heavenly bodies could be compared. On the front of the tower was a pagoda structure with five storeys, each having a door through which mannikins and jacks appeared ringing bells and gongs and holding tablets to indicate the hours and other special times of the day and night. Inside the tower was the motive source, a great scoop-wheel using water and turning all the shafts working the various devices. The wheel was checked by an escapement consisting of a sort of weigh-bridge which prevented the fall of a scoop until full, and a trip-lever and parallel linkage system which arrested the forward motion of the wheel at a further point and allowed it to settle back and bring the next scoop into position on the weigh-bridge. One must imagine this giant structure going off at full-cock every quarter of an hour with a great sound of creaking and splashing, clanging and ringing; it must have been impressive, and we know that it was actually built and made to work for many years before being carried away into exile.

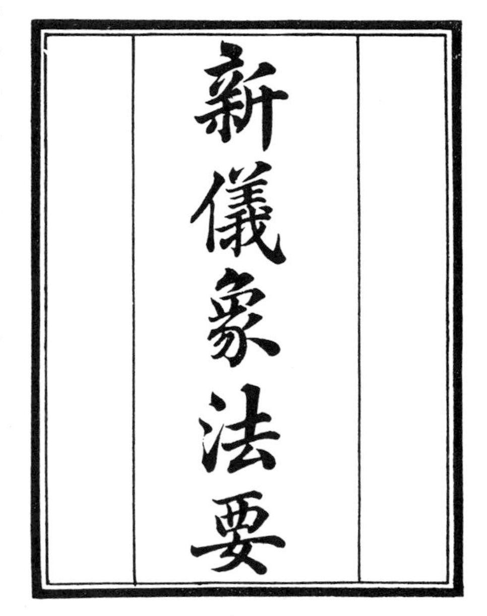

Fig. 3. Title-page of the *Hsin I Hsiang Fa Yao* (New Design for a (Mechanised) Armillary (Sphere) and (Celestial) Globe), written by Su Sung in A.D. 1090.

In what follows we shall examine all that is known of Su Sung and of his clock, and then trace the history of the Chinese tradition backwards and forwards from this date. Later we shall examine the relations between these clocks and those of Islam and Europe in an effort to see whether this early use of an escapement device in China might have had some effect on progress elsewhere in the world. The pursuit of these clocks and their makers through the pages of medieval Chinese history gave us a period of great excitement, picking up trails, losing them and then unexpectedly finding them again, while one discovery followed rapidly upon another. We feel privileged and happy to throw a spotlight on Su Sung and his associates and to restore them to a place of great honour in history.

Our work was begun in 1954 and had reached a stage sufficiently advanced to be communicated in a lecture to the (British) Antiquarian Horological Society on 11 January 1956 (cf. Ward (2)). It was then published in the form of a preliminary note in *Nature* on 31 March. Not until later in the summer of that year did we find, to our great pleasure, that parallel work on the history of horological engineering in China had been proceeding at Peking. There the well-known historian of technology Dr Liu Hsien-Chou, Vice-President of Ch'ing-Hua University, had published in October 1953 an interesting paper on Chinese inventions in power-source engineering (1), followed by a second in July 1954 on Chinese inventions in power transmission (2).[1] Reference was made to Su Sung's astronomical clock in both these works, the former dealing particularly with its water-wheel, the latter with the various forms of gearing which it contained. In the second paper the mechanism of the escapement was explained, but the historical significance of any such device occurring at that date (the late eleventh century) was not discussed. Then, at the International Congress of the History of Science at Florence in September 1956, we met Dr Liu himself,[2] and had the satisfaction of learning from his communication and from many personal talks that he and his assistants had reached conclusions substantially identical with our own.[3] At the same time we received copies of a third paper published by him, on Chinese inventions in horological engineering (3), in August. Investigations proceeding in parallel without mutual knowledge in this way led to mutually confirmatory results.

We wish to thank Dr Liu Hsien-Chou for the enjoyable discussions which took place in the inspiring atmosphere of the city and countryside of Leonardo and Galileo. And we may also take this opportunity of offering our best thanks to Dr Chang Shu-I, Librarian of the National Library at Peking, for his kindness in responding to our questions about a possible Sung edition of Su Sung's book, and for sending us spontaneously photostats of variant diagrams in an eighteenth-century manuscript copy of it which is preserved there.

[1] So far as we know, this journal was not at that time available in Western countries.

[2] The delegation included also other old friends, Dr Chu K'o-Chen, the eminent historian of astronomy and meteorology, a Vice-President of Academia Sinica, and Dr Li Nien, the eminent historian of mathematics.

[3] For particulars of the only significant variation see below, p. 57.

I

BIOGRAPHY OF SU SUNG

BEFORE proceeding further, something must be said of the life of that remarkable man whose monograph has preserved for us in such unexpected detail the construction of a mechanical clock more than three centuries before the first appearance of such clocks in Europe. Though primarily an eminent civil servant, in uprightness and ability worthy of the highest traditions of the millennial Chinese mandarinate, he was evidently one of those (by no means few in medieval China) who mastered the scientific knowledge of his time, and found opportunities for applying it in the service of the State.

Su Sung [450], whose 'style', or public name[1] (*tzu* [900]), was Tzu-Jung [451], was born in A.D. 1020 at Nan-an [850] in the prefecture of Ch'üan-chow [851] in Fukien province. In the course of his career he held many offices, advancing gradually in the bureaucratic hierarchy.[2] His first position was that of Drafting Secretary in the College of All Sages [901], one of the learned organisations of the government which, like the Han-Lin Academy, was responsible for preparing the imperial edicts and for advising the emperor and his highest ministers on all kinds of matters which might arise.[3] It embodied a great library. Under the emperor Ying Tsung (A.D. 1064–7) Su Sung became Staff Supervisor of the Ministry of Finance [902]. At about this period he received the Lien-T'ui [903] decoration, an order bestowed on incorruptible officials, in company with Han Ch'i [452], who, a prominent member of the conservative party, was driven out of public life by the reformer Wang An-Shih [453] in 1069 and died in 1075. Another member of the conservative party, Fu Pi [454], at one time prime minister, and a great enemy of Wang An-Shih, praised Su Sung (some time before his death in A.D. 1083) as a 'gentleman of the good old sort' [904]. From these indications it would seem that Su Sung was associated distinctly with the old-fashioned Confucian conservative party, but though many, perhaps most, of his friends were among its adherents, and though (as we shall see in the sequel, p. 116)

[1] See note 3 on p. 10.

[2] A good account of this in Sung times has recently been given by Kracke (1).

[3] Cf. Bazin (1); des Rotours (1), vol. 1, p. 192; Kracke (1), p. 45.

he himself was regarded as connected with it, we are told that with lofty ideals he stood rather alone, winning general respect thereby.[1]

In common with other officials of similar rank, Su Sung received foreign as well as home assignments. In 1077 he was despatched as a diplomatic envoy to the Liao kingdom of the Ch'i-tan people in the north,[2] and unexpectedly there came the opportunity for him to utilise the astronomical and calendrical knowledge which, though not professionally a member of the Bureau especially concerned with these sciences, he had long been acquiring. How this arose is recounted, not in the official history of the Sung dynasty, but (in two versions) in a book of miscellaneous reminiscences and accounts written early in the following century. We give the story in the words of its author, Yeh Mêng-Tê.

Shih-Lin Yen Yü [700], ch. 9, pp. 7*a* ff., by YEH MÊNG-TÊ [455], *c.* A.D. 1130

When Su Tzu-Jung [451] (Su Sung) was taking the provincial examinations (in his youth), it happened that an essay was set on the general principles of the heavens and the earth as manifested in the (structure of the) calendar. He came out top of the list, and ever afterwards he was particularly interested in (astronomy and) calendrical science. (Later on) in the Yuan-Fêng reign-period (A.D. 1078–85) he was sent as an ambassador (from the Sung empire) to the (Liao) barbarians. He happened to be in their country (North China) at the winter solstice, which occurred at the time predicted in their calendar (but one day later than the indication of the current Sung calendar); so he hastened to the court to present congratulations one day too early. As the (Liao) barbarians had no restrictions on astronomical and calendrical study,[3] their experts in these subjects were generally better (than those of the Sung), and in fact their

[1] We do not give a translation of the biography of Su Sung in *Sung Shih* [705], ch. 340, pp. 22*a* ff., for most of it is of political interest only.

[2] Fig. 4, showing the reception of a Liao embassy at the Chinese court in A.D. 1004, gives a good idea of the ceremonial customary on such occasions.

[3] In ancient and medieval China the promulgation of the calendar by the emperor was a right corresponding to the issuing of minted coins, with image and superscription, in Western countries. It had always been one of the most important duties of the ruler of the vast agrarian culture-area of the 'black-haired people'. Acceptance of the calendar was equivalent to recognition of imperial authority. Owing to this close association between the calendar and State power, any imperial bureaucracy was likely to view with alarm the activities of independent investigators of the stars, or writers about them, since they might secretly be engaged upon calendrical calculations which could be of use to rebels planning to set up a new dynasty. New dynasties always overhauled the calendar, and issued one with a new name, and this might happen even in successive reign-periods under the same emperor. These facts explain why Matteo Ricci's mathematical books were confiscated when he was on the way to the capital in A.D. 1600 (d'Elia (1), vol. 2, p. 122; Trigault (1), p. 370). Fortunately, they were returned to him, by mistake, before he went on to Peking in the following year. From early times, Chinese astronomy had benefited by State support, but the disadvantage of this was the semi-secrecy which it involved. Some realisation of this was expressed from time to time by Chinese historians—for example, in the *Chin Shu* [775] (ch. 11, p. 5*a*) we read: 'Thus astronomical instruments have been in use from very ancient days; handed down from one dynasty to another, and closely guarded by official astronomers. Scholars have therefore had little opportunity to examine them, and this is the reason why unorthodox cosmological theories were able to spread and flourish.'

PLATE I

Fig. 4. Audience of a Liao (Ch'i-tan) ambassador at the Sung court in A.D. 1004. Su Sung's astronomical and calendrical knowledge served him in good stead while he was Sung ambassador to the Liao kingdom (A.D. 1077–8).

calendar was correct. Of course, Su Sung was unable to accept it,[1] but he calmly engaged in wide-ranging discussions on calendrical science, quoting many authorities, which puzzled the (Liao) barbarian (astronomers)[2] who all listened with surprise and appreciation. Finally he said that after all, the discrepancy was a small matter, for a difference of only a quarter of an hour would make a difference of one day if the solstice occurred around midnight, and that is considered much only because of convention.[3] The (Liao) barbarian (astronomers) had no answer to this, so he was allowed to carry out his mission (on the earlier of the two days). But when he returned home, he reported to the emperor Shen Tsung [951], who was very pleased at his success and at once asked which of the two calendars was right. Su Sung told him the truth, with the result that the officials of the Astronomical Bureau were all punished and fined.[4]

At the beginning of the Yuan-Yu reign-period (A.D. 1086) the emperor ordered Su Sung to reconstruct the armillary (clock) [150], and it exceeded by far all previous instruments in elaboration. A summary of data concerning it was handed down to Yuan Wei-Chi [456], Director of Astronomical Observations (Northern Region) [924]. The original model was due to Han Kung-Lien [457],[5] a first-class clerk in the Ministry of Personnel, who was a very ingenious man. By that time Su Sung had become Vice-President[6] (of the Chancellery Secretariat) and simply gave the ideas to him. He could always carry them out, so that the instrument was wonderfully elaborate and precise. When the (Chin) barbarians took the capital[7] they destroyed the Observatory (or Astronomical Clock Tower) [117] and took away with them the armillary (clock) [150]. Now it is said that the design is no longer known, even to the descendants of Su Sung himself.

Shih-Lin Yen Yü, ch. 3, p. 14*b*, by YEH MÊNG-TÊ, *c.* A.D. 1130

The calendar of the Ch'i-tan (Liao) people was different by one day from that of our own dynasty (the Sung). In the Hsi-Ning reign-period (A.D. 1068–77)[8] Su Tzu-Jung [451], (Su

[1] As a high official of the Sung dynasty, he naturally had to adhere to the Sung calendar.

[2] These men were of course also Chinese, but they had taken service with the northern 'barbarian' dynasty, the ruling house of which was that of the Ch'i-tan tribal people.

[3] Cf. Maspero (1), p. 258. Owing to an interpolation method then used for plotting the variation of gnomon shadow lengths, Ho Ch'êng-T'ien [505] missed a solstice in A.D. 436 because it occurred half-an-hour before midnight; but in 440 he got the correct day because it occurred three hours after midnight. To appreciate Su Sung's point one must remember that the quarter was the smallest time division in common non-astronomical use.

[4] In view of what is here suggested about the superiority of the astronomers serving the Liao dynasty in the north it is interesting that the *Liao Shih* [809] (ch. 44, pp. 39*a*, *b*) has extremely little to say about armillary spheres (and nothing about astronomical clocks); indeed it indicates that the Ch'i-tan people never had any at all. It does however say: 'For firmness there is nothing better than metal, and for use there is nothing more profitable than water. With the fashioning of metal and the flowing of water one can know the Tao (i.e. the way and order) of the Heavens without stirring outside one's house.' There is also a reference to the astronomical clock of I-Hsing (see p. 74), so no doubt they knew something of that. Cf. Wittfogel & Fêng (1), p. 467.

[5] The text mistakenly says Chang Shih-Lien [458].

[6] Or Principal Executive Officer.

[7] In A.D. 1126. This statement is found also in the *Sung Shih*, ch. 48, p. 18*a*. The Chin people (Jurchen Tartars) took away not only the clock of Su Sung, as we shall see fully later (p. 132), but also all the five astronomical clocks proposed in the Hsüan-Ho reign-period (A.D. 1119–25) which were then under construction. Cf. the passage translated from the *Chin Shih* [810], p. 131 below.

[8] A slight discrepancy will be noticed here. But presumably Su Sung left on his diplomatic mission in A.D. 1077 and returned in 1078.

Sung) was sent as ambassador to offer congratulations (to the Liao emperor) for his birthday, which happened to fall on the winter solstice. Our (Sung) calendar was ahead of that of the Ch'i-tan (Liao) kingdom by one day, and thus the assistant envoy considered that the congratulations should be offered on the earlier of the two days. But the secretary of protocol in the Ch'i-tan (Liao) Foreign Office declined to receive them on that day. Su Sung then spoke quietly and tactfully saying that after all the calendrical schools were not in agreement, some being ahead of others, and slight differences could not be avoided. As one could not hope to unify all opinions, the best way was that each should follow their own choice of date in celebrating festivals. The Ch'i-tan (Liao) people could not reject this argument, and Su Sung was permitted to offer congratulations on the day desired (by the mission). Upon his return he reported to the emperor Shen Tsung, who was pleased and said that nothing could have been more embarrassing. Since later on (foreign) ambassadors might be repeatedly refused reception because someone (at the Sung capital) did not know about the differences in the beginnings of months, and how the Sung envoys had been allowed to have their way, the emperor decided that (mutual tolerance of calendars) should be observed for the honour of the empire.

Astronomy and calendrical science, however, were not the only sciences in which Su Sung was expert. About A.D. 1070 he had produced, no doubt with a number of assistants, the best work of his time on pharmaceutical botany, zoology and mineralogy, the *Pên Ts'ao T'u Ching* [701] (Illustrated Pharmacopoeia). Still today this treatise contains precious information on subjects such as the metallurgy of iron and steel in the eleventh century, or the therapeutic use of drugs such as ephedrine. It was incorporated in the Taoist Patrology (*Tao Tsang*, no. 761), so greatly was it appreciated by the adepts, and extensively quoted in most of the subsequent compilations of the same class in successive dynasties.

Some twelve years after his diplomatic mission to the north, Su Sung was promoted to the position of Right Vice-Minister of the Ministry of Personnel [905] and concurrently a Senior Executive of the Imperial Chancellery and Secretariat [906]. It is recorded that in these responsible posts he paid special attention to the recruitment and quality of the civil service, striving to ensure that promotions were given only on merit, not at random or by means of private influence, and warning military commanders against making trouble on the frontiers in order to gain renown. But this was also the period of his scientific activity which most concerns us here. In 1086 the emperor issued an order for the examination of the existing astronomical equipment, and for the construction of some kind of astronomical clock which should equal, or if possible excel, those which had been built at the beginning of the dynasty and still earlier during the T'ang. In the memorial of Su Sung to the emperor, a document of extraordinary interest (which we give in translation below, pp. 16 ff.), he describes how he chose his associates for the work, the most important mathematician and engineer, Han Kung-Lien, being not an official of the Bureau of Astronomy but a minor functionary in his own Ministry of Personnel. By 1088 the complete working wooden pilot model was set up for imperial approval in the Palace at K'ai-fêng, and two years later the metal parts, principally the armillary sphere and celestial globe, cast in bronze. By 1092 the writing of the explanatory monograph, the *Hsin I Hsiang Fa Yao* [702], must have been well advanced, and in 1094 (the first year of the new Shao-Shêng reign-period) this 'New Design for an Astronomical Clock' was finished and presented.

By this time Su Sung had reached the age of 75. He bore several titles, such as that of the Second Titular Rank [907], Grand Protector of the Army [908], and K'ai-Kuo Marquis of Wu-kung [909]. He was one of the Deputy Preceptors of the Heir Apparent [910]. When in A.D. 1101 he died, he left behind him a collection of literary works, as well as the two books already mentioned, both of which have circulated widely and long. He did not live to witness the fall of the capital two decades later and the retirement of the Sung empire to the southern provinces. We may think of him as one of the most outstanding of the scientifically-minded officials of the Northern Sung.

II

TRANSMISSION OF THE TEXT OF SU SUNG'S BOOK

IN A.D. 1772 the Ch'ien-Lung emperor of the Ch'ing dynasty set up by decree an organisation for the discovery and preservation of all rare books in the Chinese empire. A complete collection of the best works in the language was to be made, primarily in manuscript form, for the imperial libraries and six branch libraries situated in various parts of the country. This was the *Ssu K'u Ch'üan Shu* (Complete Collection of the Four Treasuries), so called because the books were divided into the time-honoured categories of Classics, History, Philosophy and Literature. The editorial board was led by Chi Yün [459] and Lu Hsi-Hsiung [460]; it collected and catalogued no less than 10,254 works, only 3461 of which, however, were considered worthy of incorporation into the imperial manuscript library. This was not the first time in history that a similar plan had been carried out; in the early days of the previous dynasty (the Ming), in 1408, the manuscript *Yung-Lo Ta Tien* [703] (Great Encyclopaedia of the Yung-Lo reign-period) had been completed, and 369 books were now incorporated in the new collection from that source. But the labours of Chi Yün's Commission, successfully completed in A.D. 1782, constituted the greatest bibliographical achievement of a nation of scholars. Their *catalogue raisonné*, the *Ssu K'u Ch'üan Shu Tsung Mu T'i Yao* [704], included a notice of Su Sung's book on astronomical clockwork, and we must hear their views on it.

Ssu K'u Ch'üan Shu Tsung Mu T'i Yao,[1] ed. by CHI YÜN *et al.*, A.D. 1781, ch. 106

The *Hsin I Hsiang Fa Yao*[2] in three chapters was written by Su Sung [450] whose *tzu*[3] was Tzu-Jung [451]. He was born at Nan-an[4] [850] but moved to Tan-t'u[5] [852]. He obtained his

[1] 'Analytical Catalogue of the Complete Library of the Four Categories.' For further information on this, see Mayers (1); Têng & Biggerstaff (1), pp. 27 ff.

[2] 'New Design for an Astronomical Clock'; lit. 'Essentials of a New Method for (Mechanising the Rotation of) an (Armillary) Sphere and a (Celestial) Globe'.

[3] Every educated Chinese in traditional China had a *tzu* name or 'style' given to him by his father when he came of age, in addition to his *ming*, the name which he had been given by his family at birth. The *tzu* was for use in formal documents and polite public address. Most Chinese also had (and still have) a *hao* or literary name, given to them by their friends. Other types of names were common, but these are the three chief kinds.

[4] A town in Fukien province, some miles up river from the port of Ch'üan-chow [851] (Marco Polo's Zayton).

[5] A town on the Yangtse river some distance east of Nanking. Now called Chen-chiang.

Chin-Shih [911] degree (in the state examinations) in A.D. 1042 and later became a high-ranking official; Right-hand Vice-President of the Ministry of Personnel, Vice-President of the Imperial Secretariat and Chancery, with the title of Duke of Chao Commandery. Details about him will be found in his biography in the *Sung Shih* [705] (History of the Sung Dynasty, ch. 340). The present book was written about the making of a new armillary clock (lit. sphere [150]). This occurred in the Yuan-Yu reign-period (A.D. 1086–93). But the *Sui Ch'u T'ang Shu Mu* [706] (Bibliography of the Sui Ch'u Studio) by Yu Mou [461] (Sung, fl. A.D. 1148) calls it the *Shao-Shêng I Hsiang Fa Yao* [707],[1] and the bibliographical chapter of the *Sung Shih* records (ch. 206, p. 10*b*)[2] '*(Hsin) I Hsiang Fa Yao*, in one chapter', to which the commentary adds that it was compiled in the Shao-Shêng reign-period. This is because the book was completed at the beginning of the Shao-Shêng reign-period (A.D. 1094–7).

According to the biography of Su Sung, a new armillary sphere [150] was to be built at that time (A.D. 1086) and Su Sung was asked to suggest some persons suitable for the job. He, being himself good at mathematics and calendrical science, put forward a name, that of Han Kung-Lien [457] a first-class clerk in the Ministry of Personnel, because of the latter's ingenuity. Han had studied the ancient methods. A tower of three floors was built, having an armillary sphere [150] at the top, a celestial globe [152] in the middle, and wheels with jacks attached to the same machine, all made to turn by water-power without human intervention. When the (double-) hours and quarters arrived, the jacks came out to report them, and thus the positions and degrees of the stars and constellations (read from) the instrument showed no error of even a fraction of time. Day and night, dawn or dusk, could all be seen and predicted. There had not been quite such a thing before.[3] The *Shih Lin Yen Yü* [700] of Yeh Mêng-Tê [455] (*c.* A.D. 1130) also says[4] that what was made by Su Sung exceeded by far all previous instruments in elaboration. Yeh goes on to say that 'a summary of data concerning it was handed down to Yuan Wei-Chi [456], Director of Astronomical Observations (Northern Region) [924]. But now[5] not even the descendants of Su Sung have received any instruction (in these matters).' We[6] find that the name of Yuan Wei-Chi appears indeed in the lists of Student-Astronomers [912] of the Bureau of Astronomy and Calendar. This agrees with what Yeh Mêng-Tê said, so that his words are trustworthy. From this we know that much attention was paid to it (the clock) during the Sung.

The book opens with a letter (from Su Sung) to the emperor. The first chapter comprises seventeen illustrations, from the armillary sphere to the water-level base [122]. The second

[1] 'Shao-Shêng reign-period Design for an Astronomical Clock.'

[2] It is of much interest to note that the bibliographical chapter of the *Sung Shih* also records (on the same page) another book on astronomical clockwork, the *Shui Yün Hun T'ien Chi Yao* [708] (Essentials of the (Technique of) making Astronomical Apparatus revolve by Water-Power), written by Juan T'ai-Fa [462]. Unforunately we have not been able to find out anything about this author, his work, or his date. If he did not write before the end of the Northern Sung (A.D. 1126), he must have written at the end of the Southern Sung (thirteenth century), i.e. after Su Sung's technique and book had been recovered.

[3] The words from the beginning of the paragraph to this point are an almost verbatim quotation from the *Sung Shih* ch. 340, p. 30*a* (Su Sung's biography).

[4] Cf. p. 7 above.

[5] I.e. some five years after the fall of the capital, K'ai-fêng [862], the writer being with the Sung in the south.

[6] I.e. the editors of the *Ssu K'u Ch'üan Shu*, in the course of their bibliographical researches.

chapter has eighteen illustrations, from the celestial globe to the diagrams of transits of constellations at dawn on the winter solstice. The third chapter adds twenty-five illustrations, from the tower of the armillary sphere and celestial globe to the sun-shadow scale of the gnomon device. Each illustration is fully explained. This shows how carefully the author presented the system of construction and method of use of his astronomical equipment.

After the Southern Sung (period)[1] there were only very few copies of this book remaining. The edition we now have[2] follows the text of that in the possession of Ch'ien Tsêng [463] of the Ming dynasty.[3] At the back of this book there were the two following lines, 'Edition of Shih Yuan-Chih [464] of Wu-hsing [853][4] at the San Ch'ü Tso Hsiao Studio [913], the 9th day of the 9th month of the 8th year of the Ch'ien-Tao reign-period (A.D. 1172)'. This shows that the present edition is a true copy of the text of the Sung edition. Shih Yuan-Chih's *tzu* was Tê-Ch'u [465]. He held office up to Policy Adviser in the Secretariat of the Censorate [914], and his works include a commentary on the poems of Su[5] which is still current. At the end of the book, there are 4 (final) illustrations, the Celestial Globe Drive Wheel [81] etc. with their explanations, and added commentary on various items (in the middle of the book), all provided by Shih Yuan-Chih on the basis of another edition or other versions. The editing work and commentary was very well done and the copying of Ch'ien Tsêng (of the Ming) was extremely skilful. He wrote the *Tu Shu Min Ch'iu Chi* [710] (On the Study of Old Books and their Meanings)[6] and included a discussion of this book in it. He himself said that 'all the illustrations with their lines and details (followed the original copy) without a hair's-breadth of difference. The copying work took several months to complete, and the paper and the ink were so excellent and indeed matchless, that the result was in no way inferior to the Sung edition itself.' These words were no idle boast.

The dynasty of your Imperial Majesty now has instruments the excellence and precision of which far exceed all those made during the past thousand years.[7] Of course, the invention of Su Sung is not to be compared with them. However, we may have something to learn by paying attention to these old matters, for they show that the people of that time were also interested in new inventions.[8] Furthermore, it is remarkable that such a confidential volume

[1] I.e. after about A.D. 1280.

[2] I.e. in A.D. 1781.

[3] Although Ch'ien Tsêng is here spoken of as a Ming scholar, most of his life was passed under the Ch'ing. Ch'ien was a famous bibliophile, born at Ch'ang-shu in Chiangsu province in A.D. 1629. The catalogue of his library was printed in 1669 as the *Yeh Shih Yuan Shu Mu* [709]. He died in A.D. 1699, but it was not until 1726 approximately that his *Tu Shu Min Ch'iu Chi* (On the Study of Old Books and their Meanings) was printed.

[4] A town in Chiangsu province, just to the south of the beautiful Lake T'ai-hu.

[5] Presumably Su Shih (Su Tung-P'o), the famous poet [466] (A.D. 1036–1101), who not only bore the same family name as Su Sung, but was almost his exact contemporary in life-span, and studied under the same master, the great scholar Ouyang Hsiu [467]. Both men were also affiliated with the conservative party.

[6] This book was described by the *Ssu K'u Ch'üan Shu* editors in their bibliography (ch. 87), but it did not win the honour of being transcribed into the imperial manuscript collection.

[7] The reference here is not only to the mechanical clocks introduced and maintained by the eighteenth-century Jesuits, but in particular to those which were just at this time being imported by western commercial firms such as Cox & Beale. It is remarkable that many of these wonderful clocks, designed especially for the China trade, were carrying back to China features which had originated there. Cf. p. 151 below.

[8] These words, remarkable as written by scholars of mainly literary interests in the Chinese eighteenth century, are commended to the attention of historians of science and technology.

was handed down through several hundred years and yet after repeated copying even the illustrations are as fresh as when it was first written.[1] Thus it should be considered as something indeed valuable and precious.

This of course is not the whole story of the transmission of the text of Su Sung's book. It is interesting that the *Sung Shih* records it as having had only one chapter, since it now has three. From evidence which we shall present in due order (cf. pp. 131 ff. below) we know that at the fall of the Sung capital in A.D. 1126 the astronomical clock of Su Sung was taken away to the north by the Chin Tartars. Presumably its maintenance engineers went with it, presumably also the relevant documents in manuscript. During the following fifty years the Sung people in the south tried vainly to recover Su Sung's designs, but his own family could not produce them, and other scientific men were obliged to return to the earlier methods which involved only clepsydra floats. Of particular interest is the fact that the great philosopher Chu Hsi, living in Su Sung's own province, Fukien, made efforts to continue his work, but failed since only one part (our present second chapter), on the globe and its markings, was available. The Sung archives evidently knew only of this section. Eventually, in 1172, the full text came to light, and was printed for the first time. It is not at all unlikely that it was then brought down by someone from the Chin empire in the north, for although there was always fighting on the borders, communications between north and south were hardly ever entirely interrupted. Moreover, it is not impossible that the Sung edition of Shih Yuan-Chih, or fragments of it, might still come to light, since several hundred Sung editions are preserved in the National Library at Peking.[2]

Lastly, a word about the edition which we have used. The *Ssu K'u Ch'üan Shu*

[1] The point of this remark is the danger to which books were subjected when copied only in manuscript or printed only for restricted circulation and deposition in government archives. This was another aspect of the price paid by Chinese astronomy for government support. The periodical upheavals at times of dynastic change or temporary political confusion took a heavy toll of official libraries, and there can be no doubt that this is one of the chief reasons why the Chinese literature available to us now on subjects such as astronomy and calendrical science is much less than it ought to be. For every book which is accessible to us today we know the titles of a dozen more which have perished irretrievably.

[2] The Librarian of the National Library in Peking, Dr Chang Shu-I, informs us (letter of 26 October 1956) that there is no copy of a Sung edition of the *Hsin I Hsiang Fa Yao* among the early printed books conserved there. However the library does possess an interesting MS. copy of the work, written in the Ch'ien-Lung reign-period (A.D. 1736–95), i.e. about the same time as the report of the editors of the imperial collection which we have just given in translation. Two of the illustrations in this MS. differ slightly from those in the printed editions of the nineteenth century—notably the diagram of the whole armillary sphere. Dr Chang Shu-I most kindly sent us photostats of these MS. pictures (cf. Fig. 30 below). The text, however, is identical with the printed editions, and there are no divergencies in the clockwork diagrams.

editors of 1781 certainly used the version which Ch'ien Tsêng had so carefully copied during the latter half of the previous century. We have not seen this and do not know whether any copy of it now exists. The man who perpetuated it was Chang Hai-P'êng [468] (1755–1816), a famous bibliophile of Ch'ang-shu in Chiangsu province, who incorporated it into his collection, the *Mo Hai Chin Hu* [711], printed by 1814 but not issued till 1817. As this issue was very restricted in numbers and soon became rare, most of the books in it, including Su Sung's, were incorporated into another collection, printed in a much larger edition, the famous *Shou Shan Ko Ts'ung Shu* [712] of Ch'ien Hsi-Tso [469] (1801–1844) published in 1844. This collection was reprinted in 1889 from a tracing, and again in 1922. The latter contains the text which we have used.

III

SU SUNG'S MEMORIAL TO THE EMPEROR CHÊ TSUNG

AFTER we had translated and studied the main text which describes the astronomical clock of Su Sung, we turned our attention to the memorial which he addressed to the emperor in A.D. 1092 and which is printed at the beginning of his book. Great was our delight to find that he gave not only a minute account of the stages in the construction of the clock itself, together with names and details of the collaborators whom he assembled for the work, but also a historical disquisition on the instruments of similar kind which had existed in previous centuries. This enabled us to find our way to and through a whole succession of texts the significance of which could only be understood in the light of Su Sung's descriptions, and which had consequently escaped attention until now. These will follow in due order after Su Sung's memorial.

In writing his letter to the emperor, Su Sung interweaved two main subjects, the description of his new clock, and the summary of his historical investigations. For the convenience of the reader, we give the relevant analysis of the paragraphs into which the memorial divides itself, according to marginal numbering which we have introduced.

Historical investigations	*Paras.* 1–5, 10–14, 21
Principles of former instruments	2–4, 12, 20, 21
Mechanisation of instruments by clockwork or otherwise	5, 10, 11, 13, 14
Account of the building of the clock of Su Sung and his collaborators	1, 6–9, 15–18
Description of special features	15–17
Testing	19

MEMORIAL FROM SU SUNG TO THE EMPEROR (CHÊ TSUNG)[1] PRESENTING THE ASTRONOMICAL CLOCK (LIT. ARMILLARY SPHERE AND CELESTIAL GLOBE)

Hsin I Hsiang Fa Yao [702], ch. 1, pp. 1*b* ff., A.D. 1092

(1) After your Imperial Majesty ordered a decision about the old and the new armillary spheres in the winter of the first year of the Yuan-Yu reign-period (November, A.D. 1086) I called a conference of all the calendrical (and astronomical) officials and investigated all the extant literary discussions and testimonies. I also went to the Astronomical Department (T'ien Wên Yuan) [915] of the Imperial Academy [916] and to the Bureau of Astronomy and Calendar (T'ai Shih Chü) [917].[2] By comparing (the instruments and documents) in these two places,[3] I came to the conclusion that the 'New' Armillary Sphere [150] had been built and used in the Chih-Tao (A.D. 995–7) and Huang-Yu (A.D. 1049–53) reign-periods.[4] The 'Old' Armillary

[1] The reign of this emperor, the seventh of his line, Chê Tsung [950], personal name Chao Hsü [471] (A.D. 1076–1100), was a somewhat unhappy one. When he came to the throne in A.D. 1086 the reform party headed by Wang An-Shih [453] and the conservative party headed by Ssuma Kuang [472] were locked in struggle for supremacy. Bitter controversy continued throughout the reign. During the first reign-period, Yuan-Yu (1086–93), the emperor supported the conservatives, but eventually their internal dissensions became so acute that a change was inevitable. As soon as his mother (who had been leagued with the conservatives) died, the boy emperor swung back to the policy of his father Shen Tsung, and called the reformers back to power, inaugurating the Shao-Shêng reign-period (1094–7). These facts are not at all irrelevant to our history of clockwork. The emperor's youth may well have been connected with Su Sung's abundant use of jack-work and, as we shall see (pp. 116 ff. below), Su Sung's political enemies were anxious to destroy his clock as soon as they came back into power, or later on to ignore it.

[2] Cf. des Rotours (1), vol. 1, p. 208.

[3] The existence of two observatories at the capital is interesting. Light is thrown on this by an interesting passage in P'êng Ch'êng's [476] *Mo K'o Hui Hsi* [713] (Flywhisk Conversations of the Scholarly Guest), written about A.D. 1080, ch. 7, p. 8*a*. He says:

'In our (Sung) dynasty, the Astronomical Department [915] (of the Imperial Academy, the Han-Lin) was established within the Imperial Palace, with a clepsydra (room) and an observatory with a bronze armillary sphere, just the same as those in the Bureau of Astronomy and Calendar [935]. The reports from the two observatories are supposed to be compared. Every night the Astronomical Department should state whether or not there have been vapours, unusual auspicious phenomena, oppositions and conjunctions, changes of positions of the heavenly bodies etc., and present its report before the opening of the Palace gates. Then when the gates have been opened (at dawn), the results of the Bureau of Astronomy and Calendar should arrive, after which the two should be compared and checked, in order to avoid all false reports. But in recent times the officials of the two observatories secretly copied from each other before reporting, and this went on for years. Everyone knew about it, yet no one thought it strange. Besides, the astronomical officials contented themselves with copying out the positions of the sun, moon and five planets according to very rough ephemerides, and presenting the results to the emperor—never using the (astronomical equipment in the) observatory. The (two) institutions were full of officials who simply drew their salaries without doing any work. When I became Astronomer-Royal in the Hsi-Ning reign-period (A.D. 1068–77) I exposed six of these dishonest officials (and had them punished), but soon afterwards the deceptions continued as before.'

We need not conclude, however, that the establishment of two observatories was due solely to the fear of intentional deception on the part of the officials, or shortcomings in their performance of their duties. An eminently sceptical and scientific desire for doubly checked evidence may also be admitted.

[4] These periods were those of the work of the astronomers Han Hsien-Fu [473] and Shu I-Chien [474] respectively; see below, pp. 68 ff., 63 ff., and 93 ff.

Sphere was built in the Hsi-Ning reign-period (A.D. 1068–77).[1] Because the rings and (indeed all the parts of) the instrument are thin and weak, and the Water Level Base [122] too low, it is difficult to handle. Yet according to your Imperial Majesty's order to the Secretariat, what has been planned (lit. decided on) should be put to use. Now your servant's opinion is that the system and principles of the armillary sphere and the celestial globe have been handed down to us (from former times) with detailed degrees and measurements. Yet the astronomical and calendrical officials continue to differ and argue among themselves. This is because the instruments of today no longer agree with those of old, and also the technical terms are not properly defined [153]. Furthermore (the present instruments) for observations require the human hand to move them,[2] and (as the motion of the) hand is sometimes too much and sometimes too little, so the degrees of motion of the celestial bodies (seem to) vary. And each (official) gets readings larger or smaller than those of his colleagues, and there is no definite conclusion.[3]

(2) The astronomers of former times observed and investigated the heavens and their destined (movements). They had two methods.[4] The first was the (demonstrational) armillary sphere [154] (which exhibited) the round of the heavens and the square (i.e. the horizon plane with its cardinal points) of the earth [155]. The mechanism was hidden inside, and the co-ordinates graduated all round (the rings). By the (number of) degrees travelled by the sun and the stars one could study the changes of the hot and cold seasons and the years. Such were one of the astronomical instruments [156] of Chang Hêng [480], and the bronze water (-power) driven astronomical instrument [157] of the K'ai-Yuan reign-period (A.D. 713–41) (of I-Hsing [477]).

(3) The second is the (observational) armillary sphere [158] (with its sighting-tube). The present-day 'New' and 'Old' Armillary Spheres [150] used by the Astronomical Department of the Imperial (Han-Lin) Academy and the Bureau of Astronomy and Calendar both belong to this (second category).

(4) (In addition, there is a third instrument, the celestial globe.) Wang Fan [478] an official[5] of the Wu kingdom (third century A.D.) wrote: 'The armillary sphere is the old instrument of

[1] This had been the work of the distinguished scientific official Shen Kua [475], a man quite of the type of Su Sung himself; cf. Needham (1), vol. 1, pp. 135 ff. The reason why the sphere was called 'old' was presumably because it embodied principles going back to those instruments made in pre-Sung, e.g. T'ang, times.

[2] The non-observational instruments had of course been mechanised long before, as we shall see in the sequel.

[3] These sentences may perhaps be taken to imply that the mechanisation of the armillary sphere which was now being introduced, as a new development, by Su Sung into his astronomical clock, was intended, in part, as a kind of coarse adjustment for the sphere during night observations. It certainly preserved its sighting-tube. As we shall later see (p. 46) Su Sung distinctly states that this was to be used during the daylight hours as a means of following the rays of the sun during its daily movement. Since the motion of the clock was very discontinuous, the sighting-tube must have been brought round jerk by jerk, as it were, into the direct sunlight line. If a similar arrangement was intended by night, it would have acted as a crude 'clockwork drive' giving a coarse adjustment on the basis of which the observers could make their fine adjustments. The jerkiness would have precluded its use for gaining more accurate estimations of apparent inter-stellar distances, but the motion would have helped to avoid over-shooting or under-shooting the degree marks desired.

[4] The important three following paragraphs are given also in *Yü Hai* [740], ch. 4, pp. 40*a* ff.

[5] His title, 'Chung Ch'ang Shih [919]', shows that he was an aide-de-camp of the emperor. These 'gentlemen without portfolio' (des Rotours (1), vol. 1, p. 141) were attached directly to the service of the emperor's person, rode on each side of his carriage when he went out, and remained always on hand to advise on whatever minor matters might arise.

Hsi and Ho.[1] It has been handed down for generation after generation and is called the *chi hêng* [159], the rings and the sighting tube.[2] It is used to observe the sun, the moon and the stars, and to measure the equatorial constellations in degrees. But there is also the celestial globe [161] on which are marked the celestial bodies and the stars. These two are used to study the heavens, and they give approximately correct results.'[3] Thus from this we know that besides the (demonstrational) armillary sphere [154] and the bronze (observational) armillary sphere [158] they also had the celestial globe [161], three instruments altogether. During the past dynasties little has been recorded about celestial globes, but the (astronomical) chapter of the *Sui Shu* [714] (official History of the Sui Dynasty, ch. 19, p. 17*a*) says that the Secret Treasury [918][4] of the Liang dynasty (A.D. 502–57) had one. It was also said that an (earlier) one had been made in the Yuan-Chia reign-period (A.D. 424–53) of the (Liu) Sung dynasty.[5] Thus the ancient (astronomers) were able to reach the greatest perfection in observing the heavens by using these three instruments. Nowadays we have only one, so that I am afraid our results cannot be very correct.[6]

(5) The system used in Chang Hêng's instrument was not recorded in the official histories and the old instrument of the K'ai-Yuan reign-period (A.D. 713–41) (I-Hsing's) was already lost

[1] These two names have traditionally been considered those of legendary astronomers. They occur in the *Shu Ching* [811] (Historical Classic), ch. 1 (cf. Legge (1), p. 32; Karlgren (1), p. 3), in a text which may date from the seventh or eighth century B.C. There they appear to be those of six brothers, three of each name, whom the (legendary) emperor Yao stationed at various confines of the empire to observe celestial events. But in fact Hsi-Ho is a binome, originally the name of the mythological being who is sometimes the mother and sometimes the chariot-driver of the sun. The 'astronomers' were really magicians or cult-masters, despatched to control the movements of the heavenly bodies, not to observe them. But by the time of Wang Fan the traditional interpretation had long been stabilised.

[2] This is a reference to the astronomical instrument mentioned in ch. 2 of the *Shu Ching* as having been used by the legendary emperor Shun. Its name is there *hsüan chi* [162] and *yü hêng* [163], often translated 'the jade rings or disc and the jade traverse or sighting-tube'. Much evidence has been adduced by Michel (1) to show that this very ancient device was something like a nocturnal. The *hsüan chi* would thus have been a kind of circumpolar constellation template, with indentations fitted to the stars of the Great Bear, etc., while the *yü hêng* would have been the tube of jade fitting into a hole in its centre and so pointing at the pole-star. It can easily be seen how *chi*, often modified in writing [164], came to mean the rings of the armillary sphere, while *hêng* permanently retained its original meaning of sighting-tube.

[3] The quotation is from Wang Fan's *Hun T'ien Hsiang Shuo* [715] (Discourse on Uranographic Models), conflated from a number of sources in *CSHK* (*Chhüan Shang-Ku San-Tai Chhin Han San-Kuo Liu Chhao Wên*, ed. by Yen Kho-Chün; San Kuo sect.), ch. 72, pp. 1*a* ff. Partial translation, Maspero (1), pp. 332, 333; Eberhard & Müller (1).

[4] This may be only another name for the Bureau of Astronomy and Calendar, which under the T'ang was also called the Pi Shu Ko Chü [920] (des Rotours (1), vol. 1, p. 210). But most dynasties had their secret stores and treasuries, e.g. the garage in which the South-Pointing Carriage was kept (cf. Needham (1), vol. 4).

[5] This was Ch'ien Lo-Chih's (see below, pp. 94 ff.).

[6] The whole of this paragraph seems to rest upon a misapprehension of Su Sung's. Evidence which we shall present in due course (p. 23, n. 1) shows that in the early centuries (first to fifth) the term *hun hsiang* [152] meant the demonstrational armillary sphere, with a model of the earth at the centre instead of the swinging sighting-tube. It was this, and not the solid celestial globe, that Wang Fan was talking about in the third century. The true celestial globe seems to have been introduced at the end of the fourth century, after the star-positions in the ancient catalogues had been conflated into a single list, and this new instrument was called by the same name, *hun hsiang*, since it also was an image of the actual heavens. Thus Su Sung was perfectly right that three instruments had been known in ancient times, but mistaken in thinking that Wang Fan was talking about solid celestial globes; he ought to have referred to Ch'ien Lo-Chih (see p. 98 below) instead.

before the end of the T'ang dynasty. At the beginning of the T'ai-P'ing Hsing-Kuo reign-period (A.D. 976) a Szechuanese, Chang Ssu-Hsün [479], was the first to make a plan and model [165][1] and presented it to the emperor T'ai Tsung (A.D. 976–97). The emperor authorised its construction in the Palace and it was completed in the following year. The emperor then ordered it to be erected under the Eastern Drum Tower of the Wên Ming Hall [934] ⟨Now the Wên Tê Hall⟩, with an inscription 'Armillary Clock of the T'ai-P'ing reign-period [166]'. After the death of Chang Ssu-Hsün the cords [167][2] and the mechanism [164] went to rack and ruin and there was no longer anyone who could understand the system of the apparatus. (6)

When formerly I was seeking for help I met Han Kung-Lien [457], a minor official in the Ministry of Personnel, who having mastered the *Chiu Chang Suan Shu* [774] (Nine Chapters of Mathematical Art; first century B.C.),[3] often used geometry (lit. the methods of right-angled triangles) to investigate the degrees of (motion of the) celestial bodies. Thinking it over, I also became convinced that the ancient people used *Chou Pei* [716] techniques[4] in dealing with the heavens. The theory says that *pei* [168] signifies (the longer side of the right-angled triangle), and *ku* [169] again means the gnomon, *piao* [170]. The number of miles (*li* [171]) along the diameter and the circumference of the ecliptic can thus be worked out by mathematical principles. The rules for right-angled triangles [172] and for similar right-angled triangles [173] were used to calculate from (lit. make deductions [174] from)[5] the sun's shadow, the pole-star and the displacements (in right ascension and declination) [175], so that the distances (from one point to another) could all be obtained from the gnomon and the sides of the right-angled triangle.[6] This was taught to the people of the Chou dynasty, and it was therefore called *Chou Pei*.[7] If one has mastered these methods, the numerical relations (distances and movements) of the celestial bodies [176] can all be discovered.[8]

[1] I.e. of an astronomical clock powered by a water-wheel and escapement. But he was certainly not the first, for I-Hsing and Liang Ling-Tsan had done the same thing more than two centuries before him.

[2] This phrase is particularly intriguing for in the scheme of Su Sung as we have examined it there is no component part which is stated to depend on cords. Perhaps there had been some attempt to use a belt drive instead of the chain drive of the later version of Su Sung's clock. Or more probably the cords were part of the jack-work mechanisms.

[3] This is the second of the great Chinese mathematical classics. Wang & Needham (1) have recently given a particular study of one of the most interesting of its methods. See also Needham (1), vol. 3, pp. 24ff.

[4] This is the first of the great Chinese mathematical and astronomical classics. The *Chou Pei Suan Ching* (Arithmetical Classic of the Gnomon and the Circular Paths of Heaven) [716] had attained its present form before A.D. 80, and much of it probably dates from the second century B.C., but some parts, especially those which deal with astronomical questions, may go back to the time of Confucius (sixth century B.C.). For example, it contains a planisphere which has resemblances to those on Babylonian cuneiform tablets, and propounds the views of the 'Kai T'ien' or Heavenly Dome school of cosmologists (the most archaic of the three schools), for whom, indeed, it became the standard classical text. See Needham (1), vol. 3, pp. 19ff., 210ff., 256ff. A partial translation, now old but still useful, by Biot (2) may be consulted.

[5] This word is interesting and important in the Chinese terminology of scientific method. In modern times (and probably also in Su Sung's time) it meant inference or deduction; in ancient times, especially among the Mohists, it may have meant something like induction (cf. Needham (1), vol. 2, pp. 183 ff.).

[6] These methods were of course the standby of the Han surveyors (second century B.C. to second century A.D.).

[7] Note that Su Sung adopts an alternative explanation of the title of this classic.

[8] The cosmological computations of the ancient Chinese (e.g. second century B.C.) were of course vitiated by their failure to allow for the sphericity of the earth. This does not mean, however, that some of the schools

(7) (After thinking these things over,) I told (Han Kung-Lien) about the apparatus of Chang Hêng [480], I-Hsing [477] and Liang Ling-Tsan [481] and the designs of Chang-Ssu-Hsün [479], and asked him whether he could study the matter and produce similar plans.[1] Han Kung-Lien said that they could be successfully completed, if mathematical rules were followed[2] and the (remains of) the former instruments taken as a basis [177]. Afterwards he wrote a memorandum in one chapter entitled 'Verification of the Armillary Clock by the Right-Angled Triangle Method' [717][3] and he also made a wooden model of the mechanism with time-keeping wheels [178]. After studying this model [179] I formed the opinion that although it was not in complete agreement with ancient principles, yet it showed great ingenuity [180] especially with regard to the water-powered driving-wheel, and that it would be desirable to entrust him with the building of it. I therefore recommended to your Imperial Majesty that a (complete) wooden pilot model [181] should first be made and presented to you, and that some officials should be ordered to test its use. If the time-recording [182] proved to be correct, then a bronze instrument could be made. On the sixteenth day of the eighth month in the second year (of the Yuan-Yu reign-period A.D. 1087) your Imperial Majesty gave an order that my suggestion should be carried out and that a (special) bureau should be set up, officials appointed and the necessary materials prepared.

(8) I therefore recommended to your Imperial Majesty that Wang Yuan-Chih [482], Professor at the Public College of Shou-chow [854], formerly Acting Registrar of Yuan-wu [855] in Chêng-chow [856] Prefecture, should be in charge of the construction and the receipt of public materials; while Chou Jih-Yen[4] [483], Director of Astronomical Observations (Southern Region)[5] [922] of the Bureau of Astronomy and Calendar; Yü T'ai-Ku [484], Director of Astronomical Observations (Western Region) [923] of the same Bureau; Chang Chung-

of cosmologists did not affirm it. For instance, Chang Hêng [480] in his *Hun I Chu* [718] (Commentary on the Armillary Sphere), about A.D. 120, said: 'The heavens are like a hen's egg and as round as a crossbow bullet; the earth is like the yolk of the egg, and lies alone in the centre' (*K'ai-Yuan Chan Ching*, ch. 1, p. 4*b*) [719].

[1] This looks as if Su Sung and Han Kung-Lien had access to the detailed designs of Chang Ssu-Hsün if not of I-Hsing and Liang Ling-Tsan.

[2] Here is a very important instance of a conscious application of mathematical principles to the design of a machine in the Chinese middle ages.

[3] No doubt this was connected with calculating the declination of the sun throughout the year by measurements with gnomon, armillary sphere, and globe.

[4] Chou Jih-Yen had already built an armillary sphere in the Yuan-Fêng reign-period, between A.D. 1078 and 1085, which was supported by a cloud-and-tortoise column (*Hsin I Hsiang Fa Yao*, ch. 1, p. 20*a*; cf. Maspero (1), p. 320). As we shall shortly see, it was through a hollow column of this kind under the sphere that Su Sung led up his driving shaft to turn the diurnal motion gear ring (Main Text, paras. A, I and P, pp. 28, 40 and 47 below). He distinctly tells us at the first place that this device was new. Yet elsewhere (Main Text, para. F, on p. 36 below), he indicates that some elaborate horological gearing also had been set up during the Yuan-Fêng reign-period. There is the possibility therefore that a good deal of experimental work by Chou Jih-Yen and some assistants had been going on during the decade just preceding the reception of the imperial command by Su Sung. It may be significant that Chou Jih-Yen was the first technical collaborator named, along with Han Kung-Lien, whose name has come down to us more prominently.

[5] The actual Chinese titles of these regional officials embody the terms for spring, summer, autumn and winter (des Rotours (1), vol. 1, p. 213), but it is tolerably certain that these referred to the corresponding palaces of the sky (cf. de Saussure (1)), and that the officials themselves were responsible for observing phenomena, usual or unusual, in those sectors.

Hsüan [485], Director of Astronomical Observations (Northern Region) [924]; and Han Kung-Lien should be appointed to supervise the construction [183]. (I further recommended) the Assistants in the Bureau, Yuan Wei-Chi [456], Miao Ching-Chang [486], Tuan Chieh-Chi [487] and Liu Chung-Ching [488], and the Students, Hou Yung-Ho [489] and Yü T'ang-Ch'en [490], as investigators of the sun's shadow, the clepsydra and so on. (Lastly, I recommended) the Bureau of Works Foreman [925] Yin Ch'ing [491] to be Clerk of the Works.

In the fifth month of the third year of the Yuan-Yu reign-period (A.D. 1088) a small pilot model [184] was finished and at your Imperial Majesty's order presented for testing. Afterwards the full-scale machinery was built in wood [185] and completed by the twelfth month.[1] I (then) begged your Imperial Majesty to send a court official to the Bureau (of Astronomy and Calendar) to explain (the parts to the workmen) in preparation for moving the clock to the Palace for presentation....[2] In the tenth month we had sent in a request for instructions regarding the installation, and the Palace Guard Superintendent sent the Aide-de-Camp Huang Ch'ing-Tsung [492] (to look after the matter). On the second day of the twelfth month, a letter was sent up asking exactly where (the clock) was to be assembled, and your Imperial Majesty's order came to erect it in the Chi Ying Hall [926] (of the Palace). (9)

According to your servant's opinion there have been many systems and designs for astronomical instruments during past dynasties al .differing from one another in minor respects. But the principle of the use of water-power for the driving mechanism has always been the same [186]. The heavens move without ceasing but so also does water flow (and fall). Thus if the water is made to pour with perfect evenness [187] then the comparison of the rotary movements [188] (of the heavens and the machine) will show no discrepancy or contradiction [189]; for the unresting follows the unceasing. Thus it was that Chang Hêng in his *Hun T'ien* [720] (On the Celestial Sphere)[3] said[4] that (one instrument) should be set up in a closed room and rotated by water-power [190]. Those in charge of it were to report aloud from behind closed doors to those on the observation platform [191] who were examining the heavens, so that for example one star would be rising, another would be making its transit, and yet another would be going down—and all the readings of the instruments [192] would agree like the two halves of a tally.[5] (10)

In the K'ai-Yuan reign-period (A.D. 713–41), an edict was issued authorising the monk I-Hsing [477] together with Liang Ling-Tsan [481] an administrative official of the Crown Prince's Bodyguard, and many other men of art [929] (i.e. technicians), to remake and cast a bronze astronomical instrument [160]. This was formed in the image of the celestial sphere [194] and marked with the equatorial constellations and the degrees of celestial revolution. (11)

[1] There is nothing in the memorial which shows that the final clock, with all its bronze parts complete, was ever built, but we know this from other texts; see below, pp. 115, 116.

[2] There is a small lacuna here in the text.

[3] The more usual titles of this book are either *Hun I Chu* [718] (Commentary on the Armillary Sphere) or *Hun I T'u Chu* [721] (Commentary on the Diagrams of the Armillary Sphere). Su Sung was probably quoting absent-mindedly.

[4] Su Sung is here quoting from a part of Chang Hêng's treatise which has not come down to us. It cannot be found in any of the existing fragments. But the significance of his quotation is considerable, as we shall later appreciate (pp. 100 ff.).

[5] This passage, which seems to be in the words of Chang Hêng himself, is verbally identical with that in the *Sui Shu*, ch. 19, pp. 14*b*, 15*a* (astronomical chapter), the writer of which was presumably also quoting.

Water was made to fall and rotate the (driving-) wheel automatically, so that the heavens (as represented) made one rotation in one day and one night. And there were two other rings (lit. wheels) placed outside the (representation of the) heavens carrying images of the sun and the moon, and also moving round. As the (representation of the) heavens moved westward one rotation, the sun went eastward one degree and the moon 13 degrees and a fraction. After twenty-nine rotations[1] the sun and the moon met, and after 365 rotations the sun had completed one of its circuits. The whole instrument was half sunk in a wooden casing showing the horizon line. Two wooden jacks, placed in front of the 'horizon' with a bell and a drum, were made to strike the hours and quarters automatically. This was called the 'Water-powered Celestial Map (or Clock)' (lit. water-driven spherical bird's-eye view map of the heavens) [195]. After it had been completed it was set up in front of the Wu Ch'êng Hall [927] (of the Palace) to be seen by all the ministers and officials.

(12) The celestial globe of the Liang dynasty (Treasury) (A.D. 502–56) was made of wood, as round as a ball. Upon its surface were marked the twenty-eight equatorial constellations [196] and the stars of the three (fourth-century B.C.) schools of astronomers[2] ⟨i.e. Wu Hsien [495], Shih Shen [493] and Kan Tê [494]. The stars were distinguished by three colours, green, yellow and red.⟩, the ecliptic, the equator, the milky way and so on. In addition there was a horizontal annulus surrounding the outside of the middle of the globe to signify the horizon of the earth.[3]

(13) The astronomical (clock) [150] of Chang Ssu-Hsün [479] (A.D. 979) had a tower of several storeys (each) more than 10 ft. in height.[4] Inside there were wheels, shafts, posts for the escapement mechanism [197], and arrangements for turning the wheels by water-power. There were also jacks [198] which rang a small bell, beat a drum and struck a large bell. (The clock) made one complete rotation in one day and night. There were also 12 jacks [198] each in charge of one (double-) hour. When that particular hour arrived, a certain jack came out automatically to report holding a tablet, and thus they appeared one by one. The respective lengths of the day and night were fixed by the number of quarters. Since in the winter the water froze and the clock moved slowly (or stopped) water was replaced by mercury and thus no error could occur. Then there were also the sun, moon and stars arranged according to observations (lit. looking upwards). The old method was to set the indicators for the movements of the sun and moon manually every day, but now a new system was completed (by Chang Ssu-Hsün) in which the natural phenomena were followed (automatically), and therefore it was exceedingly delicate and beautiful.

[1] A lunation is 29·53 days.

[2] The stars in the medieval Chinese heavens were divided into three groups according to which of the fourth-century B.C. astronomers—Shih Shen [493], Kan Tê [494], or Wu Hsien [495]—had determined their positions. This was done, not out of any scruples concerning the history of science, but because the names of the three men were connected with three different systems of astrology. See Maspero (2); Ueta (1); and Needham (1), vol. 3, p. 263. The determinations of star positions are preserved in the *Hsing Ching* [722] (Star Manual), which constitutes the oldest extant star-catalogue of any civilisation—a list older than that of Hipparchus. Cf. Needham, *loc. cit.* pp. 197, 268.

[3] On this, see further in connection with Ch'ien Lo-Chih and Han Hsien-Fu (pp. 98 and 69 below). The full text which Su Sung here abbreviates is found in *Sui Shu*, ch. 19, pp. 17*a*, *b* (see the translation of Maspero (1), p. 353, though it is slightly incorrect).

[4] As already suggested, it is probable that the designs of Chang Ssu-Hsün's clock were accessible to Su Sung.

From what has been said above, (we can conclude that) the astronomical equipment [162] of (14) the Imperial Observatory [191] referred to by Chang Hêng included both a (demonstrational) armillary sphere [150] and an (observational) armillary sphere [158]. What was put in the closed room was the first of these [152].[1] That is why Ko Hung [496] said:[2] 'Chang P'ing-Tzu [497] (i.e. Chang Hêng, A.D. 78–139) and Lu Kung-Chi [498] (i.e. Lu Chi, third century A.D.)[3] both considered that in order to trace the paths and degrees of motion of the seven luminaries, to observe the calendrical phenomena and the times of dawn and dusk, and to collate these with the forty-eight *ch'i*;[4] to investigate the divisions of the clepsydra and to predict the lengthening and shortening of the shadow of the gnomon, and to verify these changes by phenological observations; there was no practical instrument more precise than the demonstra-

[1] The terminology here is very confusing. *Hun i* [150] has generally been understood invariably as 'armillary sphere' and assumed to be equipped with a sighting-tube for observations. But here that would leave *hou i* [158] 'observational instrument' without possible explanation. We feel quite justified, therefore, in interpreting the first term as '(demonstrational) armillary sphere' and the second as '(observational) armillary sphere'. Su Sung goes on to speak of the instrument kept below in the closed room as a *hun hsiang* [152], i.e. an instrument which has hitherto invariably been understood as a celestial globe, that is to say, a solid ball marked with the stars, constellations and great circles. We have come to the conclusion, however, that in early texts, e.g. before the third century A.D., *hun hsiang* often, if not always, meant the (demonstrational) armillary sphere, in which the centrally swinging sighting-tube was replaced by a model earth (spherical or flat and plate-like) supported on a pin from the south pole. This is proved by the statement of Wang Fan [478] about A.D. 250 (quoted in full below, p. 99), explaining his plan of removing the earth model from the centre of the armillary sphere and replacing it by a horizontal box-top outside to symbolise the earth (*CSHK*, San Kuo sect. ch. 72, p. 5*b*). It seems that the demonstrational armillary sphere half sunk in a box-like casing persisted at least until the instruments made by I-Hsing and Liang Ling-Tsan in the early eighth century (see p. 78 below). On the other hand the true celestial globe seems not to have originated before the time of Ch'ien Lo-Chih (A.D. 440; cf. pp. 96, 98 below), and it is significant that this occurred only after Ch'en Cho [499] about A.D. 310 had constructed for the first time a standard series of star-maps based on the *Hsing Ching* catalogue which had been begun in the fourth century B.C.

In any case Chang Hêng certainly had two instruments, an observational one outside on the platform, and a demonstrational (or computatory) mechanised one inside—always assuming the truth of the tradition. Su Sung, in his apparatus, mechanised both instruments, a sphere on the roof outside and a globe in the tower inside, in allusion to Chang Hêng's arrangement, but no longer with the same purpose, to make a magnificent clock. The armillary sphere, however, could still be used for observation as well.

[2] The quotation is from the *Chin Shu* [775], ch. 11, p. 3*b*, the text of which is here by preference followed. It is only part of a much larger record of Ko Hung's words. The sentences immediately following this one are of much importance, and will be given later on (see p. 100).

[3] A gloss gives the full names. In his biography (*San Kuo Chih*, ch. 57, p. 10*a*) Lu Chi [500] (fl. A.D. 220–45) is said to have written a *Hun T'ien T'u Chu* [723] (Commentary on Diagrams of the Celestial Sphere). But the fragments preserved in the *K'ai-Yuan Chan Ching*, chs. 1 and 2 (and in *CSHK*, San Kuo sect. ch. 68, pp. 6*a*, *b*, 7*a*) are entitled *Hun T'ien I Shuo* [724] (Discourse on the Armillary Sphere) and *Hun T'ien T'u* (Diagrams of the Celestial Sphere) [725]. They contain argument against the Kai T'ien (celestial dome) theory, reinforced by references to the *I Ching* [726] (Book of Changes), on which see Needham (1), vol. 2, pp. 304 ff. Lu Chi repeats Chang Hêng's comparison between the heavens as a bird's egg and the earth as the round yolk in the middle, from which we may infer that he knew the demonstrational armillary sphere with the model earth.

[4] Su Sung's text writes wrongly thirty-eight. These are the weekly periods which constituted the better-known twenty-four named fortnightly periods (*chieh ch'i* [199]); cf. *Hsiao Hsüeh Kan Chu*, ch. 1, p. 19*b*. There were also the seventy-two five-day periods called *hou* [200] (*loc. cit.* p. 20*a*). There were forty-eight indicator-rods, one for each of the forty-eight periods, used in the clepsydra each at the appropriate time of year (*loc. cit.* p. 25*a*). See Maspero (1), pp. 207 ff.

tional armillary sphere [152].' The celestial map clock of the K'ai-Yuan reign-period (I-Hsing's) was also a demonstrational armillary sphere [152]. Chang Ssu-Hsün [479], following the method of the K'ai-Yuan (instrument), indicated the North Polar Region [88] at the top and marked the degrees of celestial rotation around the equator. The instrument rotated from due east to west. These additions were a novelty of Chang's.[1]

(15) Now (in our new clock) we have adopted the theories and principles of all the schools and both instruments, the armillary sphere and the celestial globe, are included in a tower with two compartments. The armillary sphere is placed above (on the roof platform), with the celestial globe (in the upper compartment) just below it, while the machinery, with its time-keeping wheels and shafts [201] is hidden inside (the tower). The jacks which sound the bells and drums, and which report the time, are all fixed upon these wheels. There is also a wooden pagoda of five storeys with doors in front, where the jacks appear to perform their actions and then disappear. Water-power is used to drive the wheels which move the armillary sphere and the celestial globe. Thus we have a synthesis of the methods of the different schools.

(16) (Our) armillary sphere can be used to observe the degrees of the motion of the three luminaries (i.e. sun, moon and stars). The ecliptic is added in the form of a single ring.[2] The sun is seen half on each side of this ecliptic ring. The sighting-tube is made to point constantly at it so that its rays are always shining down the tube (lit. so that the body of the sun is always (sighted through) the lumen of the tube).[3] During one complete westward revolution of the heavens the sun moves eastward one degree. This device is a new invention.[4]

(17) (Our) celestial globe shows the Polar Region [88] at the north, the stars both north of (lit. within) and south of (lit. outside) the equator, the twenty-eight equatorial constellations (lunar mansions) and their extensions in degrees, the ecliptic [382], the equator [381] and the milky way, all depicted upon its surface. This system follows the statement of Wang Fan [478] and what was described in the (Astronomical) Chapter of the *Sui* (*Shu*).

(18) There are also pearls in different colours (lit. the five colours) denoting the sun, moon and five planets, threaded on silk strings attached at each end by hooks and rings to the south–north axis. Following the waxing and waning, tarrying and hurrying, stopping and retrograding, and all the motions of the seven luminaries, the pearls are made to occupy their corresponding

[1] We do not understand what Su Sung meant by this statement.

[2] This contradicts Su Sung's description of an improved split or double ecliptic ring (ch. 1, pp. 15*a*, *b*); cf. Maspero (1), p. 314, n. 5. We have no explanation of the discrepancy.

[3] This device is remarkably reminiscent of the Bumstead sun-compass, devised for use by aircraft in conditions where use of the magnetic compass is difficult. The Bumstead sun-compass was used around 1927 by Admiral Byrd in his flights from Spitzbergen to the North Pole and back, and also to and from the South Pole. It consists of a mechanical clock mounted in the equatorial plane, the dial of which carries round a perpendicular pin or gnomon the shadow of which is held continually on a translucent plate diametrically opposite. Since the time-piece, with its single pointer revolving once every 24 hr., is mounted upon a rotatable azimuth plate with graduations, the correct position of the points of the compass can easily be ascertained (see Hughes (1), pp. 117, 118). Of course, Bumstead knew no more of Su Sung than Su Sung could have known about his long-distant successor—and of course their objectives were different.

What exactly Su Sung's objective was is not too clear. Possibly he intended the procedure as a check upon the time-keeping of his clock, though he does not say so. In that case the extreme jerkiness of its movement would simply have brought the sun's rays again directly down the sighting-tube each time the machinery advanced by a distance corresponding to a single scoop on the driving-wheel.

[4] I.e. the mechanising of an *observational* armillary sphere.

positions [202].[1] They are rotated day and night following the movements of the heavens. An observer watching the pearls verifies whether the position of a luminary which they indicate agrees with what is observed and measured on the platform (above). If there is no difference (the calendar) is considered correct; (if there is a difference the calendar is adjusted). This system follows the statements of I-Hsing [477] and Chang Ssu-Hsün [479] with some modifications. Both the instruments (i.e. the globe and the sphere) are worked by water-power and not by human force. Though we do not know whether their precision is better than those of former times, yet it seems that the general principles of the instrument design (lit. specifications and calculations) agree with what has been handed down from them (I-Hsing and Chang Ssu-Hsün).

Four clepsydras[2] have been used (for checking the clock): (1) Floating Arrow (inflow) clepsydra [203].[3] (2) Steelyard, or balancing, (inflow) clepsydra [204].[4] These two are more or less the same as those now used in the Bureau of Astronomy and Calendar [917] and the Dawn Reception Hall [928] (of the Palace). (3) Sinking Arrow (outflow) clepsydra [205]. (4) Non-stop clepsydra [206].[5] For these last two instruments the designs and methods of the technical experts [930] were adopted. All four were set up in a different room. A clepsydra clerk [931] was asked to check the (mechanised) armillary sphere and celestial globe hour by hour and quarter by quarter against them. If (the mechanism) showed agreement with the movements of the stars in the heavens it could be considered correct. This was to test whether the instrument and its calculations showed any difference in comparison with the movements of the heavens, and therefore whether the seasons of heat and cold would also be correctly predicted.[6] (19)

The *Yü Shu* (in the Shun Tien, ch. 2 of the *Shu Ching* [811]) says:[7] '(The Emperor Shun) used the sighting-tube and the circumpolar constellation template [162, 163][8] to investigate the movement of the Seven Directors (sun, moon and five planets).' This means that by observing the meridian transits of forty-eight[9] stars the earlier and latter portions of the (twenty-four fortnightly) periods can all be known. The (*Shang Shu Wei*) *K'ao Ling Yao* [728][10] says: 'By observing the motions of the jade instrument [208],[11] the hours of the day and night are dictated (20)

[1] Presumably by manual setting. The principle, or something very like it, had been used long before (cf. p. 106) for studying the relations of stellar and planetary motions, i.e. the basis of the calendar.

[2] It is interesting that the technical encyclopaedia *Hsiao Hsüeh Kan Chu* [727] written by Wang Ying-Lin [501] in the latter part of the thirteenth century but not printed till A.D. 1299, where everything is listed according to numerical categories, also lists (ch. 1, p. 32*b*) four main types of clepsydra as in use in the Sung. Another list of four is given in the *Sung Shih* [705], ch. 76, p. 3*b*, in a passage which we shall discuss later (p. 90). We shall give details of the types of water-clocks at that place.

[3] See Maspero (1), pp. 185 ff.; Needham (1), vol. 3, pp. 315 ff.

[4] Emending *pai* [207] to *ch'êng* [204]. On this see p. 90 below.

[5] On this see p. 90 below.

[6] Again the calendrical preoccupation with the sidereal and tropical years and the lunations.

[7] See Karlgren (1), p. 4.

[8] See n. 2 on p. 18 above.

[9] Emending the text from forty-seven.

[10] This book, existing now only in fragmentary form, is one of the so-called 'Weft Classics' or Apocryphal Treatises which appeared in the Han period, particularly the Later Han (cf. Needham (1), vol. 2, p. 380). The best account of them in English is that of Tsêng Chu-Sên (1). They contained much material concerning prognostication and divination, mantic statements and magical lore. Naturally they also contain material of scientific interest, e.g. records of gnomon shadow lengths. The title of this particular treatise may be translated 'Apocryphal Treatise on the Historical Classic; the Investigation of the Mysterious Brightnesses'.

[11] This is a reference to the legend that armillary instruments were known at the beginning of the Chou period (*c.* eleventh century B.C.) and that in those times they were made of jade, not bronze.

by the culmination of the constellations [209].[1] If the sphere (whether mechanised or not) [162] indicates a meridian transit when the star (in question) has not yet made it (the sun's apparent position being correctly indicated), this is called "hurrying" [210]. When "hurrying" occurs, the sun oversteps its degrees, and the moon does not attain the *hsiu* (mansion) in which it should be. If a star makes its meridian transit when the sphere has not yet reached that point (the sun's apparent position being correctly indicated), this is called "dawdling" [211]. When "dawdling" occurs, the sun does not reach the degree which it ought to have reached, and the moon goes beyond its proper place into the next *hsiu*.[2] But if the stars make their meridian transits at the same moment as the sphere, this is called "harmony" [212]. Then the wind and rain will come at their proper time, plants and herbs luxuriate, the five cereals give good harvest and all things flourish.' From this we may conclude that those who make astronomical observations with instruments are not only organising a correct calendar so that good government can be carried on (i.e. the administration of an agricultural society) but also (in a sense) predicting the good and bad fortune (of the country) and studying the (reasons for) the resulting gains and losses.[3] As the Book of Changes (*I Ching*) says: 'We may be ahead of Heaven but Heaven will not violate its course; we must follow and adapt ourselves to its times and seasons' [213].[4] This is just the point of what we have been discussing. Now we have followed the example of the Monthly Ordinances (*Yüeh Ling*) [729][5] and made for the first time

[1] Here the text of Su Sung differs slightly, but not essentially, from *Ku Wei Shu* [767], ch. 2, p. 2*a*.

[2] These four sentences as given by Su Sung make the 'hurrying' and 'dawdling' refer to the astronomical instrument. This fully agrees with the quotation of the same text given in the *Sui Shu* [714], ch. 19, p. 13*b*—an authority of weight, since its astronomical chapters had been written by Li Shun-Fêng [502] as early as A.D. 656. But a diametrically opposite sense is conveyed in the version collected in *Ku Wei Shu*, ch. 2, p. 2*b*, where the 'hurrying' and 'dawdling' refer to the stars. This fragment collection was made by Sun Chio [503] in the Ming. From all the context his version must be rejected. But unfortunately it misled Maspero (1), p. 338, into supposing that the chief use of the mechanised instruments of the Chinese was for astrological purposes. This idea must also be abandoned.

What the Chinese astronomers were worrying about was not the speed of diurnal rotation of their mechanised globes or demonstrational spheres, but rather divergences or discrepancies between the indicated positions of the stars and of the sun or moon. Their instruments were probably made to rotate not exactly once in twenty-four hours, but slightly faster, so that the sun, having a slow annual motion backwards, appeared to make a daily rotation. The difference between stars and sun was, after all, the essential astronomical element of the calendar.

[3] Here Su Sung's argument is quite logical and scientific in that he is suggesting that misfortunes such as famine come from bad harvests, and these in turn may originate from ill-timed planting, brought about in its turn by incorrect calendar-making. He is really giving a rationalistic explanation of the astrological functions (which had always been primarily concerned with State, not individual, astrology) traditionally associated with the Astronomer-Royal and his department. Of course, this does not mean that in ancient and early medieval times there had not been a mass of astrological supersitition connected with the work of the astronomers—of this we can assure ourselves by reading such a text as the astronomical sections of the *Chin Shu*.

[4] This quotation from the Wên Yen Commentary on the *kua* [233] (trigram) 'Ch'ien' in the *I Ching* (Book of Changes) is really almost a pun. The correct meaning of the original words is: 'The great man may act in advance of Heaven but Heaven will not contradict him; he may follow Heaven, but his action will be in accordance with the times of Heaven.' See Legge (2), p. 417; R. Wilhelm (1) in Baynes' tr. vol. 2, p. 15. Su Sung adapts the 'scriptural' text to his purpose, availing himself of its laconic ambiguity.

[5] The 'Monthly Ordinances' is an ancient calendar which details month by month the culminating stars, seasonal 'works and days', imperial ceremonies, and associated musical notes, numbers, tastes, sacrifices, etc. It may date from about the fifth to the seventh century B.C. It was eventually incorporated into several Ch'in and Han books (cf. Needham (1), vol. 3, pp. 195 ff.).

a set of Diagrams of the Culmination of Constellations in the Four Seasons [730] with their degrees at dawn and dusk. This is placed at the end of the chapters for your Imperial Majesty's inspection.[1] These are the main uses of our astronomical clock (lit. armillary sphere and celestial globe).

Thus as we have seen, the (demonstrational) armillary sphere [154], the bronze (observational) armillary sphere [158] and the celestial globe [161] are three things different from one another. In these matters the statements of scholars of former times are not always completely to be trusted. According to Ch'en Miao [504] (third century A.D.) the instrument made by Chang Hêng included only a celestial globe [152] with the seven luminaries.[2] On the other hand, Ho Ch'êng-T'ien [505] (early fifth century A.D.)[3] was (apparently) unable to distinguish the difference between the armillary sphere [150] and the celestial globe [152].[4] In any case if we use only one name, all the marvellous uses of (the three) instruments cannot be included in its meaning. Yet since our newly-built instrument embodies two instruments but has three uses, it ought to have some (more general) name such as 'Cosmic (Engine)' ('Hun T'ien [156]'). We are humbly awaiting your Imperial Majesty's opinion and bestowal of a suitable name upon it.[5]

Presented by (Sgd.) SU SUNG [450]
Official of the Second Titular Rank (title) [907],
President of the Ministry of Personnel [932],
Imperial Tutor to the Crown Prince [933],
Grand Protector of the Army (title) [908],
K'ai-Kuo Marquis of Wu-kung (title) [909]

[1] As we shall see, this section may have been the only part of Su Sung's work which was conserved in the Sung empire after the loss of the capital in 1126 and the retreat to the southern provinces (p. 130; cf. p. 13 and Fig. 70).

[2] Was Su Sung quoting from memory from the *Sui Shu* here? For the *Sui Shu* [714], as we have it today, seems not to make Ch'en Miao (a scientific worthy of whom very little is known) say this. The quotation from him is short. The relevant text runs as follows (ch. 19, p. 17*b*): 'The Astronomer-Royal of the Wu Kingdom, Ch'en Miao, said: "The worthies of old made an instrument of wood which was called *Hun T'ien* [156] (Celestial Sphere)." Was he not referring to this? (i.e. the celestial globe). According to these words, the armillary sphere [150] and the celestial globe [152] were two different things, having no relation to each other. Thus what Chang Hêng made was a celestial globe [152] (or demonstrational armillary sphere), with the (models of the) seven luminaries. But Ho Ch'êng-T'ien [505] could not distinguish the difference between the armillary sphere and the celestial globe; thus he went astray.' Thus according to our reading the remark about Chang Hêng was due to the writer of this part of the *Sui Shu*, and was not made by Ch'en Miao.

[3] His *Lun Hun Hsiang* [731] (Discourse on the Sphere and Globe) is preserved in part in *CSHK* (Sung sect.), ch. 24, pp. 2*b* ff. It does seem to use a rather fluctuating terminology. This is interesting in view of the gradual change already mentioned (n. 1 on p. 23 above) from demonstrational armillary spheres to solid globes.

[4] These sentences of criticism are taken from the *Sui Shu*, ch. 19, p. 17*b*.

[5] Inaudible echo must have answered 'A clock!' But history does not record that the young emperor had any good ideas on the subject. The usual etymology relates the word clock to Ger. *glocke* and Fr. *cloche*, the timepiece thus taking its name from the bell or bells which it struck, and thus distinguishing itself from the medieval *horologium*, which had included all kinds of sundials and water-clocks which told the time silently. When the Jesuits reached China at the end of the sixteenth century their mechanical clocks attracted great attention (a story which we shall tell briefly in its place, pp. 142 ff. below), and a special term, i.e. *tzu ming chung* [214], self-sounding bell, was coined for them. This helped to create the illusion that China had had no mechanical clocks at all before the time of the Jesuits.

IV

TRANSLATION OF SU SUNG'S THIRD CHAPTER

THE WATER-DRIVEN ARMILLARY (SPHERE) AND CELESTIAL (GLOBE) TOWER [1]

A Ch. 3, text pp. 2*b*–3*b*, figure p. 2*a* (Fig. 5)

A square tower is built in two storeys, narrower at the top than at the bottom, and of a height appropriate to its location.[1] Large baulks of wood are used for the four corner posts, and these are linked by stout struts of sappan wood and walled around with boarding over the wooden frame. There are two lots of flooring, one for the upper platform (i.e. the roof) and one for the upper room, and around the top of the structure a balustrade [2] is set up. There are openings at the front and at the back of the upper room (i.e. to reveal the celestial globe) and two openings (for the pagoda and the constant-level tank; see Figs. 9 and 17) facing the front in the lower room.[2] ⟨A variant text says that there is one opening at the front and one on either side of the upper room, and two double-doored openings at the front of the lower room.⟩

The armillary sphere (Fig. 6) is placed on the platform above the upper room. It comprises three sets of concentric rings, the outer ones encircling the inner, as follows (cf. Fig. 24):

(i) Component of the Six Cardinal Points [3] (i.e. north, south, east, west in the horizon plane, zenith and nadir).

(ii) Component of the Three Arrangers of Time [4] (i.e. meridian, ecliptic, equator).

(iii) Component of the Four Displacements [5] (i.e. polar-mounted declination-ring, carrying the sighting-tube).

The armillary sphere is protected by a wooden roof. The component for the six cardinal points consists of a split-ring meridian circle [6] or celestial ring [7] set in the vertical plane, and a terrestrial single ring [8] making a horizon circle [9] set in the horizontal plane. Near the south (pole) of the component for the three arrangers of time is fitted the diurnal motion gear-ring [10] (see Fig. 24). ⟨This diurnal motion gear-ring is a new device.⟩

The celestial globe [11] and its wooden casing [12] (Fig. 7) are installed in the middle (i.e. upper) room of the tower. The celestial globe also has a split-ring meridian circle [13] which is fixed in the wooden housing in a vertical plane, half above the horizon (circle) and half hidden underneath it. The single-ring horizon circle [14] is fitted on the top of the wooden casing. ⟨In joining a casing to the celestial globe we now imitate the new invention given in the astronomical chapter of the *Sui Shu* (+656).⟩

[1] The dimensions cited consistently in the text indicate a height of at least 25 ft. without the armillary sphere which must add a further 10 ft. on the top of the tower.

[2] Fig. 8 shows a staircase which is needed to obtain access to the upper storey; it is not, however, mentioned in the text.

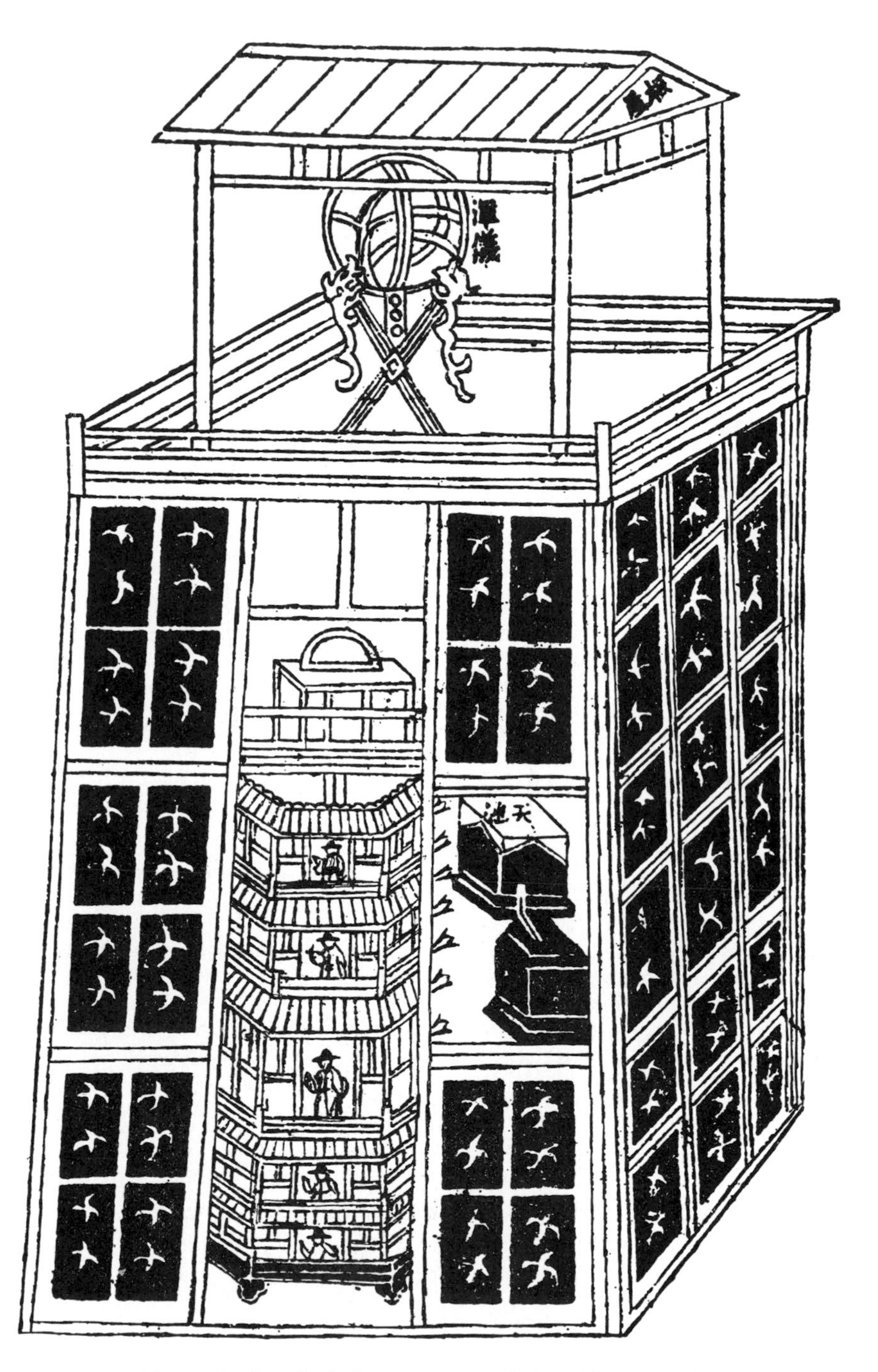

Fig. 5. Su Sung's clock-tower; general view (ch. 3, p. 2*a*).

Inside the tower, mounted one above the other on the time-keeping shaft [15] are eight wheels for time-keeping by day and night [16]. The first, which is situated above the upper bearing beam [17] (see Fig. 10) is called the celestial wheel [18] and engages with the teeth of the equatorial gear-ring [19] of the celestial globe.[1] The second wheel is for striking the (equal

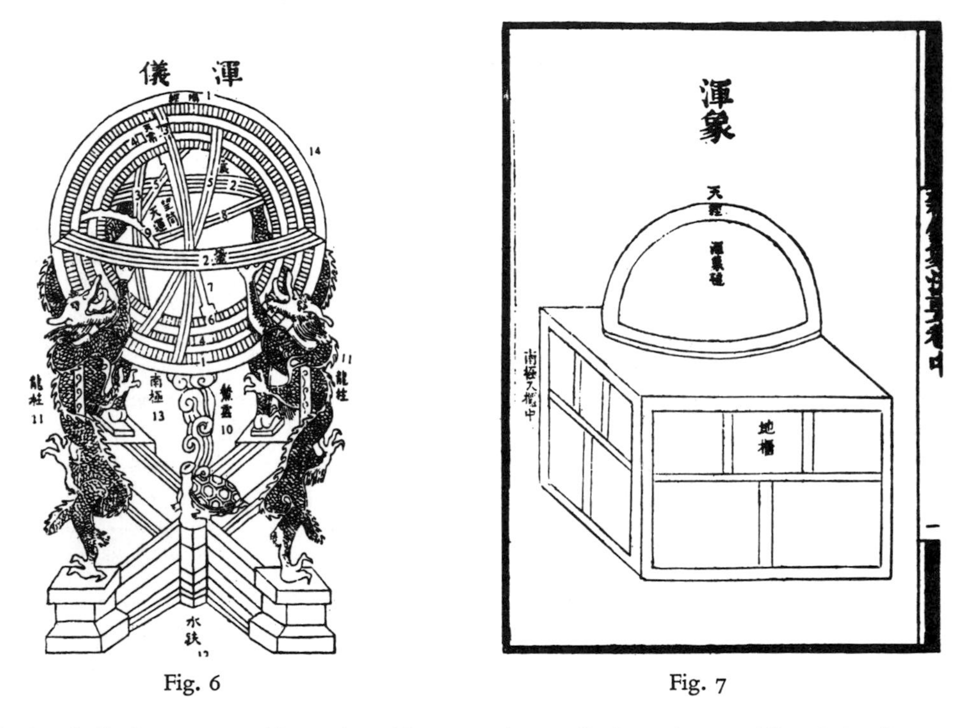

Fig. 6　　　　Fig. 7

Fig. 6. Su Sung's clock-tower; armillary sphere (ch. 1, p. 6*a*). 1, split-ring prime meridian circle of outer nest; 2, single-ring horizon circle of outer nest; 3, single-ring equatorial circle of outer nest; 4, split-ring solstitial colure circle of middle nest; 5, split-ring ecliptic circle of middle nest; 6, split-ring polar-mounted declination circle (inner nest); 7, sighting-tube; 8, diametral brace or cross-struts; 9, diurnal motion gear-ring connecting the middle nest with the power-drive; 10, cloud-and-tortoise column, concealing the transmission shaft; 11, four supporting dragon columns; 12, cross-piece of the base, incorporating water-levels; 13, south celestial pole; 14, north celestial pole.

Fig. 7. Su Sung's clock-tower; celestial globe (ch. 2, p. 1*b*).

double-) hours of the day by bells and drums [20]. The third wheel is for striking 'quarters'[2] [21] by bells and drums [22]. The fourth wheel is for the hourly jacks [23]. The fifth wheel is for the jacks exhibiting 'quarters' [24]. The sixth wheel is for striking the night-watches by gongs [25]. The seventh wheel is for the jacks exhibiting the night-watches [26]. The eighth and lowest wheel is for the float indicators of the night-watches [27].

[1] This is an alternative scheme (first variant) of the globe drive.

[2] The 'quarters' are the 100 *k'o* into which the 12 double-hours are divided; each is therefore 14 min 24 sec. in duration. See Appendix, p. 199.

All this mechanism is enclosed in an (ornamental) pagoda of five half storeys (i.e. fractions of the single lower story housing the pagoda), each one having a doorway through which the wooden jacks are seen to appear and disappear. On the first storey (from the top) a bell rings on the left, a large bell strikes on the right and a drum is beaten in the middle. On the second storey the beginnings and middles of (double-) hours are exhibited. On the third storey the 'quarters' are exhibited. On the fourth storey a gong is struck for the night-watches. On the fifth storey the night-watches are exhibited.

The great wheel [28] is mounted (in a vertical plane) behind the eight wheels mentioned above. It has seventy-two spokes [29] and thirty-six scoop-holders [30][1] mounted on the three reinforcing rings [31] of the wheel, and water is received into the thirty-six scoops [32] so supported. The hub [33] is mounted on an iron driving-shaft [34] which lies in a horizontal plane and runs in a direction from back to front of the casing. The front end of this shaft carries the earth wheel [35] which enmeshes [36] with the lower wheel [37] on the (vertical) transmission-shaft [38]. The middle (wheel) [39] rotates the time-keeping wheels [40], and hence the celestial globe, while the upper (wheel) [41] rotates the armillary sphere. ⟨A variant text says that the great wheel is fixed behind the eight (time-keeping) wheels, and that it has ninety-six spokes and forty-eight scoop-holders held between the reinforcing rings to receive water in forty-eight scoops. Through the hub goes the iron horizontal driving-shaft, running from back to front. At the middle of the front section of the shaft (i.e. between the great wheel and the earth wheel) is the lower chain wheel [104] which moves the celestial ladder [102] (chain-drive) and thus the armillary sphere above. (This is variant II of the armillary drive.) At the end (of the driving-shaft) the earth wheel moves the time-keeping gear-wheel and the celestial globe above it.⟩

To the left (the right as seen from the front) of the great wheel [37] there is situated the upper reservoir [42] and the constant-level tank [43], the latter receiving water from the former and passing it on to the scoops [44] whieh provide the motive power for the great wheel. The scoops discharge water into the sump [45] which has a hole in the bottom at the back. Through this hole the water is conveyed to the lower reservoir [46], from which it is raised by the lower noria [47] to the intermediate reservoir [48]; in this dips the upper noria [49] which turns with the manual wheel [50]. In this manner the upper and lower norias are both rotated to raise water into the upper flume [51] through which it flows finally into the upper reservoir so that the cycle can be repeated again and again.

⟨⟨A variant text says that at the south (pole) of the component for the three arrangers of time is fixed the diurnal motion gear-ring [10]. The celestial globe and its casing [12] are placed in the upper room. The celestial globe is half hidden under the horizon and there is a split-ring horizon circle round it. Beyond the body of the celestial globe is fixed the celestial globe drive-wheel. ⟨This device followed the new invention described in the *Sui Shu*.⟩ Inside the tower, the time-keeping wheels are fitted in horizontal planes.⟩⟩

The mechanism of the whole instrument, including the celestial globe, is as follows:

B Ch. 3, text pp. 4*b*–5*a*, figure p. 4*a* (Fig. 8)

Set up first of all the great wheel [28] and the eight wheels for time-keeping by day and night [40]. Four main posts [52] are erected within the tower and carry the driving-shaft [34];

[1] The great wheel consists of a pair of thirty-six spoked wheels side by side. Each pair of spokes supports a scoop-holder on which rests a scoop. The scoop-holders are not described in detail but they were probably a simple shelf fixed to a spoke on either side.

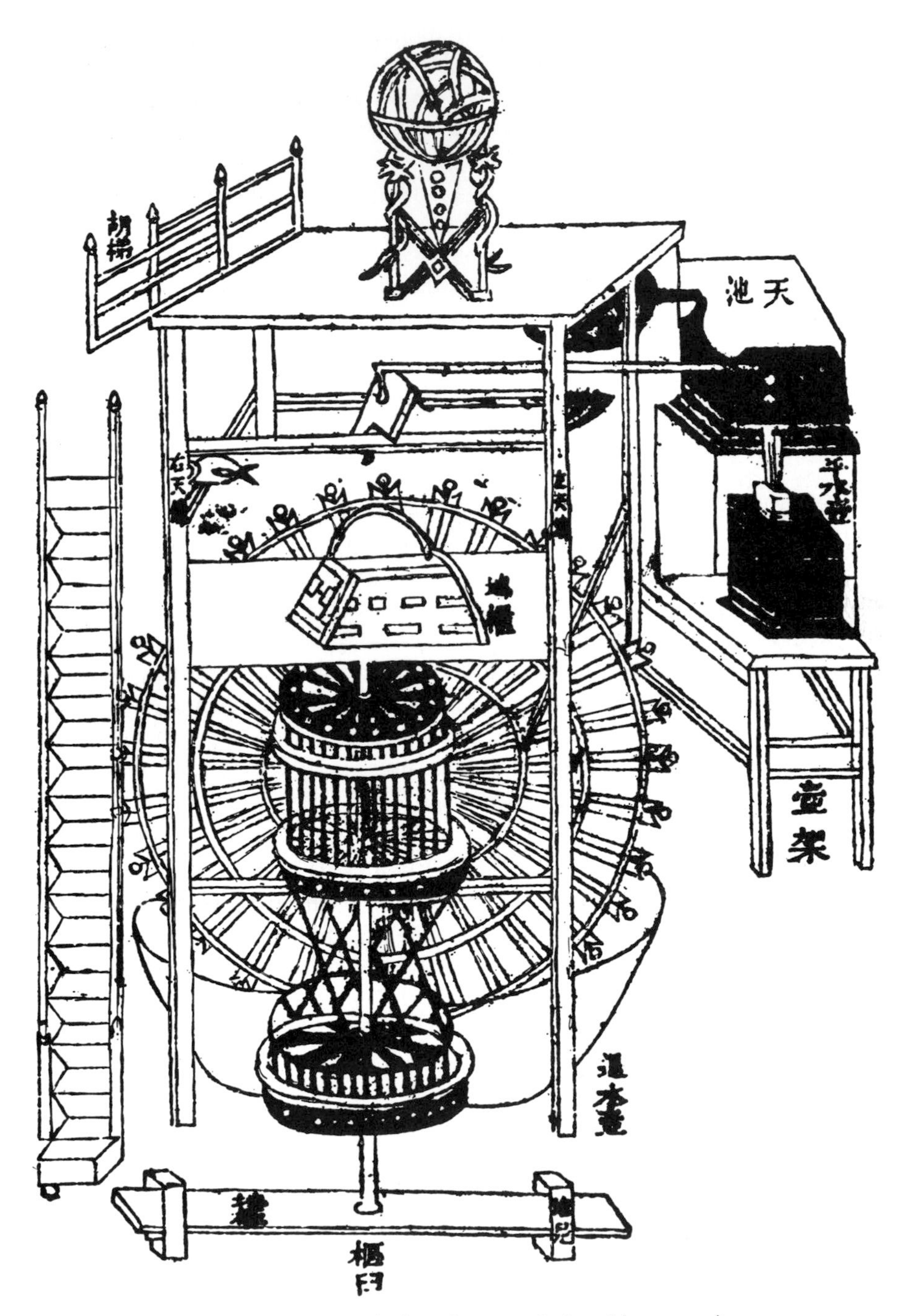

Fig. 8. Su Sung's clockwork; general view (ch. 3, p. 4*a*).

they are joined together by two main beams [53] running from left to right at the front and the back. There are also two upper beams [54] which are fixed to the posts and main beams and carry the upper stopping device [55].[1] At the left and right are fixed the celestial poles [56] (i.e. framework members), south and north, the south 'pole' being at the front of the platform between the two left-to-right beams, and the north 'pole' piercing the main post [52] below the left and right upper beams but above the main beams.[2] The time-keeping shaft [15] stands right in the middle of the tower. An upper bearing beam [17], constructed from two horizontal pieces of wood joined (side by side), has its two ends resting on the left- and right-hand celestial poles. It is placed under the celestial wheel and above the time-keeping gear-wheel [57]. A hole is made through the upper bearing beam to take the (top of) the shaft. At the bottom of the shaft there is a horizontal lower bearing beam [58] supported by two stands (one each) at its left and right ends [140]. In the centre of the lower bearing beam there is an iron mortar-shaped end-bearing [59] to take the pointed cap [60] (at the end) of the main shaft; this cap also being made of iron.[3]

The upper reservoir is to the left (right as seen from the front) of the main posts, and the constant-level tank [43] is just in front of the upper reservoir; they are both supported by wooden stands [61] (see Fig. 17). The constant-level tank is connected to the upper reservoir and has a hole in the bottom[4] through which the water flows out to the great wheel and its scoops. The sump is situated below the great wheel. The upper and lower norias with their respective tanks and the manual wheel (see Fig. 21) are all placed under the main beam and the upper beams and on horizontal bars.

The eight time-keeping wheels share the same shaft, and when the time-keeping gear-wheel is made to turn, all the other seven wheels revolve. The double great wheel [28] (see Fig. 15), which is mounted on a single hub with the thirty-six scoop-holders [30] and scoops [44] attached to its rim, receives water and is thus turned.

The upper balancing lever [62] (see Fig. 18) is fitted above the great wheel and carries the upper stop [55] at its head [63], and the upper weight [64] at its tail [65]. A connecting rod or chain [66] (hangs down) at the front of the weight (i.e. between fulcrum and weight) holding the trip lever [67] which is for setting the upper stop [68]. The lower balancing lever [69] is situated above the trip lever, and at its head is the checking fork [70], so arranged as to engage (in turn each) scoop. Furthermore, the lower weight [71] hangs at the end (of this lever) and is so adjusted as to regulate the rising and falling of the scoops. Two upper locks [72], one on either side, are put to the left and right between the main posts, lying on the beam and controlling the even movement of the great wheel.[5]

[1] See Fig. 18. In fact, the upper beams do not carry the upper stopping device itself but the upper balancing lever to to which it is affixed.

[2] The exact location and function of these 'celestial pole' peices is uncertain.

[3] Such a special heavy-duty bearing is clearly necessary to take the considerable weight—perhaps half a ton—of the time-keeping assembly.

[4] This appears to be a variant arrangement; elsewhere in the text it is stated that a siphon is used instead of a simple hole.

[5] The nature of these upper locks is not quite clear. They may be stops preventing the upper balancing lever from being displaced too high or too low.

C Ch. 3, text p. 6*a*, figure p. 5*b* (Fig. 9)

A wooden pagoda of five storeys is situated in front of the time-keeping wheels. At the first (i.e. top) storey, a wooden jack rings a bell on the left at the beginning of each (double-) hour, beats a drum at the centre at each 'quarter', and strikes a large bell on the right at the middle of each (double-) hour. At the second storey, a wooden jack appears exhibiting the beginning and middle of each double-hour. At the third storey, a wooden jack appears exhibiting the hundred 'quarters' throughout the 12 double-hours. At the fourth storey, a metal gong strikes the night watches. At the fifth storey, wooden jacks are distributed so as to exhibit the night watches by arrows (i.e. clepsydra indicator-rods).

D Ch. 3, text p. 7*a*, figure p. 6*b* (Fig. 10)

There are eight time-keeping wheels. The first (uppermost) is called the celestial wheel, and it engages [92] the toothed equatorial gear-ring [19] of the celestial globe (i.e. first variant of globe drive). The second is called the time-keeping gear-wheel [57] and (because of) the teeth [79] which it bears it follows the revolutions of the middle wheel of the transmission shaft, and so turns with it all the other seven wheels from top to bottom. The third is the wheel for striking 'quarters' by bells and drums [22]. Above this wheel are the trip-lugs [91] for striking the large bells, small bells and drums which mark the beginnings and middles of the (double-) hours and (also mark) the hundred 'quarters'. The fourth is the wheel for the hourly jacks. It contains twelve jacks [93] (which appear at the beginnings of each double-hour) and another twelve jacks (which appear at the middles of each double-hour). The fifth is the wheel for the 'quarter' jacks (i.e. those which appear at the hundred 'quarters'). The sixth is the wheel for striking the night-watches by gongs. This wheel has more trip-lugs [91] so arranged as to strike the gongs during the night. The seventh is the wheel for jacks exhibiting the night-watches. This has jacks which indicate sunrise and sunset, twilight and dawn, and the night-watches. The eighth is the wheel for the float indicator-rods keeping the night-watches; this intercepts [94] the gong [95].[1] Starting from the wheel for the float indicator-rods, there are eight wheels, one above the other on the same shaft which runs from the upper bearing beam down to the iron mortar-shaped end-bearing at the base. The five-storey pagoda conceals these wheels externally. ⟨A variant text does not give the equatorial gear-ring, but substitutes the celestial globe drive-wheel (as the form of globe drive employed).⟩

Ch. 3, text p. 8*a*, figure p. 7*b*

The time-keeping shaft [15] passes through a hole in the upper beam [17] and its lower end is supported by the mortar-shaped end-bearing [59] on the lower beam [58] (of the framework base [140]). Thus are fixed the eight time-keeping wheels [16].

E Ch. 3, text p. 8*b*, figure p. 8*a* (Fig. 11)[2]

The celestial wheel [18] is 3·8 ft. in diameter and has 600 teeth [79]; the iron shaft goes through it. This wheel and the celestial globe drive-wheel [81] engage with [82] each other (through the intermediary celestial axle) above the upper bearing beam. In front, the rim of the (celestial) wheel engages with [83] the celestial idler [84] (see Fig. 23). This (little) wheel is to move the celestial globe drive-wheel. The latter faces the South Pole in a slanting position, looking like a sloping cover. The teeth (of the celestial globe drive-wheel) enmesh with [85] the celestial idler to move the celestial globe. Hence the shaft of the time-keeping wheels is extended upwards to the wheel for the celestial movement. When the wheels are moving, the celestial wheel turns to the west (counter-clockwise as seen from above) and so the celestial idler turns to the east (clockwise as seen from the front), thus making the celestial globe drive-

[1] It is not clear how this eighth wheel functions and 'intercepts' the gong.

[2] This is an inserted passage related to **P, Q, R, S.**

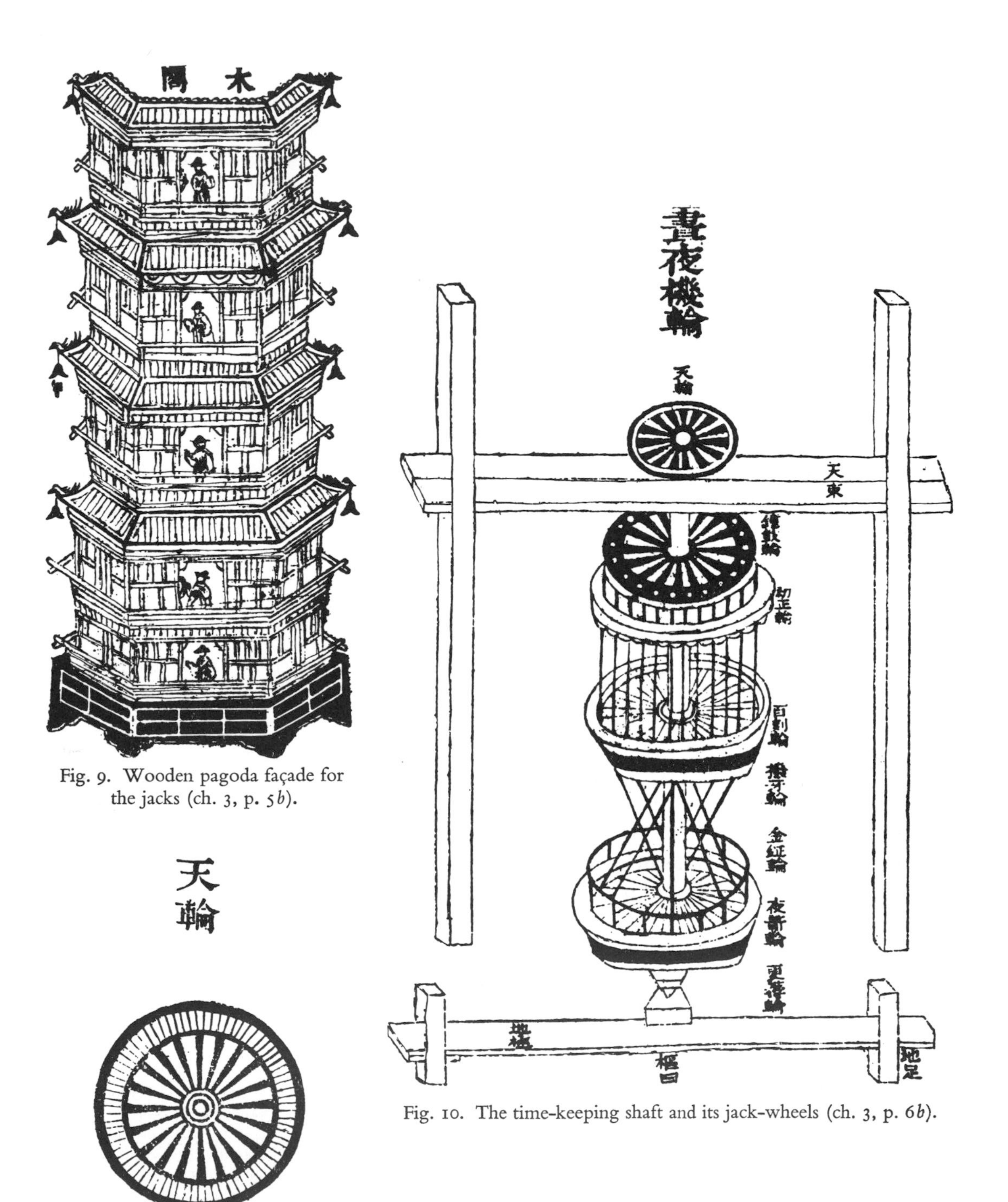

Fig. 9. Wooden pagoda façade for the jacks (ch. 3, p. 5*b*).

Fig. 10. The time-keeping shaft and its jack-wheels (ch. 3, p. 6*b*).

Fig. 11. 'Celestial wheel' at the top of the time-keeping shaft; it powers the rotation of the globe (ch. 3, p. 8*a*).

wheel and the celestial globe itself turn to the west (clockwise as seen from its North Pole) both together.

F Ch. 3, text p. 9*a*, figure p. 8*b*

The time-keeping gear-wheel follows the rotation of the middle wheel of the transmission shaft. It is situated above the wheel for striking the hours of the day by bells and drums. It has a diameter of 6·7 ft. Under this wheel are attached 600 teeth [79] which serve[1] those on the middle wheel. The middle wheel moves it six teeth for every 'quarter', fifty teeth for every (double-) hour, and the whole 600 teeth in every 12 (double-) hours. This system dates from the Yuan-Fêng reign-period (A.D. 1078–85).[2]

⟨⟨Another text says that the time-keeping gear-wheel is directly in front of the great wheel and is situated at the third floor of the pagoda. It is combined with [97] the wheel for the 'quarter' jacks, and has a diameter of 6·7 ft. Underneath are 600 teeth which carry the motion from the great wheel, each movement of which is served by[1] six spur teeth.[3]⟩⟩

G 1 Ch. 3, text p. 10*a*, figure p. 9*b*

The first (top) storey of the pagoda has three doors. At the beginning of each (double-) hour a jack wearing red rings the bell at the left-hand door. At the 'quarters' a jack wearing green beats the drum at the middle door, and at the middles of the (double-) hours one wearing purple strikes the large bell at the right-hand door.

G 2 Ch. 3, text p. 10*b*, figure p. 10*a* (Fig. 12)

The wheel for striking the (double-) hours of the day by bells and drums is situated on the first (top) storey of the pagoda. It has a diameter of 6·7 ft. and agrees in its rotation with the

[1] 'Serve' and 'served by' translate *tai* [141], which means literally 'waiting for the movement of' something else, and so moving. The term is strikingly reminiscent of that famous passage in *Chuang Tzu* [815] (*c.* 290 B.C.) where Penumbra is discussing with Shadow (ch. 2; tr. Legge (3), vol. 1, p. 196). 'Do I have to wait for something else to move, in order that my movement may come about?' Cf. the discussion in Needham (1), vol. 2, p. 51.

[2] This is a very curious remark, for it indicates that some work was going on concerning horological gear-wheels five years or so before Su Sung received the imperial command to construct a clock-tower. Was there about 1080 some important clock of which we otherwise know nothing?

[3] The only other information in the text concerning numbers of teeth on the gear-wheels is found, not in ch. 3, but in ch. 1, p. 17*a*. The text there reads as follows: 'The diurnal motion gear-ring [10] is also a new invention. It is attached to the component of the three arrangers of time and fitted well below (lit. to the south of) the ecliptic (ring). On its outside surface are 478 teeth [79] which engage with the armillary sphere shaft [77] in the tortoise-and-cloud (column). When one tooth of (the earth wheel [35] on) the driving shaft [34] turns, it moves (the lower wheel [37] on) the transmission shaft [38] by one tooth, and correspondingly the diurnal motion gear-ring by one tooth. When the latter is moved, the component of the three arrangers of time moves with it. This signifies the incessant motion of the heavens. Because the old component of the three arrangers of time had no water-power drive, this ring did not exist. It is now for the first time employed. The (number of) 478 teeth is derived from expressing the number of mean solar ($365\frac{1}{4}$) days in the year as an integral factor (*chou t'ien tu fên chih fa* [142]).

⟨⟨Another version says that the diameter of (the diurnal motion gear-ring) is 4·145 ft., its width 0·19 ft. and its thickness 0·07 ft. It is attached to the component of the three arrangers of time well south of the ecliptic. On the outer surface there are 600 teeth, etc. The (number of) 600 for the teeth was (due to) the adoption of 600 divisions in the new inflow float clepsydra of the Yuan-Fêng reign-period (A.D. 1078–85).⟩⟩'

The number 478 seems to lack any obvious relation to astronomical constants. But 487 occurs naturally as the number of gear-teeth corresponding to $365\frac{1}{4}$ days, for this is $487 \times \frac{3}{4}$. It is the smallest and most convenient whole number to give this year-length. If this explanation applies, a scribal error must be assumed at two places in the text. We shall meet again with the number 487 in connection with I-Hsing, cf. p. 78 below.

For the significance of the 600 teeth see immediately below, p. 37. The apparatus mentioned is named a float clepsydra, but one wonders whether it could have had some relation to the gearing just before Su Sung's time.

100 'quarters' and the 12 (double-) hours. At the beginning and middle of each (double-) hour and at each 'quarter' it coincides with [96][1] the corresponding (one of) the 600 teeth. On the wheel, trip-lugs are fixed so that on the 'quarters' the drum in the centre is beaten, on the (double-) hours the little bell on the left is shaken, and on the half (double-) hours the large bell on the right is struck.

The second storey of the wooden pagoda has a door opening at the centre. As the wheel for jacks exhibiting the beginnings and middles of the (double-) hours of the day and night moves with all the other wheels, jacks wearing red appear, each holding a tablet, at the beginnings of (double-) hours, and others in purple at the middles of (double-) hours.

G 3 Ch. 3, text p. 11*a*, figure p. 10*b*

The wheel for the jacks exhibiting the beginnings and middles of the (double-) hours is situated in the second storey of the wooden pagoda and has a diameter of 7·3 ft. It has twenty-four jacks in all. Twelve of them exhibit the beginnings of the (double-) hours and the other twelve the middles of the (double-) hours. At the middles and beginnings of each (double-) hour a jack appears at the middle door holding a tablet.

G 4 Ch. 3, text p. 11*b*, figure p. 11*a*

The third story of the wooden pagoda also has a door opening in the centre. As the wheel for the jacks exhibiting 'quarters' turns with the other wheels, a jack wearing green and holding a tablet appears at each 'quarter'.

G 5 Ch. 3, text p. 12*a*, figure p. 11*b*

The wheel for jacks exhibiting 'quarters' is situated in the third storey of the wooden pagoda. It has a diameter of 7·2 ft. Over it are distributed the jacks for the hundred 'quarters'. With the exception of the beginnings of (double-) hours they are called 'quarters'. There are ninety-six jacks.[2] They are arranged to correspond in timing with the sounding of 'quarters' on the bell-and-drum floor of this belfry.

G 6 Ch. 3, text p. 12*b*, figure p. 12*a* (Fig. 13)

The fourth and fifth storeys have doors opening in their centres. At sunset [133], at darkness [134], at 'waiting for dawn' [135], at dawn [136], and at sunrise [137] a wooden jack strikes a gong in correspondence with the appearance of a jack on the fifth storey. The jacks on the fifth storey appear in order to exhibit the time of night. Two and a half 'quarters' after sunset comes darkness; then comes[3] the first night-watch [138]. Every night-watch has five subdivisions [139]. When the night-watches come to an end there is the moment called 'waiting for dawn' which begins a period lasting ten 'quarters'; then, after these ten 'quarters' comes dawn. Two and a half 'quarters' after dawn there is sunrise.

G 7 Ch. 3, text p. 13*a*, figure p. 12*b*

At sunset a jack wearing red appears to report, and then after two and a half 'quarters' there comes another in green to report darkness. The night-watches each contain five subdivisions. A jack wearing red appears at the beginning of the night-watch, marking the first subdivision, while for the remaining four subdivisions the jacks are all in green. In this way there are

[1] Note that the number of teeth must therefore be divisible both by 100 and by 24, and hence 600, being the L.C.M., is the smallest number of teeth capable of such coincidence. The expression *hsiang ying* [96] is of interest, for the words are philosophical technical terms of great importance in Chinese thought—'mutual stimulus and response'. Cf. Needham (1), vol 2, p. 304.

[2] Possibly this refers to the accurate quarters of Liang Wu Ti's system which competed with the centesimal 'quarters' cited elsewhere in this text. Note that a system of 108 'quarters' to the day, which is often cited, is probably nothing more than the ninety-six accurate (modern) quarters plus the 12 (double-) hours which coincide in fact with twelve of these quarters (Maspero (1), p. 210). Cf. Needham (1), vol. 3, p. 322.

[3] There is a small lacuna in the text here; *hou*, 'afterwards', must be supplied. On night-watches see Appendix, p. 199.

twenty-five jacks for the five night-watches. When the time of waiting for dawn comes, with its ten 'quarters', a jack in green comes out to report this. Then dawn with its two and a half 'quarters' is marked by another jack wearing green, and sunrise is reported by a jack wearing red. All these jacks appear in the central doorway.

Fig. 12. Wheel with sixteen lugs for striking the (double-) hours of the day by bells and drums (ch. 3, p. 10*a*).

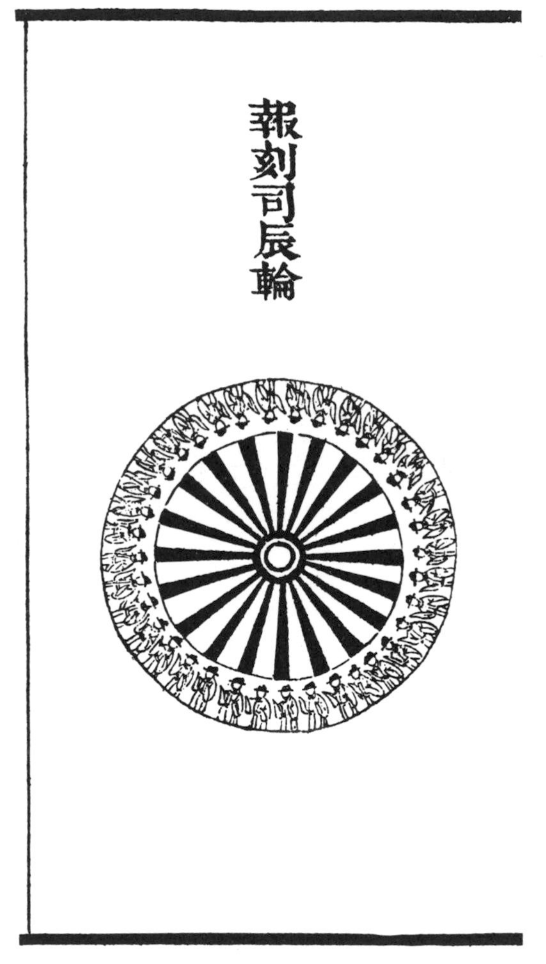

Fig. 13. Wheel for the jacks exhibiting the quarter (-hours) (ch. 3, p. 12*a*). Only thirty-six are shown.

G 8 Ch. 3, text p. 14*a*, figure p. 13*b*

The wheel for striking the night watches by gongs is situated on the fourth storey (from the top) of the pagoda. It has a diameter of 6·7 ft. To it is attached the wheel for the float indicator-rods showing the night watches. At the whole and half double-hour, a trip-lug is fixed so that it strikes the gong at these times and at sunrise and sunset.

G 9 Ch. 3, text p. 14*b*, figure p. 14*a* (Fig. 14)

The wheel for jacks exhibiting the watches of the night is situated on the fifth storey (from the top) of the wooden pagoda. It has a diameter of 8 ft. It is combined with [97] the wheel for the float indicator-rods showing the night-watches. At the sunrise and sunset, twilight and dawn, and night and early morning watches a jack with a tablet appears in the middle of the doorway to report. The wheel for the float indicator-rods has a diameter of 6·7 ft. Because the lengths of day-time and night-time vary from winter to summer there should not be a fixed division; instead there are sixty-one (different kinds of) float indicator-rods in one year. The rods differ from one another in length and they are changed (i.e. every six days) according to the seasons. In this way, throughout the four seasons there will be no errors in the lengths of night and day.

Fig. 14. Wheel for the jacks exhibiting the watches of the night; they are set round 285° of the circle (ch. 3, p. 14*a*). Only twenty-four are shown.

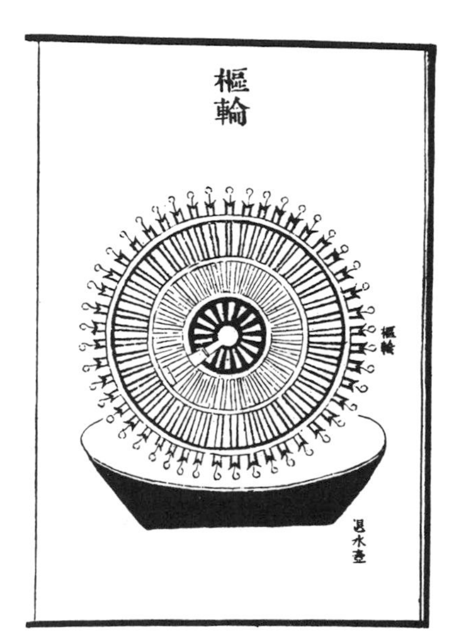

Fig. 15. Driving-wheel and sump (ch. 3, p. 15*a*).

H Ch. 3, text p. 15*b*, figure p. 15*a* (Fig. 15)

The great wheel and the sump. The great wheel has a diameter of 11 ft. and has seventy-two spokes. ⟨A variant text (as noted earlier) says ninety-six (spokes).⟩ It is a double wheel on one hub, with thirty-six scoop-holders. ⟨A variant text says forty-eight (scoop-holders).⟩ The wheel has three reinforcing rings. Each holder grasps [98] a scoop, of which there are thirty-six in all, each being 1 ft. long, 0·5 ft. wide and 0·4 ft. deep.[1] At the side of (each) scoop an iron pin [75] projects so as to actuate the trip-lever of the balancing mechanism (see Figs. 15, 26). Through the hub of the great wheel goes the iron driving shaft, stretching from the front to the back. It is from the front end that the celestial globe and armillary sphere are moved. The sump is 11·4 ft. long, 1·9 ft. wide and 3·2 ft. high at the right, 2·55 ft. high at the left and 1·55 ft. high

[1] Each scoop therefore has a capacity of 0·2 cu. ft. and holds approximately 12 lb. of water; the cycle of thirty-six scoops which are filled in 9 hours must therefore consume about half a ton of water.

in the middle.[1] It is placed under the great wheel to receive the water each time that a scoop delivers it. The sump has a hole in the bottom (perhaps not right at the bottom) on the far side, and water flows from this into the lower reservoir.

I Ch. 3, text p. 16*b*, figure p. 16*a* (Fig. 16)

The driving shaft [34] is 5·9 ft. long and 0·18 ft. square (in cross-section). It is put through the hub of the great wheel in the direction running from the back to the front. It has two cylindrical necks [99] which revolve in iron crescent (-shaped) bearings [100] on the top of the main beam [53]. This shaft is attached at its front end to the earth wheel which engages with [92] the lower wheel of the transmission shaft, which it turns. ⟨Sticking out before and behind they enmesh and one follows the other.⟩

The transmission shaft is 19·5 ft. long and made of wood;[2] it is concealed at the top by the tortoise-and-cloud (decorative central column). The upper wheel of this shaft moves the gear-wheels of the armillary sphere shaft [77]; the middle wheel of the transmission shaft moves the mechanism of the eight time-keeping wheels; the lower wheel is moved by the earth wheel which is connected to the great wheel.

The armillary sphere shaft is placed within the meridian circle [101] of the armillary sphere. It is supported by crescent (-shaped) (bearings). The back gear-wheel is moved by the upper wheel of the transmission shaft. The front gear-wheel engages with the diurnal motion gear-ring; they (i.e. the front and back gear-wheels) share the same shaft. When the latter is moved, the front gear-wheel turns too, and thus the diurnal motion gear-ring also moves.

⟨⟨A variant text omits mention of the transmission shaft, the armillary sphere shaft and its gear-wheels, giving instead the chain-drive [102] and its gear-box [103]. Yet another variant text has the driving-shaft on its crescent (bearings) engaging at its front end directly with the time-keeping gear-wheel without any transmission shaft, and with the chain-drive attached to a lower chain wheel [104] (on the driving-shaft).⟩⟩

J Ch. 3, text p. 18*a*, figure p. 17*b* (Fig. 17)

The upper reservoir and the constant-level tank. The latter has a water-level arrow (i.e. marker) [105].[3] The manual wheel forces the water to flow into the upper flume and so to the

[1] It is carefully calculated for placing to the left of the centre below the great wheel so that it projects more on the side where water is being received into the scoops. The capacity of the sump is about 33 cu. ft., sufficient, if it were full, to provide water for about 40 hours of continuous operation.

[2] Such a long shaft, subject to considerable torque, would be mechanically unsound if constructed of wood. This is probably the reason why the transmission shaft seems to have been abandoned in favour of the celestial ladder chain-drive. In fact, the change appears to have proceeded in two stages; first the upper portion only of the transmission shaft was dispensed with (variant I), and then the lower portion as well was replaced (variant II) so that the earth wheel meshed directly with the time-keeping gear-wheel. See inset, Fig. 23.

[3] The actual mechanism of this constant-level tank, though not described in any detail, is nevertheless crucial for the time-keeping properties of the clock. Nowhere is there any mention of an overflow pipe, yet with only one compensating tank the alternative system of a number of intermediate vessels (cf. p. 87 below) seems excluded. Both types were well known in Su Sung's time, but this seems something different. The floating arrow here mentioned must be considered a philological relative of the floating indicator-rod (always so named) of the inflow receiver of the classical Chinese clepsydra. Some late Japanese drawings actually show it as an arrow with feathers (Takabayashi (1), figs. 11 and 14). The most straightforward explanation of Su Sung's floating arrow is that it merely indicated the height of water in the tank so that attendants could watch it, making adjustments periodically to maintain the head. Alternatively, it could have been an automatic device for regulating the water-level by stopping the supply (something like a ball-valve). An apparatus of this kind seems to be described in the *Kuan Shu K'o Lo T'u* [757] written by Wang P'u [547] about A.D. 1135, which is quoted in a Sung edition of the *Liu Ching T'u* [758], of which the Cambridge University Library possesses a microfilm copy. See Fig. 39.

upper reservoir. As the water enters the reservoir at a non-uniform rate, the water-level tank is made to adjust (i.e. the pressure head). After escaping, the water flows to the scoops of the great wheel. Since the water remains uniformly (in motion) throughout the day and night, time can thus be correctly kept.

K Ch. 3, text p. 19*a*, figure p. 18*b* (Fig. 18)

The upper balancing lever [62] which is placed above the driving-shaft. At the fulcrum is the iron shutting axle [111]. On the horizontal cross-bar [112] between the right-hand posts, a camel back [113] is made with two iron cheeks [114] (i.e. a pair of concave bearings) so as to

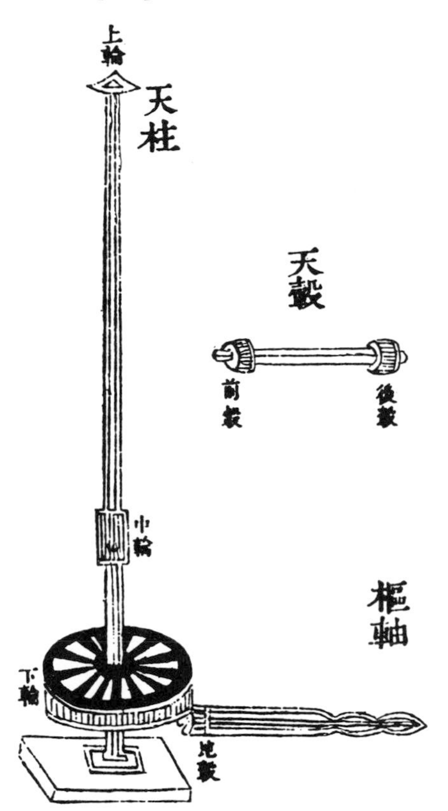

Fig. 16. Main driving-shaft and main vertical transmission-shaft (ch. 3, p. 16*a*). Inset is shown the armillary sphere shaft which connects with the main vertical transmission-shaft.

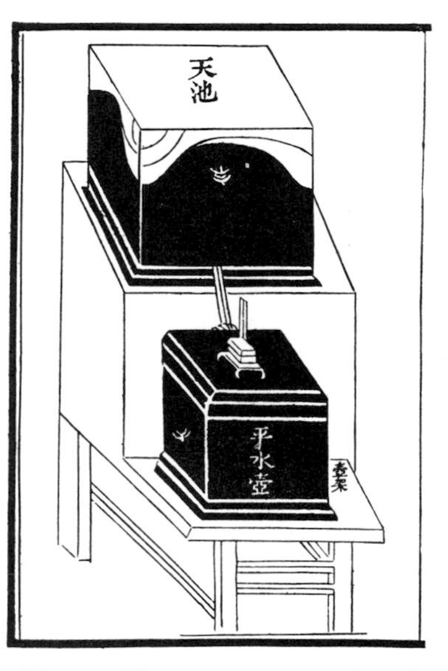

Fig. 17. Upper water reservoir and constant-level tank (ch. 3, p. 17*b*).

support the (iron shutting) axle while it moves. At the tail of the upper balancing lever there hangs the upper weight [64]; at the head [63] hangs the upper stop [55]. At a point on the right of the fulcrum, between it and the upper weight there hangs an (iron) connecting chain [66]. ⟨This is what is called an iron 'crane bird's knee' [115];[1] its length accords with the

[1] I.e. a chain (perhaps of linked rods) attached to rods or levers. We shall return to this technical term below, p. 56.

height of the great wheel.⟩ The trip-lever has its own iron shutting axle fitted above the bar running from back to front in the frame supporting the constant-level tank, and it can thus move freely.[1]

The head [116] of the trip-lever is fastened to the connecting chain; when the trip-lever moves, the upper stop acts. There are two upper locks [72], one at the left, another at the right; at the end of each a shutting axle [111] is attached. These two shutting axles are attached to the main posts (of the framework) at either side. They face each other on the left and right to stop [74*a*] the scoop-holders of the great wheel. Attached above the trip-lever is a (lower balancing) lever [69] and a weight [71]. Its shutting axle [111] is supported by two concave cheeks at the two horizontal cross-bars; one at the front and the other at the back of the constant-level tank. The axle is able to move repeatedly. At the head [116] of the (lower balancing) lever, a checking fork [117] is made to resist [74] the movement of the scoops on the driving-wheel. The weight hanging on the right of the lower balancing lever goes up and down when the scoops are emptied and filled.

L Ch. 3, text p. 20*a*, figure p. 19*b* (Fig. 19)

The upper and lower norias each have a diameter of 5·6 ft. The upper one shares its axle with the manual wheel. The front of the axle rests on the middle of the cross-bar under the upper beam, and the back of it rests on the cross-bar at the middle of the tower and the wooden pagoda, supported by a crutched post [106]. The axle of the lower noria rests at the front on the centre of the cross-bar of the main beam and is supported at the back by another crutched post abutting on to the boards at the back of the tower. ⟨The wooden pagoda, which is 7·1 ft. high, 7·3 ft. deep and 2·5 ft. wide, is walled with boarding. At the front, two wooden posts are erected supporting cross-beams at the back on which men can stand to turn the manual wheel.⟩

The intermediate and lower reservoirs.[2] The intermediate one is 7·4 ft. long, 0·95 ft. wide, 2·3 ft. high at either end and 1·5 ft. high at the centre. The lower one is 7·2 ft. long, 1·6 ft. wide, and 2·1 ft. high; they are under their respective wheels. The upper flume is situated at the top of the upper noria to receive water from it. In front of the lower reservoir a hole is made communicating with the hole in the sump. When the manual wheel is turned the upper and lower norias are turned too. Both the manual wheel and the upper noria turn towards the right, hence the lower noria turns in the contrary direction [107],[3] i.e. to the left. The latter takes up water from the lower reservoir on the right-hand side and delivers it to the intermediate reservoir, while the upper noria takes up water from the upper tank on the left-hand side and delivers it to the upper flume and so to the upper reservoir.

M Ch. 3, text p. 21*a*, figure p. 20*b* (Fig. 20)

The manual wheel and the upper flume. The manual wheel has a diameter of 4·8 ft. The upper flume is 3·8 ft. long, 0·7 ft. wide and 0·6 ft. high. On its right is a hole which is connected to the upper reservoir. The manual wheel has sixteen leaf teeth [91] which engage with sixteen peg teeth [108] on the lower noria. Eight handles [109] are fitted on the manual wheel for turning it round. The rims of the norias carry buckets [110], twenty-four in all, attached in

[1] The levers are supported by these shutting axles rather than simple fulcrums probably because of the lateral strain they must take when destroying the angular momentum of the great wheel in the course of action of the escapement.

[2] Slight confusion is caused in the text by the use of the terms 'upper and lower water-raising tank' for 'intermediate and lower reservoir' respectively.

[3] The term used, *ni hsing*, is of philosophical interest (cf. Needham (1), vol. 7). *Ni* is a word used for all forces in the universe which oppose organic pattern in a whole or its constituent wholes.

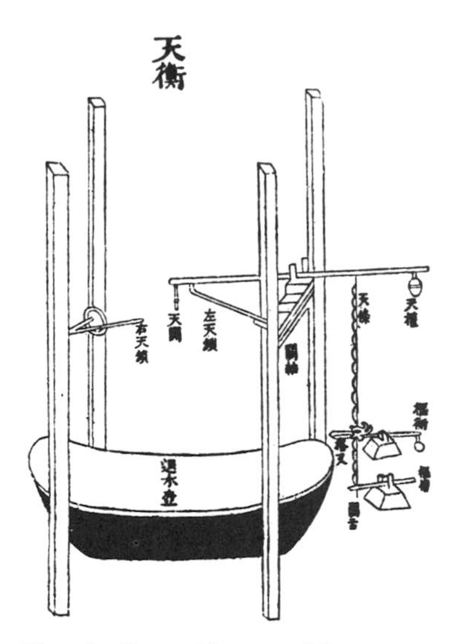

Fig. 18. General layout of the escapement stops and levers (ch. 3, p. 18*b*).

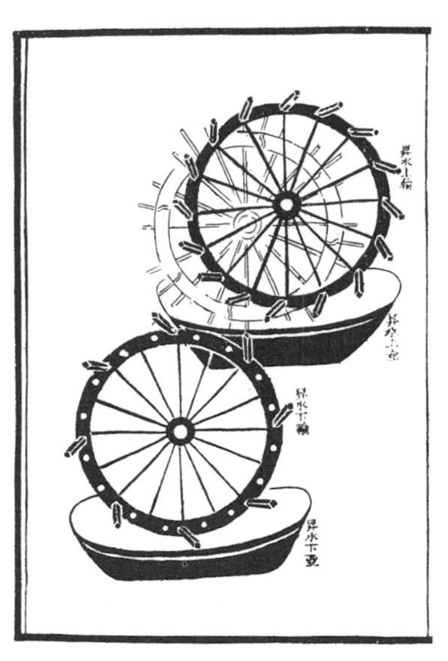

Fig. 19. Upper and lower norias and manual wheel for operating them (ch. 3, p. 19*b*).

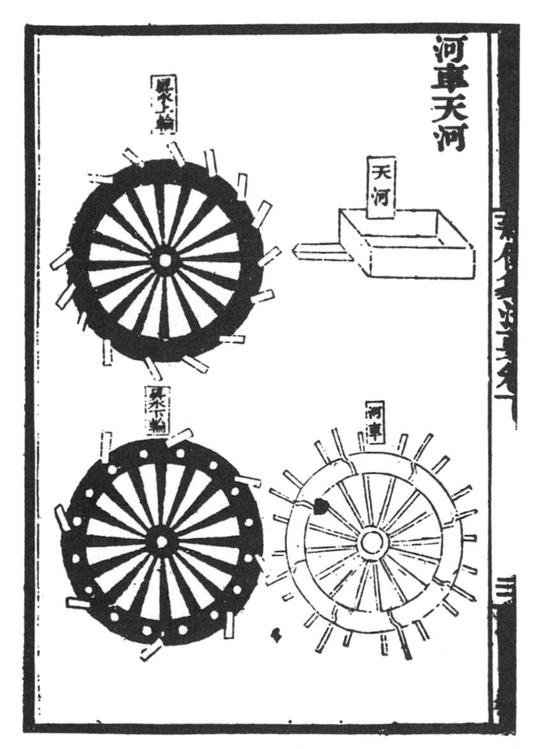

Fig. 20. Component parts of the noria system, with the upper flume (ch. 3, p. 20*b*).

a slanting position; sixteen on the upper noria and eight on the lower.[1] When the manual wheel turns it rotates the two wheels with their water-buckets and thus carries water into the upper flume from which it flows into the upper reservoir.

HOW TO OPERATE THE WATER-DRIVEN ARMILLARY (SPHERE) AND CELESTIAL (GLOBE)

N Ch. 3, text pp. 21*a* ff.

The system of water movement starts from the lower reservoir. First of all the lower reservoir is filled with water. When it is full the eight handles of the manual wheel are operated, so rotating the upper and lower norias. The lower noria raises water in its eight buckets up to the intermediate reservoir, and from there the upper noria takes it in its sixteen buckets up to the upper flume. From here it runs into the upper reservoir and thence out at the front through a siphon (cf. the simple hole mentioned previously) [73] into the constant-level tank and then again through another siphon into the scoops of the great wheel.

The scoops meet on the right the checking fork [70] of the iron lower balancing lever [69]. The checking fork resists [74] the scoops (one after the other). Each scoop when empty is caught by the checking fork to enable it to be filled with water. When it is full the checking fork can no longer support its weight so that it falls and an iron pin [75] on the side of the scoop strikes the trip-lever and moves the connecting chain. When this is moved, the upper balancing lever rises (on the left) and moves the upper stop [68]. The left upper lock [72] then opens to let one of the spokes [29] of the great wheel pass.

This means that the driving-shaft rotates, turning two instruments, the armillary sphere and the celestial globe, as follows: The armillary sphere is moved because the great wheel turns the earth wheel, thus moving the lower wheel of the transmission shaft and hence the back gear-wheel [76] of the armillary sphere shaft [77] and so on to its front gear-wheel [78] and thence to the diurnal motion gear-ring [10] and finally the component for the three arrangers of time (to which this gear-ring is fixed). Such is the manner in which the great wheel moves the armillary sphere.

The time-keeping wheels are moved as follows: the great wheel turns the earth wheel which turns the lower wheel of the transmission shaft. In its turn the middle wheel of the transmission shaft drives all the time-keeping wheels on the time-keeping shaft. Such is the manner in which the great wheel moves the time-keeping wheels. The time-keeping wheels have four components: (i) the celestial wheel moving the celestial globe, (ii) the bell and drum wheels, (iii) the wheel for jacks marking the (double-) hours, (iv) the wheel for jacks marking the quarters.[2]

The celestial globe moves as follows: the time-keeping wheels move and carry with them the celestial wheel which thus turns and rotates the celestial globe (i.e. direct drive variant). Such is the manner in which the celestial wheel moves the celestial globe.

The bell and drum wheel moves as follows: when the time-keeping shaft turns, this time-keeping wheel also turns. At the beginning of each (double-) hour, trip-lugs or pins [79] on

[1] The upper noria has more buckets than the lower so that water is carried away from the intermediate reservoir more readily than it can be put in; an overflow is thus prevented.

[2] The scheme here described is considerably simpler than that given elsewhere in the text. It may be either an earlier unelaborated version, or a later simplification.

the wheel poke the stick held by the jack on the left and make him ring the small bell. At the middle of each (double-) hour the pins poke the stick held by the jack on the right and make him strike the large bell. At the 'quarters' they poke the stick held by the jack in the centre which makes him beat the drum. These three jacks are all situated on the first (top) storey of the wooden pagoda and in the doorway, to the left, right and centre respectively. Such is the manner in which the wheels move the bells and drum.

The jacks exhibiting the hours of day and night. When the time-keeping wheels turn they carry with them that for these jacks. At the whole and half (double-) hours one of the wooden jacks on the wheel appears in the doorway on the second storey (from the top) of the wooden pagoda holding a tablet in his hand. Such is the manner in which the jacks exhibit (the hours, etc.).

The jacks exhibiting the 'quarters'. When the time-keeping wheels turn they carry this jack wheel with them. When one of the 'quarters' is due, one of the jacks (appears) in the doorway on the third storey of the pagoda to exhibit it. Such is the manner in which the wheels move the jacks for the 'quarters'.[1]

After each spoke of the great wheel has passed, the left upper lock and the upper stop close, and when this has happened one scoop falls into the sump.[2] After each scoop has so fallen, the stopping and locking device resists once again the next scoop, but since this might cause the wheel to recoil counter-clockwise [80], the right-hand (left as seen from front) upper lock (i.e. the ratchet action) acts to prevent any such movement.

The great wheel. Each time a scoop passes, the water falls into the sump and then out through the hole towards the back into the lower reservoir, and in such a manner the whole cycle repeats itself from the beginning to the end and from the end to the beginning.

(This is probably the end of the original text; later modifications now follow.)

The gnomon shadow scale [128] attached to the base of the armillary sphere. According to the old method, the gnomon shadow scale and the armillary sphere were quite separate from each other so that the armillary sphere could not be used to measure the length of the sun's shadow, and the gnomon shadow scale could not be used to study the degrees of motion of the seven luminaries. Now the two are combined in the following manner:

O Ch. 3, text p. 23*b*, figure p. 23*a*[3]

Under the armillary sphere is placed the gnomon shadow scale [129]. Its surface meets the centre of the water-level base [122] which has water channels cut in it for levelling. The gnomon shadow scale is 13 ft. long and shows the movement of the sun's shadow in the north-south direction. The surface of the scale carries divisions in feet and 'inches' (tenths). Its two sides are marked with the twenty-four fortnightly periods [130]. The distance from the surface of the scale to the surface of the horizon circle [131] or the mid-point of the sighting-tube

[1] It is interesting that the simplified variant of the time-keeping system omits the paraphernalia previously described which are concerned with the striking and exhibiting of the unequal hours or night-watches.

[2] It is not made clear whether the scoops become submerged under the water-level in the sump or whether they merely tip their contents into it.

[3] This page is significantly different in in style and subject-matter from all other sections of the chaper; it is almost certainly a later addition to the text.

[132] is just the height of the (standard) gnomon, i.e. 8 ft.[1] Thus the distance from the surface of the horizon circle or the mid-point of the sighting-tube to the lowest point of the tortoise-and-cloud (moulding) is also 8 ft. At noon, the sighting-tube is pointed at the sun and the light goes through the hole of the tube and falls upon the surface of the scale at a certain division which may be taken as a standard reading. The sighting-tube is used for observing the movements of the five planets, whether going forward or retrograding, tarrying or hurrying, and thus for investigating their times of passage. It is also used for studying the obliquity of the ecliptic, and the various polar distances of the stars. Thus, while the sighting-tube determines the changes occurring in the heavens, the (gnomon) scale measures the twenty-four fortnightly periods by the varying lengths of the sun's shadow, and thus determines the changes occurring on the earth.[2] If both are combined in one instrument the phenomena in heaven and earth may be synthesised. In such a way the calendar can be regulated without errors or inaccuracies.[3]

⟨The sections (remaining of this chapter) are from another recension.⟩

P Ch. 3, text p. 24*b*, figure p. 24*a* (Fig. 21)

The celestial globe drive-wheel. The body of the celestial globe is spherical like a ball and has a diameter of 4·565 ft.[4] On its surface, the circumference is marked with 365 and a fraction degrees [86], and the constellations and stars both north and south of the equator are marked; there being 246 names (of constellations) and a total of 1281 stars. The polar region [88] is situated at the northern part with thirty-seven names of groups and a total of 183 stars. The sum total is therefore 1464 stars. It (the globe) is circumscribed by the ecliptic and equator. The twenty-eight lunar mansions [89] are shown in their succession, and also the path where the sun, moon and five planets move (i.e. the $\pm 5^\circ$ band of the zodiac). In the middle a shaft is put through from north to south. At the end of this shaft is the celestial globe drive-wheel which

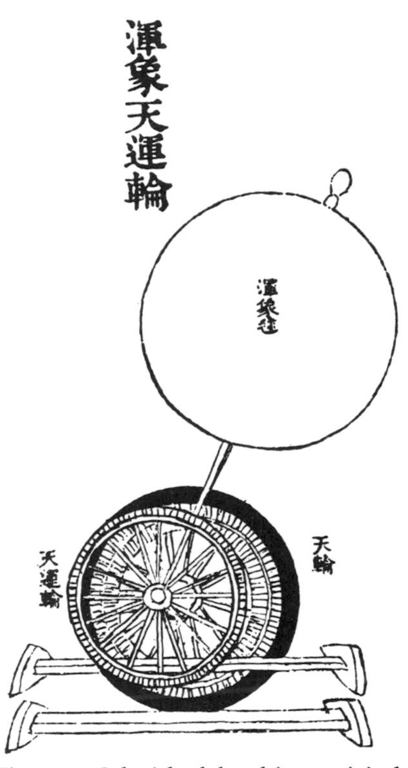

Fig. 21. Celestial globe drive, original system with rotating polar axis (ch. 3, p. 24*a*).

[1] This standard presumably dates from a time when the foot was so defined that 8 ft. was the average height of a man. Cf. the 6 ft. gnomon length in the Anglo-Saxon usage of Bede.

[2] It is remarkable that the gnomon is used here purely for calendrical rather than horary determinations.

[3] In connection with the possibility of error, it should be noted here that the clock mechanism is so arranged as to provide an intermittent rotation of 3·6° every 'quarter' hour. This is ideal for the working of jacks but it is ill-suited to driving an observing armillary. Unless observations are made only on exact 'quarter' hours, as indicated by the appropriate jack-work, they will be subject to an intolerable error.

[4] The dimension of 45·65 'inches', given to such unaccustomed accuracy, is presumably intended as an exact eighth of 365·2, the number of days in the year and the number of degrees of the equatorial circle. Other Chinese texts confirm that it was usual to construct globes so that they were to a scale of some round number of inches to the 'degree'. In this case the simplicity seems to have been lost by calculating the diameter rather than the circumference as such a round number. Cf. Needham (1), vol. 3, pp. 383 ff.

engages with [90] the celestial idler and the teeth [79] of the celestial wheel [18]. When the celestial wheel moves it turns the celestial globe drive-wheel and (hence) the celestial globe on which the degrees, positions of stars and constellations, ecliptic, equator, and the track (i.e. zodiac) of the sun, moon and five planets are all marked with their appropriate names and degrees.

Q Ch. 3, text p. 25*b* figure p. 25*a*

The iron celestial idler is placed by the base of the wooden housing of the celestial globe. The two ends of the idler are situated on the compartmental framework in the tower, underneath the cross-bar stretching from right to left between the celestial wheel and the celestial globe drive-wheel. The idler engages with [90] the teeth [91] of the two wheels. As the celestial wheel turns to the west, the idler rotates to the east, and hence the celestial globe follows this rotation. This is to show the heavens turning westwards (instead of eastwards, as they would without the idler).

R Ch. 3, text p. 26*a*, figure p. 25*b* (Fig. 22)

The celestial ladder (i.e. chain-drive) is 19·5 ft. long (as was the transmission shaft it replaced). The system is as follows: an iron chain with its links joined together to form an endless circuit [118] hangs down from the upper chain wheel of the celestial ladder [119] which is concealed by the tortoise-and-cloud (moulding), and it passes round the lower chain wheel of the celestial ladder [104] which is fixed to the driving shaft. Whenever one link [120] moves, it moves forward one tooth [108] of the diurnal motion gear-ring and rotates the component of the three arrangers of time [4] thus following the motion of the heavens.

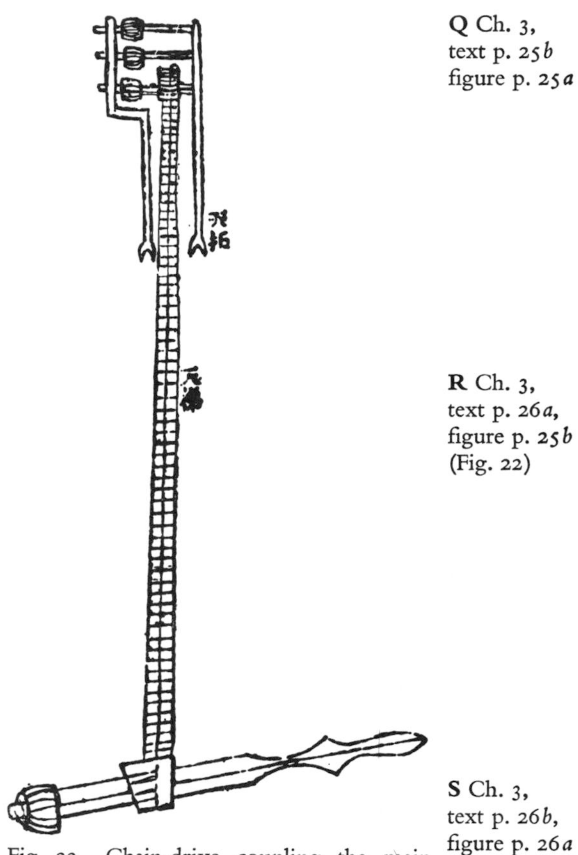

Fig. 22. Chain-drive coupling the main driving-shaft to the armillary sphere gear-box, in the first and second modifications (ch. 3, p. 25*b*).

S Ch. 3, text p. 26*b*, figure p. 26*a*

The celestial ladder gear-case is concealed by the tortoise-and-cloud (moulding). Each side of it is 3·7 ft. long and has a fork [121] at the bottom which fits into the centre of the water-level base [122]. The distance between the two sides is 0·31 ft. The front side is bent [123] at a point one-quarter of the total length from the top. The distance between the two sides at the top is 0·45 ft. ⟨This is because of the bend.⟩ Above this, three holes are bored through each side to carry the three axles with four gear-wheels. The top gear-wheel is called the upper pinion [124] and it engages with [125] the diurnal motion gear-ring in the armillary sphere. The next below it is called the middle pinion [126] and it engages with the upper one. Next is the lower pinion [127], and it engages with the middle one. The next one (i.e. the fourth) is called the upper chain wheel of the celestial ladder [119] and it is mounted behind the lower celestial pinion and on the same axle with it. From this upper chain wheel hangs the celestial ladder.

V

EXPLANATION OF SU SUNG'S CLOCKWORK

THUS ends Su Sung's description of his astronomical clock tower. It may be that the reader who has followed it with the help of the original illustrations which we have reproduced will have gained a tolerably clear idea of the way in which it worked. But in case some further explanation should be needed we shall now give a running commentary on four diagrams which constitute a reconstruction of the whole mechanism. Fig. 23 shows the general scheme of the power-source and its transmitted drives. Insets give details of the two alternative forms of shafting and endless chain mentioned in the text (p. 40 above), and of the alternative globe-rotating gear. Fig. 24 illustrates the component parts of the armillary sphere and shows the different types of teeth required for the various gear-wheels. The general layout of the escapement is shown in Fig. 25 and the successive stages of the escapement cycle in Fig. 26.

In Fig. 5 we have seen the general external appearance of the 'Combined Tower' [117] or 'Tower for the Water-powered Sphere and Globe' [1]. The armillary sphere [150] is on the platform at the top, the celestial globe [11] is in the upper chamber of the tower, half-sunk in its wooden casing; and below this stands the pagoda-like façade [144] with its five superimposed storeys and doors at which the time-announcing jack figures appeared. On the right the housing is partly removed to show the water storage tanks [42, 43]. The scale of the whole is clearly deducible from the internal evidence of the text; the height must have been between 30 and 40 ft. in all, as is seen in the imaginative reconstruction given in the frontispiece.[1]

The modern drawing (Fig. 23) should now be compared with Su Sung's own general diagram (Fig. 8). This latter sees the structure from the south or front,[2] but in the reconstruction we view it from the east side. The great driving-wheel

[1] We are greatly indebted to our friend Mr John Christiansen for his generous gift of this graphic reconstitution and for the great pains with which he applied his artistic talent in the preparation of it.

[2] Su Sung, as was usual with medieval Chinese engineers, always describes components of machinery by the cardinal points, upper pieces being named 'heavenly' and lower ones 'earthly', while the points of the compass replaced 'left' and 'right'.

[28], 11 ft. in diameter, carries thirty-six scoops [32, 44] on its circumference, into each of which in turn water pours at uniform rate from the constant-level tank [43]. The whole cycle of scoops must have filled every nine hours, consuming in

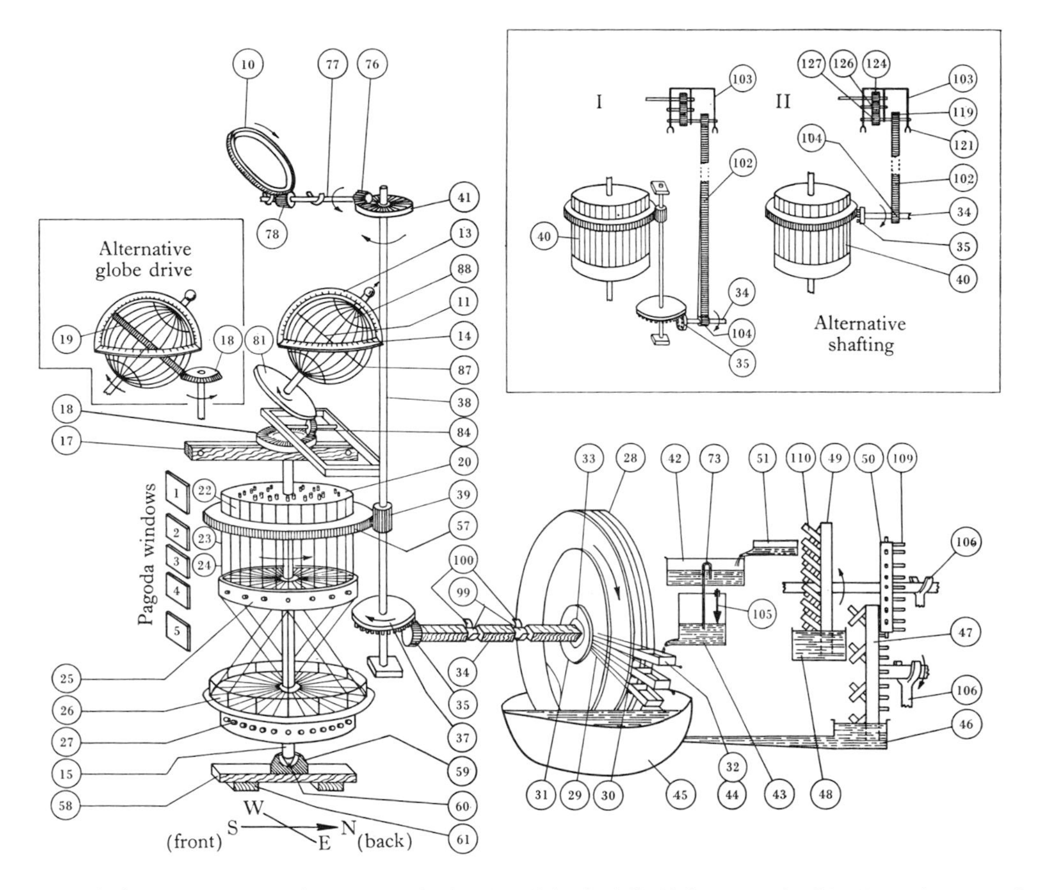

Fig. 23. 'The Water-Powered Armillary (Sphere) and (Celestial) Globe Tower' of Su Sung (A.D. 1090); detailed reconstruction of the clockwork mechanism. The insets give the alternative forms (later modifications) of shafting and chain-drives. The numbers refer to the glossary of technical terms in the main text.

the process about half a ton of water. The main driving-shaft of iron [34], with its cylindrical necks [99] supported on iron crescent-shaped bearings [100], ends in a pinion [35] which engages [92] with a gear-wheel at the lower end of the main vertical transmission-shaft [38]. This drives two components. A suitably placed pinion [39] connects it with the time-keeping gear-wheel [57] which rotates the whole of the jack-work borne on the time-keeping shaft [15]. This consists of half a dozen superimposed horizontal wheels carrying round the figures or jacks

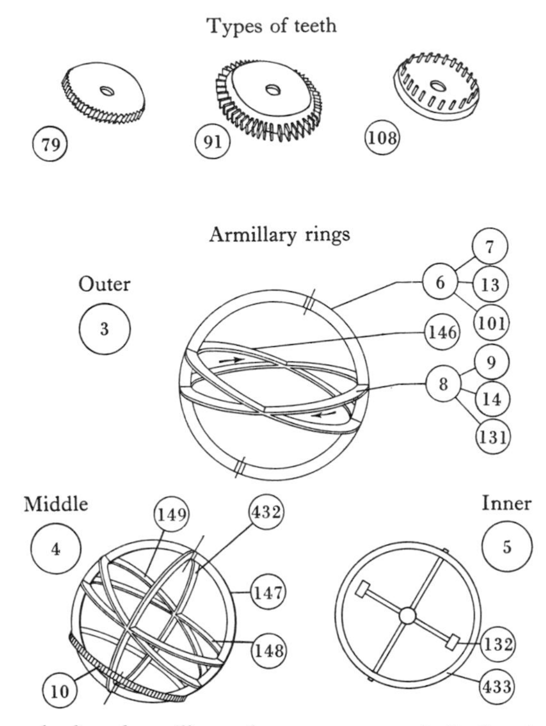

Fig. 24. Gear-wheels and armillary sphere components in Su Sung's clock-tower.

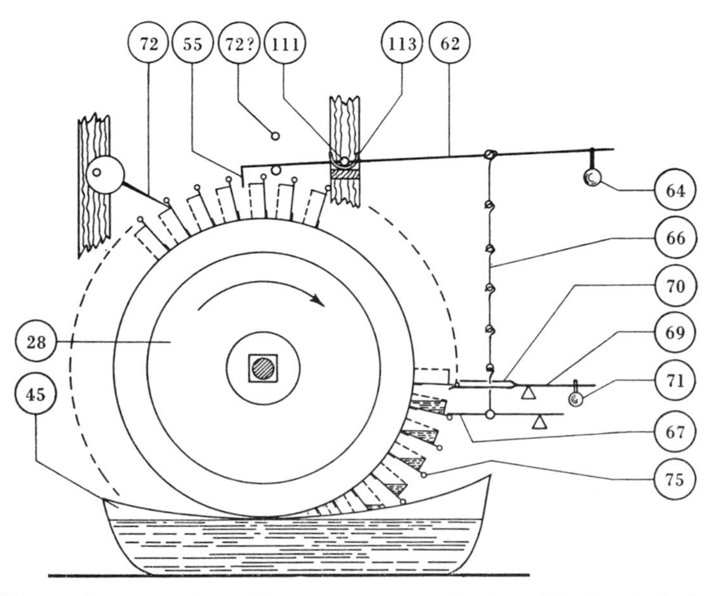

Fig. 25. Reconstruction of the escapement mechanism of Su Sung's clock.

[145]. Since each of these wheels is from 6 to 8 ft. in diameter, the total weight must have been very considerable, so the base of their shaft is fitted with a pointed cap [60] of metal, and supported in an iron mortar-shaped end-bearing [59]. As we have seen, the jack-work wheels performed a variety of functions in the visual and auditory reporting of time. The quarters, the night-watches and their divi-

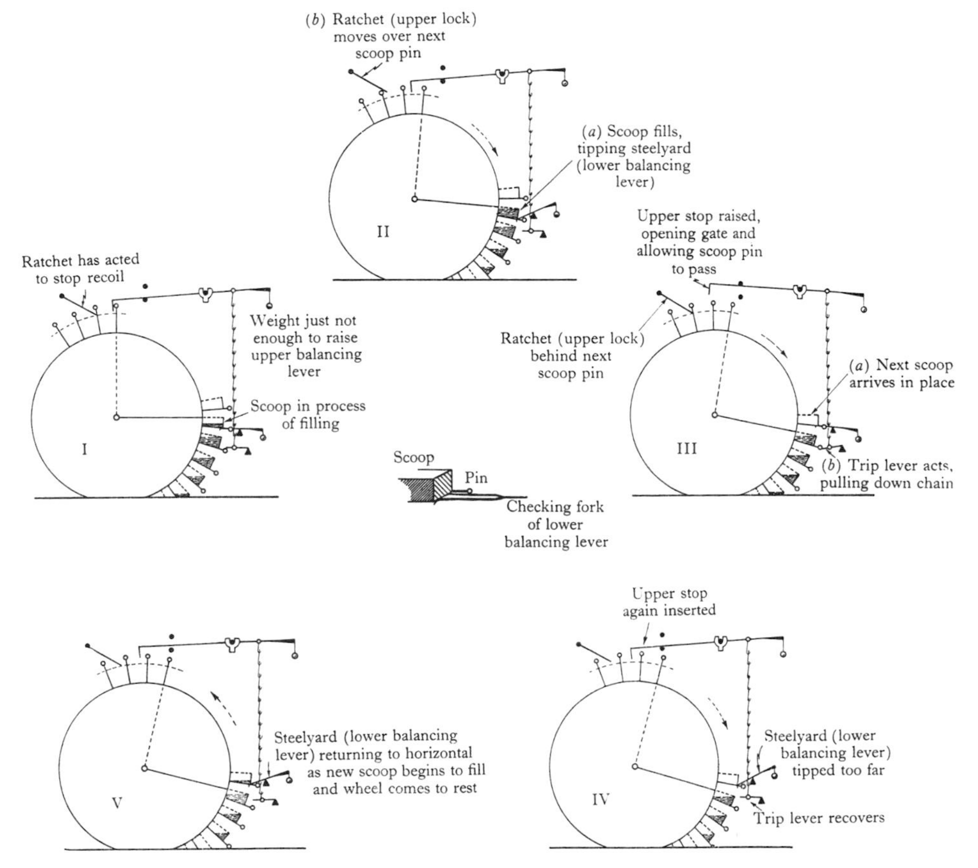

Fig. 26. Stages in the escapement cycle of Su Sung's clock.

sions were announced by placards, and the quarters as well as the beginnings and middles of double-hours were struck on bells, gongs or drums.

Evidence will be brought forward presently (see p. 78 below) that horizontally mounted jack-wheels of this kind go back to the seventh century A.D. in China. They certainly lasted through many centuries, as may be seen from Fig. 27 which shows a sixteenth-century European Renaissance clock by Isaac Habrecht. Most of the monumental clocks of major importance in Europe have one or more horizontally rotating jack-wheels. One may mention, for example, the Electors of the Holy Roman Empire who still today circulate at noon high above the Nürnberg market-place on the tower of the Marienkirche.

The rotation of the jack-wheels was not the only duty, however, of the time-keeping shaft, for at its upper end it engages by means of oblique gearing [18, 84] and an intermediate idling pinion with a gear-wheel [81] on the polar axis of the celestial globe. This can hardly have been bevel gearing in the modern sense, but closer description is impossible, for Su Sung does not explain it in sufficient detail. The angle of these gears corresponded of course with the polar altitude at K'ai-fêng. Now we have noted that the text contains a number of additions which record improvements in the clock mechanism, probably dating from the last years of the eleventh century. In these an alternative globe drive is given, the uppermost gear-wheel [18] enmeshing directly with an equatorial gear-ring [19] passing round the globe (Fig. 28). Presumably the original gearing proved difficult to maintain.

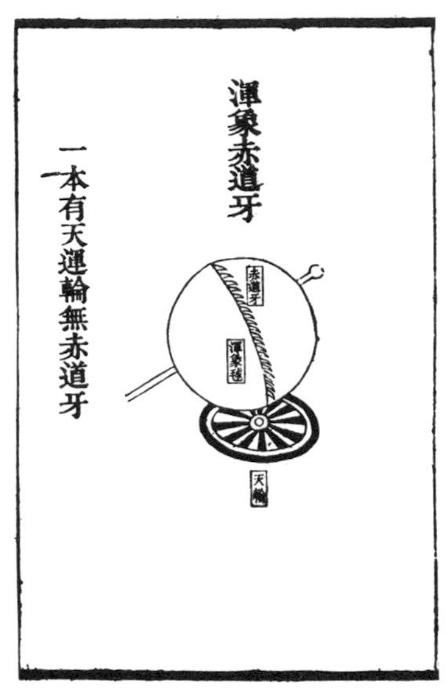

Fig. 28. Su Sung's clockwork; celestial globe drive. Modified system in which the uppermost gear-wheel of the time-keeping shaft enmeshes directly with an equatorial gear-ring (ch. 2, p. 4*a*).

We now return to the main vertical transmission-shaft and the second component which it drives. Its uppermost end provides the power for the rotation of the armillary sphere. This is effected by right-angled gears and oblique gears connected by a short idling shaft. The oblique engagement is made with a toothed ring called the diurnal motion gear-ring [10] fitted round the intermediate nest or shell of rings of the armillary sphere, not equatorially but along a declination parallel near the southern pole.[1] The three nests of rings or circles which were in use in Chinese armillary spheres of this period[2] will be better appreciated from Fig. 24. The outer nest, called the Component of the Six Cardinal Points [3], was fixed to the framework of the instrument, and comprised a split-ring meridian circle [6, 7, 13, 101], a single-ring horizon circle [8, 9, 14, 131], and a single-ring equatorial circle [146]. The middle or inter-

[1] In our diagram (Fig. 23) the length of the idling shaft [77] is exaggerated. The main vertical transmission shaft came right up through the central supporting column under the armillary sphere.

[2] For a full discussion of the history of armillary spheres and their construction in ancient and medieval China see Maspero (1) and more recently Needham (1), vol. 3, pp. 339ff.

PLATE II

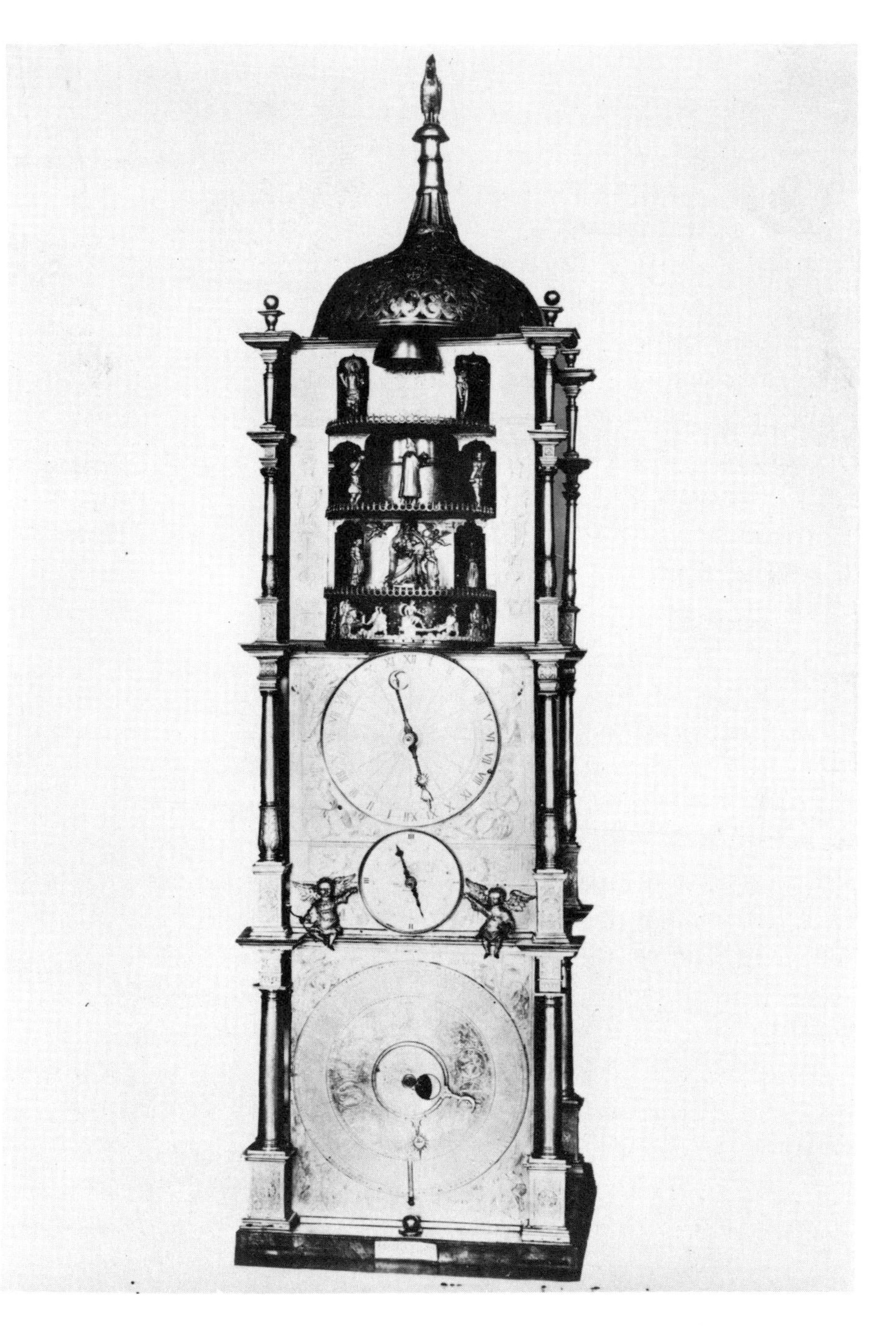

Fig. 27. A sixteenth-century European clock by Isaac Habrecht which includes horizontally rotating jack-wheels

mediate nest, called the Component of the Three Arrangers of Time [4], had a split-ring solstitial colure circle [147], a single-ring equator circle [148], a split-ring ecliptic circle [149], a single-ring equinoctial colure circle [432], and of course the diurnal motion gear-ring [10] already mentioned. Lastly, the innermost nest was represented by one circle only, the split-ring polar-mounted declination circle [433] which carried the sighting-tube [132] on a cross-strut arrangement (Fig. 29), the whole being termed the Component of the Four Displacements [5].[1] There can be no doubt that Su Sung's armillary sphere was intended for practical observations of star positions. The fact, therefore, that it was coupled to a time-keeping mechanism confers upon it a great importance in the history of astronomical science, since the apparatus was nothing less than the first of all 'clock drives'. As is well known, equatorial telescopes in modern times are always provided with a clockwork mechanism which compensates for the motion of the earth during a period of observations so that the instrument can remain fixed upon a certain point or region of the heavens. This invention (or re-invention) belongs to the early years of the nineteenth century. What exactly Su Sung's purpose was in embodying an observational instrument in the structure of his astronomical clock raises a number of interesting points with which there is no room to deal in this place, and the reader is referred to a full discussion of the matter which is given elsewhere.[2]

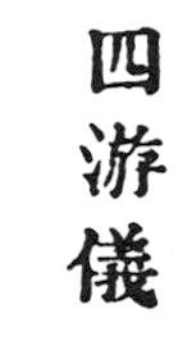

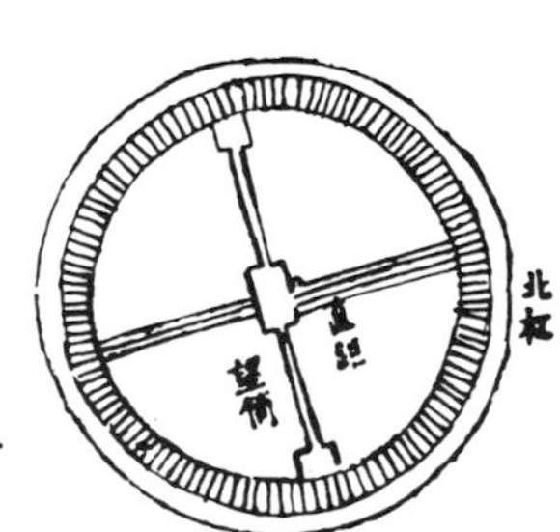

Fig. 29. Su Sung's armillary sphere; the Component of the Four Displacements, with its split-ring polar-mounted declination circle, sighting-tube and cross-struts (ch. 1, p. 9*b*).

In due course the original model of the armillary sphere drive proved unsatisfactory, as had been the case with the drive of the globe, and improvements were made as time went on. We know that the main vertical transmission-shaft was made of wood and nearly 20 ft. long. This must soon have shown itself to

[1] The disposition of the three nests may be seen rather well in the diagram given in the eighteenth-century MS. copy of Su Sung's book preserved in the National Library in Peking. This we reproduce in Fig. 30, with thanks to the Librarian, Dr Chang Shu-I, to whom we owe our knowledge of it. Though at first sight somewhat cruder than that which is found in the printed version (given already in Fig. 6 above), in fact it shows the three component nests rather more clearly. There are only a few defects in the drawing: (*a*) the ecliptic ring on the far side of the instrument is drawn straight and not curved; (*b*) the same ring on the near side does not pass clearly in front of the sighting-tube strut; and (*c*) the sighting-tube itself, which is labelled 'the jade traverse for observing the stars and constellations', is drawn in three staggered portions instead of one continuous straight line.

[2] See Needham (1), vol. 3, pp. 359ff.

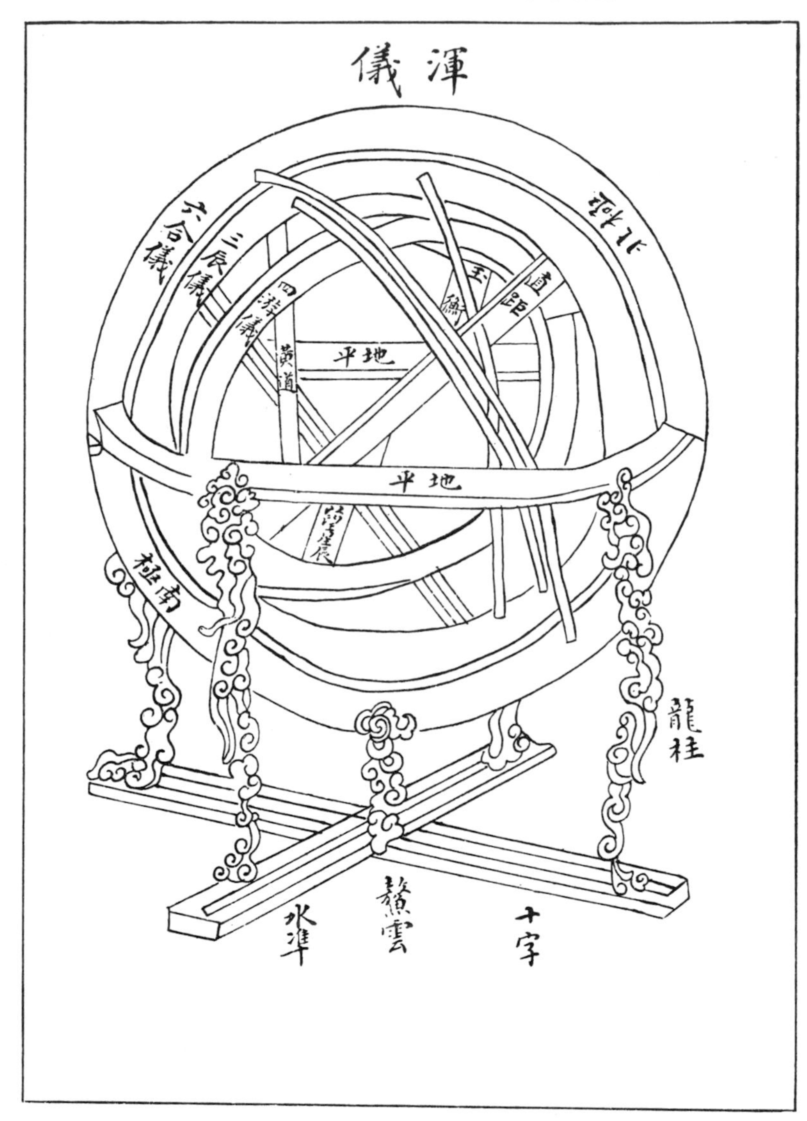

Fig. 30. Su Sung's armillary sphere as drawn in the eighteenth-century MS. of the *Hsin I Hsiang Fa Yao* preserved in the Peking National Library (courtesy of Dr Chang Shu-I). For comparison with Fig. 6.

be mechanically unsound, and in the later variants, probably about A.D. 1100, it was first shortened, and finally abolished altogether. The later designs are shown in the inset in Fig. 23. In the first modification the vertical shaft had no other duty than to turn the chief time-keeping gear-wheel [57] and its associated wheels [40], while in the second the 'earth-wheel' driving pinion [35] connected the main driving-shaft [34] directly with the time-keeping gear-wheel itself, so that the vertical shaft was suppressed altogether. But in both cases the motive power was conveyed to the armillary sphere on the upper platform by means of an endless chain drive or 'celestial ladder' [102], rotating three small pinions [124, 126, 127], presumably of different sizes, in a gear-box [103]. The final design shortened the chain, thus making it more efficient.[1]

Next must come a word or two about the water-power circuit. Water stored in the upper reservoir [42] is delivered into the constant-level tank [43] by a siphon [73] and so passes to the scoops [32, 44] of the driving-wheel [28], each of which has a capacity of 0·2 cu. ft. (about 12 lb. of water). As each scoop in turn descends the water is delivered into a sump [45]. Apparently the clock was never so located as to be able to profit by a continuous water-supply and instead of this the water was raised by hand-operated norias [47, 49] in two stages to the upper reservoir. Presumably maintenance mechanics from the Ministry of Works came daily to do this, for the dimensions given allow for rather more than 24 hours running. The bearings of these norias were supported on crutched columns [106], and the water reached the upper reservoir along an upper flume [51]. The fact that the whole power-source was thus self-contained within the housing of the clock-tower may have had far-reaching and quite unintended implications. At a later stage, indeed, we shall venture the suggestion that it may have been responsible for inducing, by the repercussions of a misapprehension, the strange idea of the possibility of perpetual motion—an idea which antiquity never knew, either in east or west, but which haunted the later Middle Ages and even the Renaissance (cf. p. 192 below).

We can now examine what Basserman-Jordan has called the soul of any time-keeping machine, namely the escapement. All that Su Sung's draftsman could depict of it we have already seen in Fig. 18 but from the elaborate description in the text above (pp. 41 ff.) it is possible to offer the reconstruction in Figs. 25 and 26. The whole mechanism was called the 'celestial balance' [62], and it did indeed depend upon two steelyards or weigh-bridges acting on each of the scoops in turn. The larger of these steelyards bore the same name as the mechanism as a whole (*t'ien hêng*); we refer to it as the upper balancing lever. The operation of the whole device was as follows. The lower balancing lever [69], suitably weighted at its further end [71], prevented the fall of each successive scoop until full or nearly

[1] This feature of the clock may perhaps be considered the most remarkable of all for its time. We shall dwell upon its great importance for the history of technology at a later point (p. 73 below), in connection with the possibility of its first development a century or so before the time of Su Sung.

full, by means of a 'checking fork' [70]. The adjustability of this weight is an important feature. For although the main part of the time-keeping depended upon the constancy of flow of the water, as if from a clepsydra, the adjustment of the weight on the lower balancing lever could permit the scoops to descend when less than full, and thus allow of the regulation of the time-keeping, to a limited extent, by purely mechanical means. We shall return to this point below (p. 113).

The specifications suggest that release could hardly have occurred more often than once every 10 min., and probably only once every quarter of an hour.[1] But then the descending scoop and its projecting pin had to trip another lever, the 'stopping tongue' [67], which was connected by means of a chain forming a parallel linkage system (the 'iron crane bird's knee' [115])[2] with another weigh-bridge, the upper balancing lever [62], carrying a suitable 'upper weight' [64] at its further

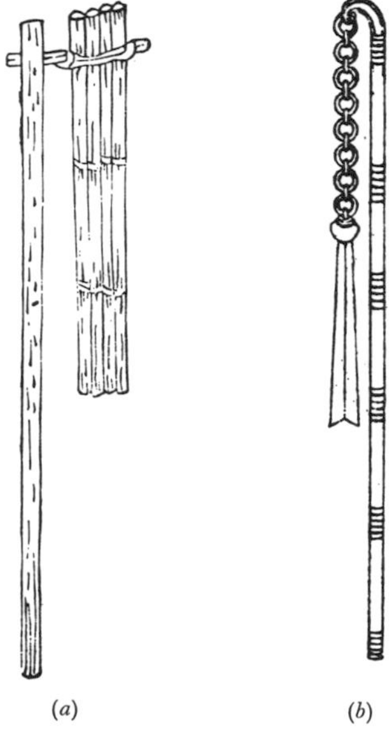

Fig. 31. The origin of a technical term in eleventh-century mechanical engineering. A farmer's flail from the *Nung Shu* of A.D. 1313 compared with an iron war-flail (the 'crane bird's knee') from the *Wu Ching Tsung Yao* of A.D. 1044. The phrase came to be applied to combinations of rods and chains in linkwork.

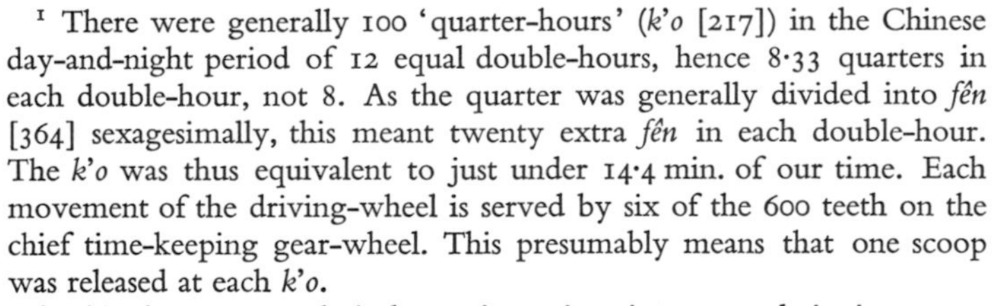

[1] There were generally 100 'quarter-hours' (*k'o* [217]) in the Chinese day-and-night period of 12 equal double-hours, hence 8·33 quarters in each double-hour, not 8. As the quarter was generally divided into *fên* [364] sexagesimally, this meant twenty extra *fên* in each double-hour. The *k'o* was thus equivalent to just under 14·4 min. of our time. Each movement of the driving-wheel is served by six of the 600 teeth on the chief time-keeping gear-wheel. This presumably means that one scoop was released at each *k'o*.

[2] This phrase, as a technical term in engineering, strangely invites one to inquire into its origin. Literary encyclopaedias cite as the first appearance of the expression a passage in an ode by the third-century poet Tso Ssu [685]. In his poem on the capital of the State of Wu (*Wu Tu Fu* [826]), speaking of the defence of the country, he says that 'every household has its "crane bird's knee"' (see *Wên Hsüan*, ch. 5, p. 10*a* [773]). The commentator, Liu Yuan-Lin [686], says: 'The "crane bird's knee" is a sort of *mao*. Its leg is like the calf of the crane's leg, large above and small below; that is why it is called the "crane bird's knee".' We do not know the exact date of this commentary, but it must be before A.D. 530, the time of the compilation of the *Wên Hsüan*, and after A.D. 270. The commonest meaning of *mao* is a lance or spear, but as *ch'ui mao* [436] it means a war-mace or war-flail, and that is how it must be taken here. For indeed the farmer's jointed flail (*lien chia* [437]), shown in Fig. 31 (*a*) from the *Nung Shu* [827] of A.D. 1313 (ch. 14, p. 28*b*), had a military analogue in which the loose end-piece did not simply revolve on an axle but was attached to the handle of the weapon by a chain. This is shown in Fig. 31 (*b*) from the *Wu Ching Tsung Yao* [828] of A.D. 1044 (ch. 13, p. 14*a*), where it is called *t'ieh lien chia pang* [438]. The description says: 'The war-flail with a chain originated with the western barbarian tribes who used it on horseback in combat with Chinese foot-soldiers; it is like the flail of the farmers but more elegant because made of iron. It is in fact useful for striking from above, so Chinese soldiers learnt well how to use it, and employ it effectively against the tribal warriors.' With this background gained, it is easy to see how naturally it would have occurred to artisans making a machine of linked rods and chains to use the expression 'iron crane bird's knee'. Other technicians also made use of

end.[1] This lever was fitted at its fulcrum with a crosswise axle [111] moving in a special concave bearing (a 'camel back' [113] with two 'iron cheeks' [114]), and ended directly above the driving-wheel in an 'upper stop' [55, 68]. As the scoop fell, its pin depressed the stopping tongue, which pulled down the connecting chain [66] and the right-hand end of the upper balancing lever, thereby raising the left-hand end and withdrawing the upper stop from between the empty scoops and pins at the top of the wheel.[2] At the same time as this gate was opened, recoil was prevented by upper locks [72] which inserted themselves in ratchet

derived phrases—thus a poem of which the sound of the words at the ends of the first and third lines did not rhyme was said to have a 'crane's knee'. And Dr Lu Gwei-Djen informs us that in medicine too *ho hsi fêng* [439] was a technical term for rheumatism of the knee.

[1] It will be understood that the left-hand arm of the upper balancing lever is the heavier of the two, so that its normal position is one of interference with the rotation of the driving-wheel. The upper weight would be so adjusted as not quite to compensate for the weight of the left-hand arm, but to facilitate the action of the stopping tongue trip-lever below.

[2] In our first reconstruction of the escapement mechanism (Needham, Wang & Price (1)), following the text as closely as we could, we visualised (and depicted) a different, and opposite, action. We supposed that the 'celestial rod' (*t'ien t'iao* [66]) was a solid rod, and assumed that the fulcrum of the stopping tongue trip-lever [67] was on the side nearest to the wheel, not, as in Fig. 25, on the farther side. We thought therefore that when the scoop pin depressed the stopping tongue, the rod was pushed upwards, thus lowering the left-hand end of the upper balancing lever, and *inserting* the upper stop [55] among the scoops and pins at the top of the wheel. This indeed involved moving the position of the fulcrum from its position in Su Sung's drawing (Fig. 18), but *t'iao* is not one of the more usual words for chain, and seemed to imply a rigid rod. On the other hand, the connection does look more like a chain than a rod in this picture. The first mentions of it in the main text, in sections B and K (pp. 33 and 41 above) are quite ambiguous. Section N is also indecisive. It does not actually say (p. 44 above) which end of the upper balancing lever rises when the lever below is tripped. Later on (p. 45 above) it does not say what each spoke of the great wheel has passed when the upper lock and the upper stop close.

In our discussions with Dr Liu Hsien-Chou at Florence, however, we found that he was convinced that the 'celestial rod' was really a chain (though it may well have been a chain of linked short rod lengths). In his publication too (3), p. 9, he adopted this interpretation. We now believe that he was right in this, and have adjusted our description accordingly. But apart from the weight which one may or may not be disposed to put upon the exact detail of the drawing in Fig. 18 itself, we can find only one piece of positive evidence in favour of the downward-pulled chain. Section K of the main text says distinctly that the 'head' [116] of the stopping tongue trip-lever is attached to the 'celestial rod' or chain (p. 42 above). Since in several other places, the ends of the levers nearest to the great wheel are termed 'head' or 'brain' while the further ends are termed 'tail', this constitutes rather good ground for assuming that the fulcrum must be farther away from the wheel than is the connecting chain, and therefore that this chain must be pulled down when the lever is tripped.

Thus our original reconstruction envisaged the upper stop of the upper balancing lever as inserted only for a short time to assist the settling of the next empty scoop on the lower balancing lever. But the view now generally accepted is that the upper stop was kept inserted most of the time, and only opened briefly like a gate to allow one scoop to pass through.

The suggestion may be worth making that the chain originated as a manually operated device for use with a series of balancing clepsydras mounted on the periphery of a wheel. This will be better understood in the light of the discussion which we give below (p. 85) on the history of the different types of clepsydras in ancient and medieval China. If some transition of this kind were to be imagined, or even substantiated from texts not yet discovered, it would offer a curious parallel with the story of the automation of valve action in the steam-engine.

manner behind each passing scoop.[1] These motions had the effect of bringing the next scoop to rest under the constant stream of water. Clearly the torque on the driving-wheel and hence the whole duty of rotating all the parts of the clock depended on the weight of water in the scoops of the quarter periphery between the lower balancing lever weigh-bridge and their point of discharge into the sump. The work to be done would have required nice adjustment with this. The whole cycle of retention and release is reconstructed as nearly as we can visualise it[2] in Fig. 26.

It will at once be seen that the whole design is reminiscent of the anchor escapement of the late seventeenth century, since the driving-wheel is a scape-wheel and the 'pallets' are inserted alternately at two points on its circumference separated by 90° or less rather than the 180° of the crown wheel. In Fig. 32 we have attempted to indicate graphically the course of events in the escapement cycle, plotting the angle through which the driving-wheel moves forward or backward against time. Obviously the alternation of the 'pallets' occupies only a relatively short time in the cycle, taking place during the period of forward motion of the driving-wheel. Most of the cycle is spent with both 'pallets' inserted, while the scoop is filling. This, however, was something which the water-wheel linkwork escapement had in common with most subsequent escapements, namely that the time it spent in motion was much less than the time it spent at rest. In a modern watch, as Michel (4) has pointed out, the wheels are motionless for $\frac{19}{20}$ of each second, so that they move for only one hour out of every twenty, and in precision clocks a much lesser proportion. Hence the long periods which can be worked by horological machinery without serious wear. It is interesting in this connection to note that Su Sung's clock ran successfully from A.D. 1092 until 1126, and then again for some years

[1] The text talks about two upper locks, a left-hand one and a right-hand one. In Fig. 25 we show but one of these anti-recoil pins or ratchets, on the left at the top of the wheel. Su Sung's diagram (Fig. 18) only vaguely indicates the position of the right-hand ratchet pin, but it probably worked on the upper right-hand quadrant of the great wheel, and for simplicity we have omitted it. Alternatively, the right upper lock might have been a check for the excursion of the upper balancing lever (see Fig. 25).

[2] No doubt the more subtle aspects of Su Sung's mechanism will not reveal themselves until a working model has actually been constructed. We may hope that such a task will be undertaken by one of the great museums of science and technology. Moreover, we should not like to claim that his book has yet given up all its secrets. In particular it will be seen from Fig. 18 that some kind of star-shaped gadget is indicated at the point where the connecting chain [66] crosses the checking fork of the lower balancing lever [69]. We have had much discussion with Dr Liu Hsien-Chou on the possible function of any direct connection between these two parts of the mechanism. Although he was inclined to look favourably upon it, there is no mention of it in the printed text of Su Sung, and none of us has been able to see what function it could possibly have performed.

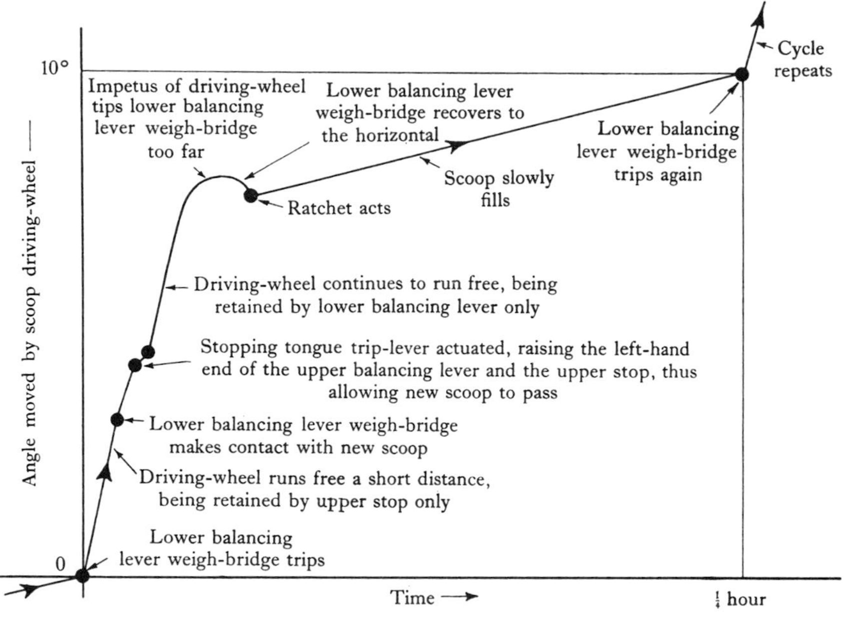

Fig. 32. A diagrammatic time-chart to illustrate the course of events in the escapement action of Su Sung's clock.

at least after its removal to Peking—or so at least the historical records imply (see Ch. VII below).

Of course the solution of the problem of the escapement by chain and link work naturally had a certain medieval awkwardness, but this can hardly be matter of blame for engineers of the eighth century, the real time of origin (as we shall see, p. 78 below) of the device. It differs from the types of escapement of the later West because the action of the arrest is brought about not by any mechanical oscillation but by gravity exerted periodically as a continuous steady flow of liquid fills containers of limited size. The peculiar interest of the water-wheel linkwork escapement, hitherto unknown to historians of technology either in East or West, lies in the fact that it constitutes an intermediate stage or 'missing link' between the time-measuring properties of liquid flow and those of mechanical oscillation. It thus unites, under the significant aegis of the millwright's art, the clepsydra and the mechanical clock in one continuous line of evolution.

VI

THE HISTORICAL BACKGROUND OF SU SUNG'S CLOCK

THE next task which presents itself is to survey the general history of the mechanisation of astronomical instruments—that is to say, the wider setting of the invention of the mechanical clock with its escapement—within the Chinese culture-area before and after the time of Su Sung. The most convenient plan will be to work backwards first, starting from the eleventh century and tracing our way step by step until we reach the first. With the knowledge we shall then have gained we shall be able to make another start from the eleventh century, following the fate of Su Sung's clock itself, and by extension all the events in the world of clock-making (so far as we can assemble them) which took place between the twelfth century and the time of the coming of the Jesuits late in the sixteenth.[1]

[1] There have been few previous attempts to follow this story, and they have seen it very much 'through a glass darkly'. A hundred years ago, a medical missionary in Ningpo, D. J. McGowan, to whom we are indebted for many records of interest on Chinese science and technology, submitted to the Commissioner of Patents of his country, the United States, a memorandum on Chinese horology. The principal object of this was to describe to American clock-makers the kind of clock which would be likely to sell well in the Chinese market, but he prefaced his remarks by an introduction on the history of time-keeping in China which aroused such interest that it was reprinted at least twice in the monthly journals of the day.

McGowan's account of ancient Chinese clocks gives the impression of having been derived from some encyclopaedia such as the *Yü Hai*, relevant parts having been read out to him by a Chinese friend. Lacking detail and documentation it could hardly be very convincing. McGowan knew of an 'orrery' made by 'Tsianghung' (Chang Hêng) between A.D. 126 and 145, and of another in the third century. He stated quite rightly that in the sixth century time was measured by the *weight* of water which fell into a vessel on a balance, and that mercury was also used for this. Though he knew of the sinking clepsydras used by Buddhist monks, he gave but a passing mention of I-Hsing's 'striking clepsydra', praising however as a 'remarkable specimen of art' the later 'orrery', with complex machinery, made by another 'Tsiang' (i.e. Chang Ssu-Hsün) about 980. Finally he gave a garbled version of Chu Pien's hearsay account (see p. 116 below) of Su Sung's astronomical clock, and described vaguely the clock with automata of the Mongol emperor 'Shung-tsing' (i.e. Shun Ti) (see p. 140 below). McGowan added two pictures of clepsydras crudely redrawn from the *Liu Ching T'u* (cf. p. 91).

In recent years the eminent Japanese scholar Shiratori Kurakichi stumbled upon the history of clockwork in the course of his exhaustive studies of Chinese relations with Byzantium. It became clear to him that water-power clocks had been used in medieval China, but as he lacked engineering knowledge his translation of the passages concerning I-Hsing and Liang Ling-Tsan (with *shui yün hun t'ien i* [cf. 195] rendered as 'water-moved astrolabe' for example) is not helpful. Shiratori acutely noted, however, the similarity of the balance-arm clepsydra at Antioch (see p. 88 below) to the Chinese steelyard clepsydras, and even went so far as to suggest

I ARMILLARIES BEFORE SU SUNG

The first text which we may notice is an account given by Shen Kua [475] about A.D. 1086 of the astronomical instruments which he constructed in the Hsi-Ning reign-period about fifteen years before. He thus wrote only a few years before Su Sung criticised his armillary sphere in the opening of the memorial to the emperor which we have just read. It seems clear enough that none of Shen Kua's instruments were mechanised, but that he knew of earlier ones which had been.

Mêng Ch'i Pi T'an [738], ch. 7, p. 8*a*, by SHEN KUA, A.D. 1086

Astronomers have (two main kinds of instruments; first) the armillary sphere [150], which is an apparatus for measuring (positions and distances in) the heavens. It is erected upon some lofty platform for making observations of the starry patterns manifested on high. This is what the ancients knew as the *chi hêng* [159].[1]

(Secondly), there is the celestial globe (or demonstrational armillary) [152], which is essentially a model of the heavens (or, an image of the heavens). It is (often) moved round by falling water, or rotated by mercury [260],[2] and being placed in a closed chamber, it will keep pace with the motion of the heavens like (the two halves of) a tally.[3] This is what was made by Chang Hêng [480] and Lu Chi [500][4] (in the second and third centuries A.D.), and the instrument which was placed in the Wu Ch'êng Hall [927] in the K'ai-Yuan reign-period (of the T'ang dynasty) was nothing dissimilar (eighth century).[5]

Mêng Ch'i Pi T'an [738], ch. 8, p. 5*a*, by SHEN KUA, A.D. 1086

The armillary sphere in the Bureau of Astronomy and Calendar (Ssu T'ien Chien [935]) was made by Han Hsien-Fu [473], who was a Calendar Official in the Ching-Tê reign-period (A.D. 1004–7).[6] It was based on the methods in use at the time of Liu Yao [506][7] (d. A.D. 328) and K'ung T'ing [507] (fl. A.D. 304–29), Ch'ao Ch'ung [508] (fl. A.D. 398–405), and Hsieh Lan

that it might have been Chinese-made. As this type of water-clock was well known among the Arabs, such an origin is very improbable. But conversely Shiratori noticed that Sulaimān al-Tājir, travelling in China in A.D. 851, reported clepsydras with weights in city clock towers.

[1] It will be seen that Shen Kua accepted the usual identification of the armillary sphere with the *hsüan chi* and *yü hêng* of the *Shu Ching* (cf. p. 18). It is interesting that some writers denied it; for example, Chou Mi [540] in his *Ch'i Tung Yeh Yü* [739], about two centuries later (ch. 15, p. 5*b*). In the light of what we now know of its probable nature, they were quite right.

[2] A reference at least to the clock of Chang Ssu-Hsün [479], on whom see p. 70 below.

[3] Cf. p. 101 below.

[4] This is the only text we have noted which attributes a mechanised instrument to Lu Chi (fl. A.D. 220–45), but he is often said to have constructed spheres.

[5] A reference to the clock of I-Hsing [477], on whom see p. 78 below.

[6] Note that Su Sung said nothing about this astronomer's work. We shall see below (p. 68) that he was rather aside from the main line of tradition, and his instruments were certainly not water-powered.

[7] Last emperor of the Former Chao or Northern Han dynasty, a very short one (A.D. 304–29) in the Shih-liu Kuo (Sixteen Kingdoms) part of the Six Dynasties period. He was the patron of K'ung T'ing.

[509] (fl. A.D. 415).[1] It was therefore much too simple [215]. The armillary sphere in the Department of Astronomy (of the Han-Lin Academy) (T'ien Wên Yuan [915]), on the other hand, was made by Shu I-Chien [474] who was Director of Astronomical Observations (Northern Region) in the Huang-Yu reign-period (A.D. 1049–53).[2] This adopted the methods of Liang Ling-Tsan [481] and the monk I-Hsing [477] in the T'ang (eighth century A.D.). It is very elaborate and complicated but difficult to use [216]. So in the Hsi-Ning reign-period [A.D. 1068–77)[3] I myself constructed an armillary sphere,[4] and also invented[5] a jade clepsydra with floating indicator, and a bronze gnomon, all of which were placed in the Bureau of Astronomy and Calendar, under the care of special officials. Meanwhile the old bronze armillary[6] of the Bureau of Astronomy and Calendar was sent to the imperial storehouse of vestments and objects for further reference and study.

The following passage, taken from the *History of the Sung Dynasty*, introduces us to some of the basic history of the armillary sphere in China, and shows us the active interest taken in such instruments by imperial majesty itself. But it also indicates that during the prosperous days of Jen Tsung's reign no particular need was felt for mechanical clocks such as had previously existed, and such as that which Su Sung was presently to build again.

Sung Shih, ch. 76, pp. 1*a* ff., by TOKTAGA [630] & OUYANG HSÜAN [631], *c.* A.D. 1345

The Armillary Sphere of the Huang-Yu reign-period (A.D. 1049–53)

The emperor Yao ordered his ministers Hsi and Ho[7] to make the sighting-tube (lit. horizontal traverse [218]) to observe the degrees of the positions of the stars. The rings and tube [159] were made of hard jade[8] because they wanted the apparatus to endure all weathers and to be always

[1] All these astronomers were of the Liu Ch'ao (Six Dynasties) period. Their designs were a good deal simpler than those of the T'ang, when Li Shun-Fêng in the seventh century and then I-Hsing in the eighth introduced important developments.

[2] According to a passage in the *Yü Hai*, ch. 4, p. 38*a*, this instrument had gear-teeth of some kind, so that it may have been mechanised.

[3] The emperor for whom Shen Kua was working was Shen Tsung [951], personal name Chao Hsü [511] (A.D. 1048–85, reigned 1067 onwards). A ruler of merit, but somewhat disposed to fruitless military operations. Inclined to the Reform party, and selected Wang An-Shih as minister. Served also by many other thinkers and technicians of eminence—Chou Tun-I, Chang Tsai, the Ch'êng brothers, and Ssuma Kuang.

[4] Here Shen Kua mentions nothing resembling gear-wheels. But in his memorial to the emperor [951] of 1074, given in *Yü Hai*, ch. 4, pp. 35*b* ff., where he enumerates the special features of his armillary sphere, there is a curious passage. The ninth special feature is described as follows: 'In the instrument now at the Bureau of Astronomy teeth are arranged around the back of the ring of the components of the Three Arrangers of Time [217, 4], so that they do not meet with the sighting-tube [218, 132]; they ought to be moved and placed on each side (as is done in my new instrument).' The word for teeth here [219] is not very usual in the sense of gear-teeth, and perhaps we may suggest that what he was talking about were studs alongside the degree graduations permitting of taking readings by touch only during night observations. Later instruments certainly had them.

[5] Actually in A.D. 1074.

[6] Presumably Shu I-Chien's.

[7] Legendary characters. See n. 1 on p. 18 above.

[8] Also a legend. See n. 2 on p. 18 above.

movable, not decaying with age. In later generations, bronze was cast to make the armillary ring [220] in representation of the (real) celestial sphere. From the time of Lohsia Hung [512] who made the T'ai-Ch'u [936] calendar (104 B.C.) and used an armillary instrument, down to the time of the emperor Ho Ti of the Eastern Han (reigned A.D. 89–105), there was only the equatorial (ring).[1] But the projections of the *hsiu* (equatorial lunar mansion) extensions on the ecliptic varied somewhat with the seasons.[2] So when the emperor asked the Imperial Astronomical Counsellor Yao Ch'ung [513] (fl. A.D. 92) and others the reason, they all answered: 'The pattern of the stars (the constellations) has its (own) standard path to follow, but the motions of the sun and the moon take the ecliptic path. It is because we have had no corresponding ring in the instrument that errors have arisen.' The ecliptic ring was thus first established in the fifteenth year of the Yung-Yuan reign-period (A.D. 103) by Chia K'uei [514] (A.D. 30–101).[3] In the seventh year of the Yen-Hsi reign-period (A.D. 164),[4] Chang Hêng [480] altered the system by adopting a scale of four-tenths of an inch to the degree. Afterwards Lu Chi [500] (fl. A.D. 220–45), Wang Fan [478] (A.D. 219–57), K'ung T'ing [507] (fl. A.D. 304–29), Hsieh Lan [509] (fl. A.D. 415), Liang Ling-Tsan [481] (fl. A.D. 710–30) and Li Shun-Fêng [502] (fl. A.D. 627–80) each made (an armillary sphere). Since the great disturbances of the Five Dynasties period (A.D. 907–60), the methods handed down have largely been lost.[5] In the beginning of the Hsiang-Fu reign-period (A.D. 1008–16), Han Hsien-Fu [473] made (1010) an armillary sphere in which only the polar-mounted split declination ring [221] turned with the sighting-tube, the ecliptic and equatorial rings being fixed and immovable.

In the beginning of the Huang-Yu reign-period (A.D. 1049) the emperor ordered the calendar experts, Shu I-Chien [474], Yü Yuan [515] and Chou Ts'ung [516] to consult with others, and they adopted some of the systems and methods of Li Shun-Fêng and Liang Ling-Tsan.

[1] This historical account is quite reasonable. From the time of Shih Shen and Kan Tê in the fourth century B.C. down to the end of the first century B.C., the astronomers may well have used only a single graduated ring (with some kind of alidade or fiducial points), set up in the equatorial or meridian planes as desired. Measurements on the former gave them the position in the *hsiu* (lunar mansions), which was their equivalent of right ascension, and measurements on the latter gave them the north polar distances, which was their equivalent of declination. It is worth observing, in parenthesis, that the Chinese system of star co-ordinates was from the beginning essentially the same as the modern system—unlike the Greek, which was based on ecliptic co-ordinates. Cf. Needham (1), vol. 3, p. 266.

[2] This little problem in spherical geometry was a constant source of trouble for the ancient Chinese astronomers, and we shall see later (p. 108) how Chang Hêng solved it graphically with the aid of a spherical ruler.

[3] The discrepancy in this date arises from the fact that Chia K'uei's proposal for the addition of a permanently fixed intersecting ecliptic ring was not adopted until just after his death.

[4] This date is impossible, for it is agreed on all hands that Chang Hêng died between A.D. 138 and 140. The actual time must have been between 110 and 130. This mistake occurs in many of the accounts (cf. pp. 101 ff.); the only explanation we can offer for it is that there was another reign-period of similar name, the Yen-Kuang reign-period, which lasted from A.D. 122 to 125. But it did not have a seventh year, it ended at the fourth. There was also the Yen-P'ing reign-period, of but one year, A.D. 106. Confusion among reign-periods is a not uncommon source of error.

[5] This text illustrates how every historian who was responsible for the astronomical chapters in the dynastic histories gave short general summaries of the history of astronomy and astronomical instruments, often several times over. Recorded speeches, too, continually go over the same ground. The texts are consequently very repetitive. But the details vary a little, and it would be most desirable to have a precise and thorough study of this corpus of the history of science, with all variant statements—a task which has only been attempted once before in a western language, in the work of Gaubil (1, 2) in 1732.

Accordingly an armillary sphere with an ecliptic mounting [222] was reconstructed and cast. Besides this, clepsydras and gnomon scales [223] were also made. The emperor ordered the Academician Ch'ien Ming-I [517] to record the methods in detail and one of the Palace stewards Mai Yün-Yen [518] to take charge of the construction work. After completion, the armillary sphere was put in the Observatory [224] of the Department of Astronomy (of the Academy) [915], the clepsydra in the Bell-and-Drum Tower of the Wên Tê Hall [937] (of the Palace) and the Gnomon Scale in the Bureau of Astronomy and Calendar [935]. The emperor himself[1] wrote a book entitled *Hun I Tsung Yao* (Brief Account of the Armillary Sphere) [732] in ten chapters, discussing the merits and shortcomings of the (instruments of) the previous dynasties, but afterwards this book was kept in (the Palace) and not circulated. We shall record in what follows the design of the ecliptically mounted sighting-tube and celestial latitude ring [225]....

2 DIAL-FACES AND ANAPHORIC CLOCKS

It will have been noted that the description of Su Sung's clock contains nothing resembling a dial. Although the stationary dial-face with moving pointer is a development associated with the first European mechanical clocks of the fourteenth century, the rotating dial-face had been used in Hellenistic times. The anaphoric clock,[2] described about 30 B.C. by Vitruvius (IX, 8), consisted of a bronze disc with a planispheric projection of the stars of the northern hemisphere, and as many as are found between the equator and the tropic of Capricorn which formed the rim of the disc. The circle representing the zodiac was provided with 365 (or 182 or 91) small holes, into which was plugged from day to day a little stud representing the sun. The disc was made to rotate by the simple mechanism of a clepsydra containing a float attached to a cord terminating in a counterweight and wound round a drum on the horizontal axis bearing the planispheric

[1] This emperor was the fourth of the House of Sung. His personal name was Chao Chên [519] and his temple name (that by which he is usually known) Jen Tsung [946] (A.D. 1010–63). Succeeding to the throne in 1022, he ruled for forty years as a most enlightened monarch. Hostilities with the Hsi-Hsia State or with Korea were avoided or minimised; great literary enterprises such as the *Hsin T'ang Shu* [761] (New History of the T'ang Dynasty) (1061) were successfully brought to completion. In the ranks of his civil servants and scholars were all four founders of the great philosophical school of Neo-Confucianism. Important innovations included the issue of paper money in Szechuan in 1023. Chao Chên was long remembered for his humane and courageous actions during a plague at the capital in 1054. He was fortunate in that the violent quarrels between the conservative and reforming parties had hardly begun in his time. He was by no means the only emperor to take an interest in scientific or proto-scientific matters. The Cambridge University Library possesses an illustrated manuscript study of meteorological phenomena (e.g. parhelia) written by the Ming emperor (Chu Kao-Chih [661] (also, by coincidence, Jen Tsung [946]) in A.D. 1425. The decision that Chao Chên's book on armillary spheres should be 'on the secret list' and not printed, was equivalent to a death-sentence upon it—much to our loss.

[2] So called because it portrayed the successive rising (ἀναφορά) of constellations above the eastern horizon during the night.

disc.[1] In what follows we shall suggest that this mechanism may also have been used in China at some stages.[2]

The disc of the Hellenistic anaphoric clock was separated from the spectator by an immobile network of bronze wires, a vertical wire representing the meridian, three concentric wires the equator and the tropics, and between the tropics other wires indicating the zodiacal months. Across all the circles was an arc representing the horizon of the place, and there were crosswires dividing the concentric circles into 12 day hours and 12 night hours, above and below the horizon respectively. Neugebauer (1) and, more completely, Drachmann (1) have shown that this arrangement was the forerunner of the astrolabe of medieval times with its *rete*. We do not have to follow this development further here since the astrolabe was not known or used in Chinese civilisation. But the question does arise whether any similar kind of planispheric clock was employed there. Apart from the Vitruvian description, actual bronze engraved discs from the Roman empire have survived, notably one of about A.D. 250 found at Salzburg and described by Benndorf, Weiss & Rehm (1). No such objects have yet come to light in China, or at least none such has so far been recognised. However, there are certain

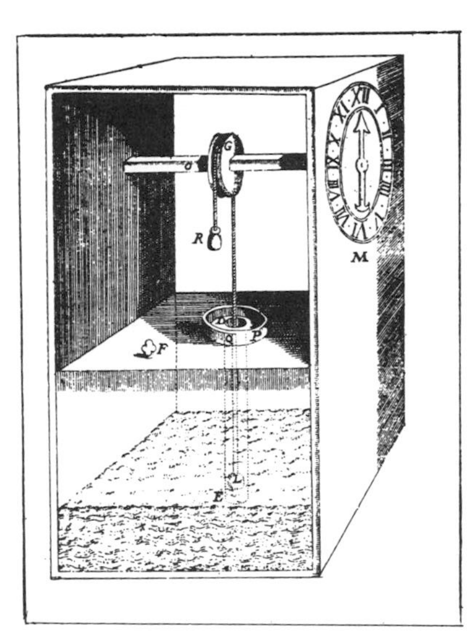

Fig. 33 (*a*). The anaphoric clock of Hellenistic times surviving in an illustration of A.D. 1644 by Isaac de Caus. Anciently the whole dial rotated, but by this time a stationary dial-face with mobile pointer had come into use.

[1] Diels (1), p. 213, Price (1), Usher (1), p. 97, 2nd ed. p. 145. The anaphoric clock lasted in active use until at least the end of the seventeenth century, and we illustrate it from the book on hydraulics published by Isaac de Caus in 1644 (Fig. 33*a*). By this time the dial had become stationary, with a pointer travelling over it. Isaac was the son of that Solomon de Caus whose name is associated with the early history of the steam-engine and the rolling-mill.

[2] We do not know of any Chinese illustration of it until the seventeenth century, when Johann Schreck and Wang Chêng gave one in their *Ch'i Ch'i T'u Shuo* [835] of 1627 (ch. 3, pp. 53*b* ff.). This is reproduced in Fig. 33*b*. The late Professor Fritz Jäger of Hamburg identified without difficulty the original of this picture in the 'Machinae Novae' of Fausto Veranzio (Venice, 1616), as may be seen from his unpublished notes. Later on we shall refer more fully to the Jesuit missionary scientist Schreck and to his Chinese collaborator (p. 145 below). The fact that they illustrated an anaphoric clock as something new can hardly be taken to imply that the principle had never been known or used in ancient or medieval China. Many centuries had passed by in many provinces before their time, and all we can say is that if knowledge of it there had been, the tradition had not come down to Wang Chêng and his circle. On the comparative history of dial-faces (mobile or immobile) and pointers, more must be said in the appropriate place (see p. 159 below).

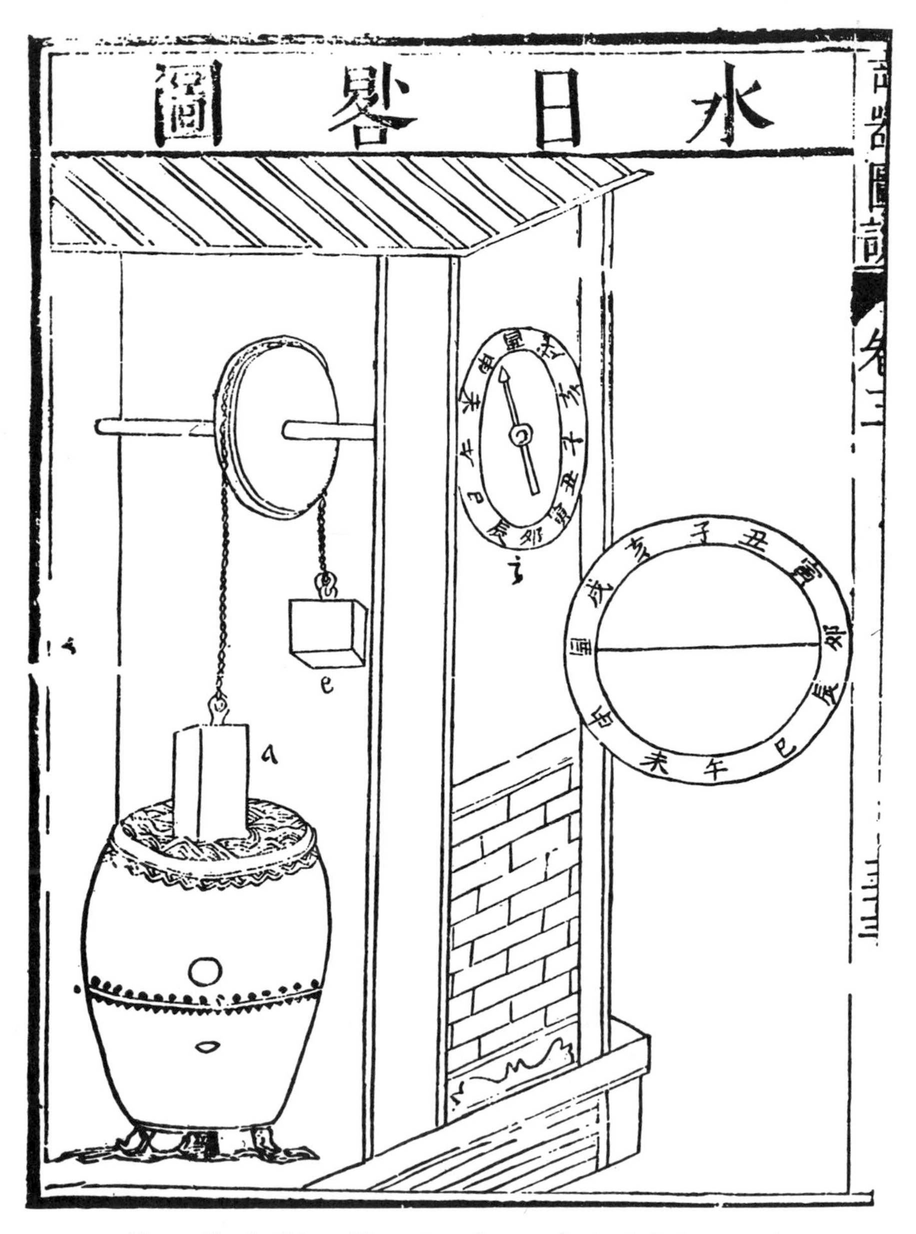

Fig. 33 (*b*). A Chinese illustration of an anaphoric clock (A.D. 1627).

literary evidences which go some way to suggest that anaphoric clocks were not unknown there. Here is one text from the thirteenth century, referring to devices which were in use in the early part of the eleventh. We introduce the subject at this point for chronological reasons but we shall refer to it again in later connections (p. 80), notably to point out that the power provided by any conceivable kind of float mechanism would have been insufficient to rotate the comparatively heavy armillary spheres and celestial globes repeatedly mentioned in the Chinese accounts.

Tung T'ien Ch'ing Lu [733], in *Shuo Fu* collection, ch. 12, p. 29*b*, by CHAO HSI-KU [521], *c.* A.D. 1235

> Fan Wên Chêng Kung (Fan Chung-Yen [522], A.D. 989–1052)[1] had in his home an ancient mirror. On the back the 12 (double-) hours were marked by hemispherical protuberances like (*wei-ch'i*) chessmen [226], and whenever one of these hours arrived, (one of them) shone like the (full) moon.[2] The whole thing rotated without ceasing. There was also a twelve-bell double-hour device [227] in another scholar's house which responded to the (double-) hours, sounding them automatically [228].[3] What magical powers these old bronze objects have![4]

A second relevant passage comes from the early tenth century and concerns some device which had been in use at the beginning of the ninth. Though it points to the existence of something like an anaphoric clock, it is not decisive evidence on the matter, since the plate may only have been a mirror-back mnemonically ornamented. We are not distinctly told that the dial turned automatically.

Ch'ing I Lu, ch. 2, p. 18*b* [736], by T'AO KU [532], between A.D. 902 and A.D. 970

> In the Palace Treasury of the T'ang Dynasty there was a yellow plate with a circumference of 3 ft. Around the disc there were designs of animals and other things. In the Yuan-Ho reign-period (A.D. 806–20) it was occasionally used to see how the symbols changed following the passing of the hours of the day. For instance, at the Ch'en hour (i.e. double-hour) (7–9 a.m.) there was a dragon playing among a decoration of flowers and herbs, but when the plate turned (automatically?) to the Ssu hour (i.e. double-hour) (9–11 a.m.) a snake appeared, while when the hour (i.e. double-hour) of Wu came (11 a.m.–1 p.m.), then the turning plate showed

[1] An eminent scholar and official in the reign of Jen Tsung. Governor of Yen-an in Shensi; strongly opposed to Buddhism. For a biography, see Fischer (1).

[2] This suggests some kind of lamp shining through a hole in the disc. Lamps lighting with the hours were a feature of many of the Arabic striking water-clocks (Wiedemann & Hauser (1); Needham (1), vol. 1, pp. 203 ff.), e.g. those described by al-Jazarī in A.D. 1206 (Fig. 34).

[3] Cf. the many examples of jack-work elsewhere described in this monograph.

[4] These remarks of Chao Hsi-Ku occur in the course of an argument that old bronze objects are good demonifuges and have other strange virtues.

a horse. (These were the correct symbolic animals for the times.) It was therefore called 'the Twelve (Double-) Hour Plate' [251]. This instrument was handed down (at the end of the dynasty) and was still in existence during the Later Liang dynasty (A.D. 907–23) of the House of Chu.

We shall return presently (p. 159) to the question of the anaphoric clock.

3 THE DECEPTIVE LANGUAGE OF HAN HSIEN-FU (ELEVENTH CENTURY)

In the next passage we catch a glimpse of instrument-making activity at the end of the tenth century A.D. under the emperors T'ai Tsung and Chen Tsung. It has nothing new to teach us about astronomical clockwork except one thing: the necessity for the closest scrutiny of the texts *as a whole*. In this description of the work of the astronomer Han Hsien-Fu there are at least four technical terms (see our annotations) which could be extremely misleading since all of them occur, *with different meanings*, in the description of Su Sung's clock. 'Horizontal wheel' is not a gear-wheel but a flat annulus round an armillary sphere, 'dragon columns' are not transmission-shafts but pedestals for the instrument, 'celestial wheels' are not gear-wheels but armillary rings, and 'slanting wheel' is not a toothed ring or pinion but just the equatorial ring. The whole description is in fact a 'cautionary tale', and illustrates well the great difficulty experienced in the Middle Ages—and by no means only in China—in coining precise new terms for new things. Perhaps Su Sung derived some of his terms from words which had previously been loosely used in the Bureau of Astronomy for earlier devices, but there seems something distinctly anomalous about Han Hsien-Fu's terminology. We should not be surprised to find other evidence that he stood in some way outside the main stream of astronomical tradition.

Sung Shih, ch. 48, pp. 4*b* ff., by TOKTAGA & OUYANG HSÜAN, *c.* A.D. 1345. Parallel text in *Yü Hai* (Encyclopaedia of WANG YING-LIN, A.D. 1267), ch. 4, pp. 30*a* ff., which is in several places better.

On a kêng-ch'en day[1] in the twelfth month of the first year of the Chih-Tao reign-period (A.D. 995) there was cast a new bronze armillary sphere.[2] Han Hsien-Fu [473], Director of

[1] This was the usual method of identifying days. The ten cyclical signs (*kan* [229]) and the twelve cyclical signs (*chih* [230]) formed by permutation and combination a regularly recurring series of sixty. This kind of day-count goes back probably to the Shang period (fourteenth to eleventh centuries B.C.). From the fourth century B.C. this cycle of sixty was applied to the years also.

[2] The *Sung Shih* has *t'ung hou i* [158]; the *Yü Hai* has *t'ung hun i* [160]. Additional reason for believing them to be the same thing.

PLATE III

Fig. 34. A striking water-clock from a MS. of al-Jazarī's treatise on mechanical contrivances (*Kitāb fī ma'rifat al-ḥiyal al handasīya*), A.D. 1206. At the top, the signs of the zodiac exhibited, then figures successively appearing and lamps successively illuminated, below that, golden balls dropped into brazen cups from the beaks of brazen falcons to strike the chime; lastly, an automaton orchestra of five musicians (after Cresswell). So far as we know, all such Arabic (and Byzantine) water-clocks worked on the anaphoric principle.

PLATE IV

Fig. 35. The Imperial Library in A.D. 1007 at the capital of the (Northern) Sung dynasty, K'ai-fêng; the emperor attending a reading.

Astronomical Observations (Western Region),[1] who was a specialist in astronomical instruments, had memorialised at the beginning of the Shun-Hua reign-period (A.D. 990) asking that this should be done. The emperor[2] had therefore ordered that funds should be made available so that (Han) could start planning and select the cleverest metal-workers for the job. Looking round at his counsellors he said that though the methods of old had been lost, he had trusted (Han) Hsien-Fu's knowledge of the Yin and Yang and of calendrical science, and so had given him the order to study them. 'Now', he went on, 'I have watched the motions of the sun and moon, dawns, dusks and seasons, and the measurements of the positions of stars and constellations shown on the instrument, and seen that checked by the sighting-tube it shows not a hair's-breadth of error. So I believe that it is verily the treasure of the Observatory.' Then he ordered the construction of a special emplacement for it in the Bureau of Astronomy and Calendar, and a report on the matter was sent to the Bureau of Historiography for recording. Fifty rolls of multi-coloured silk were bestowed upon (Han) Hsien-Fu.

(Han) Hsien-Fu's armillary had nine special features.[3] [In abbreviated form: (1) split meridian ring, (2) movable ring carrying sighting-tube, (3) axle of sighting-tube, (4) special sighting-tube.] The fifth was called the horizontal annulus[4] (lit. wheel [231]);[5] it was 6·13 ft. in diameter and 18·39 ft. in circumference,[6] and it was placed outermost above the water-levels [232]. The surface showed the eight *kua* [233],[7] the ten cyclical signs [229], the twelve cyclical signs [230],[8] the twenty-four fortnightly periods [199] and the seventy-two five-day periods [200].[9] Further in (on the annulus) there were marks which indicated the times of sunrise and sunset (at the solstices) corresponding to the four azimuth points [234],[10] the middles of the (double-) hours, and the hundred quarters of each day and night. [Abbreviation continued: (6) equatorial ring, (7) ecliptic ring.] The eighth was called the dragon columns [235],[11] of which there were four, each 5·5 ft. long, situated beneath the horizontal annulus (lit. wheel),

[1] See n. 5 on p. 20 above. A slight discrepancy in the official post with the following text will be noticed.

[2] T'ai Tsung [948] (personal name Chao Ching [524], A.D. 939–97), the second of his line. His reign was on the whole prosperous in spite of risings within and wars with the Ch'i-tan (Liao) Tartars without. A scholarly man, he paid great attention to education, finance and historical study.

[3] The Chinese terms for these will be found conveniently rehearsed in Wang Ying-Lin's *Hsiao Hsüeh Kan Chu*, ch. 1, p. 12*b*.

[4] We take this to have been a broad flat ring in the horizon plane bearing multiple graduations. As we shall see, something similar had been incorporated in the instruments of Ch'ien Lo-Chih more than 500 years earlier (see p. 97).

[5] This term *lun* (*p'ing chun lun*) is very deceptive, for at first sight it seems to be something like the horizontal wheels of Su Sung's clock. But the context shows that it was quite different.

[6] These measurements correspond to 0·5 in. to the degree.

[7] These symbols are the eight basic trigrams of the *I Ching* (Book of Changes), and had manifold meanings. See Needham (1), vol. 2, pp. 304 ff.

[8] The word here used, *ch'en*, is one of the synonyms for the twelve *chih* (cf. *Hsiao Hsüeh Kan Chu*, ch. 1, p. 21*b*).

[9] Note that these are all divisions of the year. The graduations on the annulus were no doubt a kind of tabulation for interpreting degrees on the equator.

[10] For this term *wei*, cf. *Hsiao Hsüeh Kan Chu*, ch. 2, p. 1*b*.

[11] Again this term, *lung chu*, is very deceptive, for at first sight it looks like the name of the vertical shaft in the clock of Su Sung. But here it undoubtedly means just the supporting columns of the instrument—and it will be remembered that Su Sung himself used the words in both senses.

because the dragon can transform and change itself and govern the celestial bodies. The ninth was called the water-levels [232], arranged in the form of a cross....

All this was based upon the Ch'ien-Yuan calendar (of Wu Chao-Su [525], A.D. 981). (Han) Hsien-Fu presented to the emperor a work in ten chapters entitled *Hun I Fa Yao* [734] (Brief Account of the Armillary Sphere),[1] and it was deposited in the (Imperial) Library....[2]

Yü Hai, ch. 4, pp. 32*b* ff.

On a chia-yin day in the intercalary second month of the third year of the Hsiang-Fu reign-period (A.D. 1010) the Astronomer-Royal reported that Han Hsien-Fu, Director of Astronomical Observations (Northern Region),[3] had completed the construction of (another) (observational) bronze armillary sphere [158, 160].[4] The emperor[5] ordered that it should be moved to the Lung T'u Hall [938] (of the Palace), and asked (Han) Hsien-Fu to select students worthy of receiving instruction from him in these matters.

On a wu-yin day in the eleventh month the emperor called all his ministers to assemble in the Lung T'u Hall and examine the bronze armillary sphere. Its works included two astronomical rings (lit. celestial wheels [236]),[6] one horizontal [237] (i.e. the horizon) and one slanting [238] (i.e. the equator), each having 362 (i.e. 365·2) degrees. (Within) the slanting equatorial ring (lit. wheel) and (its associated) ecliptic ring, there was the sighting-tube for making correct observations of the heavenly bodies.[7]

4 THE MERCURY DRIVE OF CHANG SSU-HSÜN (TENTH CENTURY)

With Chang Ssu-Hsün's apparatus we come again to an indubitable astronomical clock, composed mainly of an armillary sphere (demonstrational), or a celestial globe, activated by a scoop-bearing driving-wheel and gearing something like Su Sung's, and not lacking mechanical figures to sound the hours. No less than eleven technical terms are used in the description with exactly the same meanings as in Su Sung's text, without which, of course, it would not have been possible to

[1] His book is duly listed in the dynastic bibliography (*Sung Shih*, ch. 206, p. 9*b*).

[2] See Fig. 35, which depicts the emperor attending a reading in the Imperial Library in A.D. 1007, only a few years later.

[3] Cf. n. 5 on p. 20. His biography will be found in *Sung Shih*, ch. 461, pp. 6*b* ff.

[4] Here there is the difference already noted between the two texts.

[5] Chen Tsung [947] (personal name Chao Hêng [527], A.D. 968–1022). He was particularly devoted to Taoism so it is not surprising that he took a great interest in the astronomical equipment made by Han Hsien-Fu. Unbroken peace prevailed in his time, modified by the necessity of paying danegeld to the Ch'i-tan Liao, and sustained perhaps in part by elaborate Taoist mystifications designed to show that the emperor was particularly favoured by Heaven (cf. Wieger (1), vol. 2, pp. 1572 ff.).

[6] Again a very confusing piece of terminology. It is very unusual to find armillary rings called 'celestial wheels' (*t'ien lun*), and at first sight it was tempting to suppose that gear-wheels were meant, such as those which Su Sung had at the top of his vertical shafts and used for driving his globe and sphere.

[7] Still more deceptive is the term 'slanting wheel' (*ts'ê lun*) which would seem more probably connected with oblique gearing like Su Sung's than simply another name for the equatorial armillary ring. But the context, which includes nothing whatever about any driving mechanism or water-power, indicates clearly that Han Hsien-Fu was making observational armillary spheres and not astronomical clocks.

interpret the present passage. It is evident that Chang's clock was a very fine construction, particularly interesting in that mercury was employed in the closed circuit instead of water, thus enabling time to be kept even in hard freezing weather. But it must have been rather beyond the general level of the time, for Su Sung tells us (p. 19 above) that after Chang's death it very soon went out of order and there was no one who could keep it going. It is interesting that Chang Ssu-Hsün was a Szechuanese, for that populous province in the west had been the scene, just previously during the late T'ang and Five Dynasties periods, of the earliest expansion of another admirable invention, that of block printing. Liu P'ien [528] tells us how in A.D. 883, on his holiday outings, he used to examine the printed books being sold outside the city wall of Ch'êng-tu [857] (cf. Carter (1), p. 60). Notable, too, is the fact that these were mostly on proto-scientific subjects (oneiromancy, geomancy, astrology, planetary astronomy, and speculations of the Yin-Yang school).

Sung Shih, ch. 48, pp. 3*b* ff., by Toktaga & Ouyang Hsüan, *c.* 1345 A.D. Parallel text in *Yü Hai* (Encyclopaedia of Wang Ying-Lin, A.D. 1267), ch. 4, pp. 29*a* ff., which is in several places better. (Texts conflated in the translation.)

At the beginning of the T'ai-P'ing Hsing-Kuo reign-period (A.D. 976) the Szechuanese Chang Ssu-Hsün [479], a Student in the Bureau of Astronomy [939], invented an astronomical clock (lit. an armillary sphere [150])[1] and presented (the designs) to the emperor T'ai Tsung,[2] who ordered the artisans of the Imperial Workshops to construct it within the Palace. On a kuei-mao day in the first month of the fourth year (A.D. 979) the elaborate machine was completed, and the emperor caused it to be placed under the eastern drum-tower of the Wên Ming Hall [934].[3] The system of Chang Ssu-Hsün was as follows: they built a tower of three storeys (each) over 10 ft. in height, within which was concealed all the machinery. It was round (at the top to symbolise) the heavens and square (at the bottom to symbolise) the earth. Below there was set up the lower wheel [239],[4] lower shaft [240][4] and the framework base [140].[5] There were also horizontal wheels [241], (vertical) wheels fixed sideways [242], and slanting wheels [243]; bearings for fixing them in place [244]; a central stopping device [245][4] and

[1] So the text distinctly says, but we are not sure that it may not have been a globe (i.e. a solid ball portraying the constellations).

[2] Personal name Chao Ching [524]; see n. 2 on p. 69 above. Chang Ssu-Hsün was working at the beginning of his reign, Han Hsien-Fu towards the end of it. Presumably the latter rose to prominence after the death of Chang and the breakdown of his machinery.

[3] Here the *Yü Hai* text substitutes 'in the Clepsydra Room at the south-east corner of the Wên Ming Hall'. It often gives more correct information taken from older, more contemporary documents. In any case, both versions make it clear that the apparatus was under cover, and therefore not an observational armillary sphere.

[4] All these terms occur with minor variations but exactly the same meanings in the description of Su Sung's clock (see above, pp. 28 ff.).

[5] This term will be seen in Su Sung's diagrams (ch. 3, pp. 4*a*, 6*b*, 7*b*) and text (p. 4*b*). Cf. Figs. 8, 10.

a smaller stopping device [246][1] (i.e. the escapement); with a main transmission shaft [38].[1] Seven jacks rang bells on the left, struck a large bell on the right, and beat a drum in the middle to indicate clearly the passing of the quarter (-hours). Each day and night (i.e. each 24 hours) the machinery made one complete revolution, and the seven luminaries moved their positions around the ecliptic. Twelve other wooden jacks were also made to come out at each of the (double-) hours, one after the other, bearing tablets indicating the time. The length of the days and nights were determined by the (varying) numbers of the quarters (passing in light or darkness). At the upper part of the machinery there were the top piece [247], upper gear (-wheel or -wheels) [248],[1] upper stopping device [55][1] (escapement), upper (anti-recoil) ratchet pin [249], celestial (ladder) gear-case [103],[1] upper beam [17][1] (of the framework), and the upper connecting-rod [66].[1] There were also (on a celestial globe) the 365 degrees (to show the movements of) the sun, moon and five planets; as well as the Purple Palace [88] (north polar region), the lunar mansions (*hsiu*) in their ranks and the Great Bear; together with the equator and the ecliptic which indicated how the changes of the advance and regression of heat and cold depend upon the measured motions of the sun. The motive power of the clock was water, according to the method which had come down from Chang Hêng in the Han dynasty through I-Hsing and Liang Ling-Tsan in the K'ai-Yuan reign-period (A.D. 713–41). But the bronze and iron (of their clocks) had long gone to rust [250] and could no longer move automatically.[2] Moreover, as during winter the water partly froze and its flow was greatly reduced, the machinery lost its exactness, and there was no constancy between the hot and cold weather. Now, therefore, mercury was employed as a substitute, and there were no more errors....

The images of the sun and moon were also attached high up (to the globe) and according to the old method they had been moved by human hand (each day), but now success was attained in having them move automatically.[3] This was a marvellous thing. (Chang) Ssu-Hsün was considered the equal of the T'ang clock-makers and was made Assistant in charge of the Armillary Sphere (or Clock).

Fêng Ch'uang Hsiao Tu [735], ch. 1, p. 1*b* (Maple-Tree Window Memories), by a Mr Yuan [529], completed by a later writer soon after A.D. 1202

In the T'ai-P'ing Hsing-Kuo reign-period (A.D. 976–83) the Szechuanese Chang Ssu-Hsün [479] made an astronomical instrument (lit. an armillary sphere [150]) different from the older ones. It was of extremely ingenious construction and was set up in a tower of several storeys (each) more than ten feet high, having seven jacks made of wood each in charge of one of the seven luminaries (sun, moon and five planets). These automatically struck bells and beat upon drums. And there were twelve jacks in the form of immortals, each in charge of one of the (double-) hours, so that when one of these arrived, the jack came out holding a tablet to

[1] All these terms occur with minor variations but exactly the same meanings in the description of Su Sung's clock (see above, pp. 28 ff.).

[2] This had become a stereotyped phrase (cf. p. 79).

[3] Evidently some complex gearing was employed. A suggested reconstruction by Liu Hsien-Chou (3) is given in Fig. 37. For a glimpse of this a few centuries later, see also p. 121 below.

report it, and all of them did this one after the other. My great-grandfather the Venerable (Yuan) Tsan-Shan [530] once entered the clepsydra room of the Wên Ming Hall [934] (of the Palace) and saw it himself.[1]

There is one point in the description of Chang Ssu-Hsün's clock mechanism which deserves particular comment. This is the expression which we have translated 'celestial (ladder) gear-case'. If it is to be taken in exactly the same sense as in the main text of Su Sung (p. 47 above), it clearly indicates the use of a chain-drive towards the end of the tenth century A.D. Unfortunately as no further description is given, and as no illustrative diagrams have come down to us, one cannot be quite sure. But a chain-drive towards the end of the eleventh century is extraordinary enough, and some might well consider it the most remarkable of all the features of Su Sung's clock, since the linkwork escapement had originated earlier. Although an endless belt of a kind had been incorporated in the magazine arcuballista of Philon of Byzantium (second century B.C.),[2] there is no evidence that this was ever actually built, and it certainly did not transmit power. Much likelier as the source of inspiration for Chang Ssu-Hsün or Su Sung was the square-pallet chain-pump [434, 435] working in a flume; this has been widespread in the Chinese culture-area during nearly two millennia—for its origin can be traced back at least to the second century A.D. and probably to the first.[3] Of course this also was for conveying material and not for transmitting power from one shaft to another—hence the originality of Chang or Su. In fact, historians of engineering[4] mention no chain-drive in the true sense in Europe until the nineteenth century. About A.D. 1438 Jacopo Mariano figured an endless hanging chain for manual use like those for small hoists in engineering workshops today. About 1490 Leonardo da Vinci made elaborate sketches of hinged-link chains, and used them for purposes such as turning the wheel-lock of a gun. This transmitted the power of a spiral spring, but the chain was not endless. In 1588 Ramelli depicted a chain (again not endless) in oscillatory motion over the geared driving-wheel of a double-barrel pump. Not until 1832 did Galle invent a type of hinged-link chain suitable for a chain-drive, and this was put to use in 1863 by

[1] In the first chapter elsewhere the original author, Mr Yuan, mentions a date in the period 1102–1106, so that as he lived till nearly the age of 100, what he says about his great-grandfather is quite credible.

[2] See Beck (1) and Schramm (1). Dr A. G. Drachmann is convinced that with its pentagonal gear-wheels it could never have worked in practice. Nevertheless Philon seems to have seen something similar which did, probably because it had proper gears. As his device armed the arcuballista it was in part a power transmission, though not from axle to axle, but mainly it was a conveyor belt.

[3] See Needham (1), vol. 4, pt. 1. The analogous machine in the occidental world was the *sāqīyah*.

[4] Uccelli (1), p. 75; Feldhaus (2), cols. 562, 444, 445; Feldhaus (3); Matschoss & Kutzbach (1).

Aveling for cars and in 1869 by J. F. Trefz for bicycles. One can see that Su Sung, and perhaps Chang Ssu-Hsün even more so, were considerably in advance of their time.

5 I-HSING AND THE FIRST INVENTION OF THE ESCAPEMENT (EIGHTH CENTURY)

The period of Chang Ssu-Hsün's activity had been at the beginning of the long-lived Sung dynasty (the end of the tenth century). Between his time and that of the chief of the earlier mechanical clocks there had been a long interval. For in order to find it we have to continue our backward path to I-Hsing and Liang Ling-Tsan at the beginning of the eighth century. Wang Ying-Lin, the editor of the thirteenth-century *Yü Hai* encyclopaedia, must have been quite aware of this gap, for the only piece which he could find to introduce between the two groups of texts was a poetical essay by Yang Chiung [531] on the armillary sphere, which has nothing to do with clockwork, and was in any case written rather before the time of I-Hsing, i.e. in A.D. 676.

Let us now turn, therefore, to the most venerable of all escapement clocks, that constructed, in the third decade of the eighth century, by the Tantric Buddhist monk I-Hsing in collaboration with the scholar Liang Ling-Tsan who (like Han Kung-Lien afterwards) occupied a minor administrative post (in this case connected with what corresponded to the Brigade of Guards). Technical terms used in the two passages to be quoted indicate beyond doubt that the mechanism was similar to that described by Su Sung. In addition there are many other interesting circumstances.

Hsin Thang Shu [761], ch. 31, pp. 1*b* ff., by OUYANG HSIU [467] & SUNG CH'I [662], A.D. 1061

Chiu Thang Shu [760], ch. 35, pp. 1*a* ff., by LIU HSÜ [663], A.D. 945 (the better text)

Cf. *Yü Hai*, ch. 4, pp. 21*a* ff. (poor text)

At the beginning of the Chên-Kuan reign-period (A.D. 627–49) the Astronomer-Royal Li Shun-Fêng [502] memorialised the emperor[1] saying that the instruments [158] in the obser-

[1] This was T'ai Tsung [948]; personal name Li Shih-Min [533] (A.D. 597–649). Second and perhaps greatest of the emperors of the T'ang. Starting life as a military officer under the Sui, he revolted and succeeded in enthroning his father in A.D. 618 as first T'ang emperor. Having crushed all opposition, whether

vatory were all according to the methods remaining from the (Northern) Wei dynasty (ended A.D. 532); therefore they were inaccurate and difficult to use for predicting celestial movements. The emperor T'ai Tsung therefore authorised him to make and cast a new bronze armillary sphere [150], and it was completed in the seventh year (A.D. 633).[1] (Li) Shun-Fêng then wrote a book in seven chapters entitled *Fa Hsiang Chih* [737] (The Miniature Cosmos)[2] discussing the merits and shortcomings of the armillary spheres of previous dynasties. There is more about this in his biography.[3] Then T'ai Tsung praised the new armillary sphere and ordered that it should be conveyed into the Palace to the Ning Hui Hall [940] to be used for observations [252]. But some years later it got lost.

Long afterwards, when Hsüan Tsung [949][4] had ascended the throne, in the ninth year of the

from within the family or without, he took over power from his father in 626 and began a reign of great brilliance which lasted a quarter of a century. Interested not only in military matters but also in history and technology, he knew how to encourage men like Li Shun-Fêng, and welcomed Nestorian Christian clergy as well as Taoist priests and Buddhist monks. He entertained cordial diplomatic relations as far west as Byzantium, receiving in A.D. 643, for example, an embassy from the emperor Theodosius. Such missions may well have brought news of the striking water-clocks at places like Gaza and Antioch (cf. Needham (1), vol. 1, pp. 186, 193, 204). The transmission would have been in all probability a 'stimulus diffusion', for there is no reason to suppose that the Byzantine works employed anything like the same mechanism as those of I-Hsing. But the stimulus would have come just at the right time.

[1] This is fully described in *T'ang Hui Yao* [812], ch. 42, pp. 5*b* ff., and briefly in *Hsin T'ang Shu*, ch. 31, pp. 1*b* ff. See Maspero (1), pp. 321 ff.; Needham (1), vol. 3, p. 347. Its importance lay in the fact that it was the first of the Chinese armillary spheres to have three layers of rings—a practice which afterwards became standard. We may tabulate them as follows:

	Meridian circles			Equator circle	Horizon circle	Ecliptic circle
	Fixed prime meridian ring	Solstitial colure ring	Polar-mounted declination ring or celestial latitude ring			
Liu Ho I [3], fixed outer nest	+	−	−	+	+	−
San Ch'en I [4], middle nest for diurnal rotation	−	+	−	+	−	+
Ssu Yu I [5], inner ring with sighting-tube	−	−	+	−	−	−

[2] This must have got lost rather soon, for the Sung bibliography does not mention it, though it is in the T'ang lists (see *Hsin T'ang Shu*, ch. 59, p. 13*a*).

[3] In fact, there is not (cf. *Chiu T'ang Shu*, ch. 79, p. 6*b*; *Hsin T'ang Shu*, ch. 204, p. 1*a*). Of course there is a brief description of the book in both cases.

[4] Personal name Li Lung-Chi [535] (A.D. 685–762). The most unfortunate of the T'ang emperors. Succeeding to the throne in 712, he prospered for some thirty years, a great patron of music, painting and literature. The greatest of the T'ang poets all knew his court. In later life, however, growing social and

K'ai-Yuan reign-period (A.D. 721), the Astronomer-Royal repeatedly advised the emperor that there were great difficulties in the prediction of eclipses. The emperor therefore commissioned the (Buddhist) monk I-Hsing[1] [477] to prepare a new calendar. The monk said that if it was really desired to start a new epoch one ought to know the exact relations of the equator and the ecliptic (and the correspondence of their degrees), and that therefore the astronomers ought to be asked to observe and measure this. But they replied that it could not be done because all previous instruments had been on the equatorial system and had no ecliptically-mounted sighting-tubes.[2]

Just at that time Liang Ling-Tsan [481], an administrative official of the Crown Prince's Bodyguard, was acting as a secretary in the Li-Chêng Library [941] (where the texts of edicts were drafted and checked), and he (seeing the need) came forward with a wooden model of an armillary instrument including a sighting-tube in a ring (mounted in the ecliptic axis) [225]. This was a very accurate production. So I-Hsing memorialised the emperor, saying: 'Although the men of old understood the principle of this movable sighting-tube in the ecliptic axis [225], they never actually embodied one in an instrument. Indeed with the usual equatorial instruments it is difficult to follow the movements on the ecliptic as it partakes of the diurnal rotation. The men of old thought much about such a device but could never put it into practice. Now Liang Ling-Tsan has invented this design so that the path of the sun and the nodes of the moon all agree with the observed phenomena. This is an extremely important thing for accurate calendar-making.[3] An apparatus of this kind in bronze and iron ought to be made in the Library compound in order to study and check the positions of the planets without any mistake.' (So the order was given, and) in the thirteenth year (A.D. 725) this (instrument) was completed.

economic strains exposed the country to the military rebellion of An Lu-Shan [536], a Sogdian army general in the Chinese service, from which China never recovered in this dynasty (see Pulleyblank (1)). Among the incidental results was the death of the emperor's famous and beautiful concubine Yang Kuei Fei [537]. We refrain from suggesting that the astronomical clock of I-Hsing and Liang Ling-Tsan was invented to honour and amuse her with its jack-work, for she did not join the imperial entourage till 738, more than a dozen years after it was completed. It is curious, however, that her personal name Yü-Huan (538) (jade rings) could have an astronomical reference.

[1] The secular name of this Tantric Buddhist monk (A.D. 672–727) was Chang Sui [539]. For the significance of the sect to which he belonged see Needham (1), vol. 2, pp. 425 ff. Like Matteo Ricci, he astonished people by his feats of memory. Combining knowledge of Indian (and hence indirectly of Hellenistic) mathematics and astronomy with the whole range of the Chinese traditions in these subjects, he was the most outstanding physical scientist of his time. Unfortunately all the books which he wrote were afterwards lost, and only the titles remain. A full-length biography of him, with translations of all the available texts which bear on his activities, is urgently necessary. See *Ch'ou Jen Chuan* [770], chs. 14–16 (Fig. 36).

[2] This does not mean that earlier Chinese armillary spheres had had no ecliptic rings. This feature had in fact been introduced first by Chia K'uei [514] in A.D. 103, and, as we shall see, Chang Hêng [480] graduated *hsiu* extensions on it graphically (p. 108). Cf. also the text concerning Ch'ien Lo-Chih below (p. 97). But what had never been done was to mount a sighting-tube in the ecliptic axis. This was the system used by Ptolemy, and it clearly indicates Hellenistic astronomical ideas mediated through the Indian Buddhists. However, the originality and particularity of the Chinese astronomical system was not easily influenced, and the ecliptically mounted sighting-tube was not employed again until the time of the Jesuits, with the possible exception of the sphere of Shu I-Chien [474] and Chou Ts'ung [516] about A.D. 1050.

[3] What I-Hsing wanted to do was to measure celestial latitudes (i.e. positions based on the ecliptic) because, being influenced (as a Buddhist) by Indian astronomy, he saw how fundamental they were for eclipse theory. It will be remembered that in his time there were three clans of Indian astronomers resident at the capital (Ch'ang-an [858]) and working for the Bureau of Astronomy; cf. Needham (1), vol. 3, pp. 202 ff.

PLATE V

Fig. 36. Chang Sui, known in religion as I-Hsing (A.D. 682–727), the Tantric Buddhist monk who, with Liang Ling-Tsan, is responsible for the invention of the first of all escapement mechanisms about 725. An impression by a modern artist, Chiang Chao-Ho, made for a recent series of Chinese postage-stamps commemorating ancient and medieval discoverers and inventors (courtesy of Dr Kuo Mo-Jo).

[Another long speech follows, giving the history of the calendar and of astronomical instruments in general, and mentioning Li Shun-Fêng's plan for an ecliptically mounted sighting-tube. The text then continues:]

So the emperor (Hsüan Tsung, Ming Huang) himself wrote an inscription for it, and by edict (A.D. 723) entrusted I-Hsing and Liang Ling-Tsan and other capable technical men [929] to cast and make new bronze astronomical instruments [154].

One (of these) was made in the image of the round heavens [194][1] and on it were shown the lunar mansions (*hsiu*) in their order, the equator and the degrees of the heavenly circumference. Water, flowing (into scoops), turned a wheel automatically [253 *a*, *b*], rotating it one complete revolution in one day and night (24 hours) [254]. Besides this, there were two rings (lit. wheels) fitted around the celestial (sphere) outside, having the sun and moon threaded on them, and these were made to move in circling orbit [255].[2] Each day as the celestial (sphere) turned one revolution westwards, the sun made its way one degree eastwards, and the moon $13\frac{7}{19}$ degrees (eastwards). After twenty-

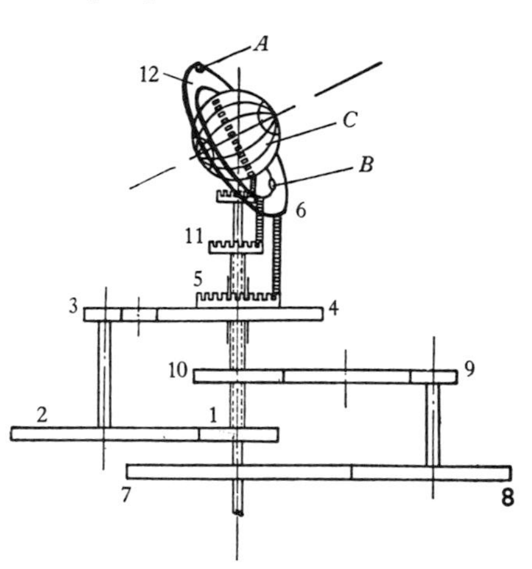

Fig. 37. Liu Hsien-Chou's reconstruction of a possible mechanism for the 'orrery' movements of medieval Chinese clocks (I-Hsing in 725, Chang Ssu-Hsün in 979, and Wang Fu in 1124, etc.). Only the sun and moon movements are shown as well as that for the celestial globe, and the design involves both concentric shafting and gear-wheels with odd numbers of teeth.

[1] This is a rather unusual expression. It is clear that Su Sung, three and a half centuries later (cf. p. 21 above), believed it to have been a solid celestial globe, but we feel that the general description concords much better with a demonstrational armillary sphere half sunk in a box-like casing the top of which represented the earth's horizon.

[2] Judging from Chang Ssu-Hsün's account (p. 72 above), this movement was manual, but the nuance of the text is perhaps rather in favour of mechanically produced motion. The former plan would have been like Su Sung's pearls of different colours threaded on strings (p. 24 above), but both Chang Ssu-Hsün and Wang Fu (p. 121 below) succeeded in building a mechanical movement of some kind. Liu Hsien-Chou (3) has now offered a hypothetical reconstruction (Fig. 37) for the devices of both I-Hsing and Wang Fu. We feel some doubt, however, that it could have been built by the former in the early eighth century A.D., since it involves not only concentric shafting but many gear-wheels cut with odd numbers of teeth. Liu Hsien-Chou's first gear-train, consisting of the wheels 1, 2, 3, 4, 5, 6 has to have the following numbers of teeth on each: 12, 60, 6, 72, 12, 73. With this the rotation of the sun is effected. His second gear-train, consisting of the wheels 7, 8, 9, 10, 11 and 12, has teeth in the following order: 127, 73, 6, 15, 6 and 114. With this the revolution of the moon is accomplished. The sphere or globe is rotated by the central spindle.

Wiedemann (1) and Price (1, 7) have drawn attention to the gearing in Arabic and European astrolabes as one of the most important elements in the prehistory of clockwork. The oldest evidence we have from these sources is textual, the MS. of al-Bīrūnī written about A.D. 1000, in which the design has the following gear-train: 40−10+7−59+24−48. But an Arabic geared astrolabe of similar pattern made in A.D. 1221 is preserved in the Museum of the History of Science at Oxford. A famous French example of *c.* 1300 is described by Price (1, 7). We have seen reason to believe (p. 36 above) that Su Sung in 1088 employed at least one odd-toothed gear-wheel, namely that of 487 teeth. If this is true, then Wang Fu in 1124 would not have found the making of such wheels too difficult. But whether or not I-Hsing and Liang Ling-Tsan could

nine rotations and a fraction[1] of a rotation (of the celestial sphere) the sun and moon met. After it made 365 rotations the sun accomplished its complete circuit. And they made a wooden casing the surface of which represented the horizon, since the instrument was half sunk in it. It permitted the exact determinations of the times of dawns and dusks, full and new moons, tarrying and hurrying. Moreover there were two wooden jacks standing on the horizon surface, having one a bell and the other a drum in front of it, the bell being struck automatically [256] to indicate the hours, and the drum being beaten automatically to indicate the quarters.[2]

All these motions were brought about (by machinery) within the casing, each depending on wheels and shafts [257], hooks, pins and interlocking rods [258], stopping devices and locks checking mutually [259] (i.e. the escapement).[3] Since the (clock) showed good agreement with the Tao of Heaven,[4] everyone at that time praised its ingenuity. When it was all completed (in A.D. 725)[5] it was called the 'Water-driven Spherical Bird's-Eye-View Map of the Heavens' [195] and set up in front of the Wu Ch'êng Hall [927] (of the Palace) to be seen by

have accomplished much in this direction as early as 725 is another matter, and proof that they did so would be of exceeding interest.

It should be added that our hesitations about the difficulties of making odd-toothed gear-wheels are not universally shared. Dr A. G. Drachmann, for instance, writes to us (letter of 26 November 1956) pointing out that after turning a blank to the diameter required it would only have been necessary to measure its circumference by means of a long strip of paper or thin brass and then to divide this line into the desired number of parts. The blank wheel could then easily be marked out following the divisions on the strip.

[1] The fraction is given by

$$\frac{365\frac{1}{4}}{13\frac{7}{19}-1} = \frac{19\times 3\times 487}{4\times 47\times 5} = 29\frac{499}{940}.$$

The fraction here is identical with that of the Ssu Fên Calendar [816] prepared by Li Fan [671] in A.D. 85. As may be seen from Maspero (1), p. 241, considerably better values were in common use by the time of I-Hsing and it is therefore remarkable that the value which had been current when Chang Hêng was working should still have been used. The retention of this constant is paralleled by the persistent use of the simplest of all values for π (see p. 96). Moreover, the number 487 appearing here was perpetuated in Su Sung's constructions, as we have already seen (p. 36). It therefore seems highly probable that the whole series of devices from Chang Hêng to Su Sung (at least) were accompanied by some kind of manuscript corpus. Such a collection, which would explain this conservatism in points of detail, must have dated from the second century A.D. The records 'in the files' must have been taken over by successive groups of artisans through the centuries.

[2] This is the earliest reference we have found to the use of jack-work in combination with an escapement.

[3] Note the similarity in technical terms to some of those used in the description of Su Sung (pp. 28 ff.)

[4] *Tao* [409] is one of the most ancient and important philosophical words in the Chinese language (cf. Needham (1), vol. 2, pp. 8, 11, 33, 36 ff., 433, 484). For the ancient Confucians it meant the right way of life within human society. For the ancient Taoists it meant the Order of Nature. Here it means just this. Since it involved the harmonious co-operation of all component organisms in the universal organism, it was not at all the same thing as, nor could it easily generate, the conception of laws of Nature.

[5] This date is taken from *Yü Hai*, ch. 4, pp. 25*a*, *b*.

the multitude of officials.[1] In A.D. 730 candidates in the imperial examinations were asked to write an essay on the new armillary (clock) (lit. armillary sphere [150]).[2]

But not very long afterwards the mechanism of bronze and iron began to corrode and rust,[3] so that the instrument could no longer rotate automatically [250]. It was therefore relegated to the (museum of the) College of All Sages (Chi Hsien Yuan [942]) and went out of use.[4]

Yü Hai [740] (Encyclopaedia of Wang Ying-Lin, A.D. 1267), ch. 4, p. 24*a*, quoting from *Chi Hsien Chu Chi* [741][5] (Records of the College of All Sages),[6] by Wei Shu [541], *c.* A.D. 750

In the twelfth year of the K'ai-Yuan reign-period (A.D. 724) the monk I-Hsing constructed an armillary sphere with an ecliptically-mounted sighting-tube [225] in the (Li-Chêng) Library [941] and presented it (to the emperor) when it was finished. Earlier, he had received an imperial order to reorganise the calendar, and had said that observations were difficult because there was no apparatus with this ecliptic component. Just at that time Liang Ling-Tsan made a small model (of an armillary sphere with an ecliptically-mounted sighting-tube) in wood and presented it. The emperor asked I-Hsing to study it, and he reported that it was highly accurate. Therefore a full-scale (sphere) in bronze and iron was made in the Library compound, taking two years to complete. Then it was presented to the emperor who praised it very much and asked (Liang) Ling-Tsan and I-Hsing to study (further) Li Shun-Fêng's *Fa Hsiang Chih* [737] (The Miniature Cosmos), so that later on they drew up complete plans of the armillary sphere [261]. And the emperor wrote an inscription in *pa-fên* [262] characters which was carved on the ecliptic ring (lit. wheel [263]) and which said:

'The moon in her waxing and waning is never at fault
Her twenty-eight stewards escort her and never go straying,
Here at last is a trustworthy mirror on earth
To show us the skies never-hasting and never-delaying.'[7]

[1] The continuity which was felt to exist between this first of the escapement clocks and the astronomical apparatus rotated by cruder methods since the time of Chang Hêng (second century A.D.) is well shown in the comment by Shen Kua [475] in the passage quoted on p. 61 above, written 350 years later.

[2] This sentence is interpolated by us from *Yü Hai*, ch. 4, p. 25*b*.

[3] Later this became a stock phrase, cf. p. 72 above. Cf. also *Liao Shih* [809], ch. 44, p. 39*b*.

[4] These two paragraphs have also been translated by Maspero (1), pp. 309, 310. We depart from his version in some slight particulars.

[5] Or *Chi Hsien Shu Yuan Chu Chi* [741].

[6] The College of All Sages seems to have been a kind of duplicate Academy like the Han-Lin. In the T'ang dynasty it also had the duty of drafting and checking the texts of edicts. But it also incorporated all kinds of experts who were available for consultation by the emperor. Entrance to it was by nomination and not (as in the case of the Han-Lin) by examination. See des Rotours (1), vol. 1, pp. 17, 19. The fact that I-Hsing was connected with it suggests that it was needed as well as the Han-Lin since it could make use of the services of scholars and experts who were not in the strict Confucian tradition.

[7] Although this poem was written on the observational armillary sphere and not on the clock, it seems singularly appropriate for the latter, and suggests that time-keeping by mechanical means, at a much higher level than anything previously known, was 'in the air'.

The scholar Lu Ch'ü-T'ai [542] received an imperial order to write an inscription containing the year and month of construction, and the names of the workers [264], underneath the plate [265].[1] The observatory used it for observations, and it is still employed nowadays.

And so in the thirteenth year (A.D. 725) there was made in the Library compound this armillary sphere [150]. After it was finished it was set up by imperial order outside the Fu Chêng Gate [943] to show the hundred officials. After I-Hsing's improvement of the ecliptic component the emperor ordered the casting of bronze to make another armillary astronomical instrument [266]. The Chief Secretary of the Left Imperial Guard Liang Ling-Tsan, and his colleague of the Right, Huan Chih-Kuei [543], took charge of plans for the separate parts, and a great (demonstrational) armillary sphere [267] was cast 10 ft. in diameter.[2] It showed the lunar mansions (*hsiu*), the equator and all the circumferential degrees. It was made to turn automatically by the force of falling water acting on a wheel [253*a*, *b*]. Discussing it, people said that (what) Chang Hêng (second century A.D.) (had described in his) *Ling Hsien* [742][3] (Spiritual Constitution of the Universe) could have been no better.

Now it is kept in the College of All Sages at the eastern capital. In the compound there is the observatory [268], which is the place where I-Hsing used to make his observations.

Such are the details of the instrument which, so far as we can see, was the first escapement clock. But between this point (A.D. 725) and the beginning of the Christian era there were many other examples of astronomical globes or spheres being slowly rotated by water-power. Since this evolution seems entirely germane to the story which we desire to unfold, we propose to give the texts concerning them in due order. In the earlier stages of our investigation we were inclined to believe that the arrangement for performing this consisted of a float in a clepsydra, bearing with it a cord in its rising or sinking which could be made to give slow rotatory movement to a piece of apparatus. But although this could have worked for the Hellenistic anaphoric clock with its simple dial vertically mounted, it became extremely difficult to see how so meagre a power could have turned round an obliquely-mounted sphere or globe, even if lightly made in wood.[4] And

[1] We do not understand this expression. It led us at first to think that the clock might have had an anaphoric dial, but the general description negatives this.

[2] This is large—nearly as large as those ascribed to Wang Fan (see p. 109 below) and Chang Hêng (see p. 100 below).

[3] This is an important statement for it shows that the rotation of a celestial globe or demonstrational armillary sphere was actually claimed by Chang Hêng himself in his own writings. Cf. pp. 21, 107.

[4] Because it is necessary to use a counterpoise for stability, the effective force which can be provided by a float is half of the weight of water it displaces. As an example of a float-driven mechanism one may consider the Salzburg anaphoric clock plate. When complete it must have weighed *c.* 35 kg. The radius of the centre axle-hole is 1·7 cm. and if this had borne a drum of about 17 cm. radius, the force required to overcome friction would be about 0·7 kg. Such force could be provided by a float having a displacement of 1·4 litres, corresponding to a cross-section 12 cm. square and a length of 10 cm. submerged.

Comparing this with the requirements of Su Sung's clock, one may calculate that if a float had been used instead of a water-wheel, it would have needed a displacement of about 360 litres, corresponding to a float

the texts say frequently enough that bronze instruments were so turned. We therefore adopted the view that the mechanism consisted of a vertical water-wheel with cups like a noria, attached to a shaft with one trip-lug. Clepsydra drip into the cups would accumulate periodically the torque necessary to turn the lug against the resistance of a peg-tooth wheel, either itself forming the equatorial ring, or attached to a polar axis shaft. The sphere would thus move on by one tooth each time. Needless to say, the time-keeping properties of such an arrangement would be extremely poor, but the men of those early centuries were doubtless satisfied with very rough approximations.[1] This introduction may serve, then, for the next group of texts.

Before proceeding on our way, however, we shall insert one quotation which demonstrates the use of some kind of jack-work a few decades earlier than in the clock of I-Hsing. Presumably it was operated in the way just described.

of 70 cm. cube. The mechanical problems involved in handling great caisson-like floats of this kind displacing tons of water would have been even greater than those which had to be solved in any water-wheel design. This is assuming that Su Sung used bronze sheet of about 3 mm. thickness (as in the Salzburg plate) for his celestial globe—a view which is consistent with the information we have (cf. p. 127) on the weight of bronze employed for the whole instrument. A globe of solid wood of the same dimensions ($4\frac{1}{2}$ ft. diameter) would have been more than three times as heavy, and one of solid bronze would have been out of the question, both on account of the availability of the metal and the possibility of applying the power required.

These considerations may be borne in mind later on when we are considering the apparatus of Chang Hêng (p. 109).

While not entirely adverse to these considerations, our friend Dr A. G. Drachmann of Copenhagen (letter of 26 November 1956) finds the greatest advantage of the water-wheel over the float in the fact that it never needed to be stopped even for a few minutes while filling or emptying of tanks was carried out.

[1] It will be seen that this interpretation has also the further advantage that it envisages a certain continuity between the early rough attempts at applying water-power to the rotation of an astronomical instrument, and the invention and perfection of the escapement. The judgment of Henri Maspero was for once at fault when he decided not to study the third chapter of Su Sung's book. He wrote (1), p. 335: 'We know from Chang Hêng himself that his instrument was made to rotate by hydraulic means and to follow precisely the diurnal revolution of the heavens by the regulation of a clepsydra. Chang Hêng wrote a special work on the motive power of his apparatus, but only a few lines of this are now extant, and these concern only the clepsydra itself. Su Sung, whom I have already so much leant upon, gives a very detailed account of the methods used by him in applying water-power to turn an armillary sphere, a celestial globe, and a *clock* [italics ours], combined in a single apparatus, but his designs breathe forth an engineering art so much more advanced than that of Han times that it seems to me unprofitable to translate this part of his book. It would tell us nothing about what Chang Hêng did. All one can say is that Chang succeeded in making his sphere rotate automatically in some way which his contemporaries thought wonderful....' Maspero's monograph was of course consecrated primarily to the astronomical instruments used in Han times. But it seems that he did not realise the extraordinary nature of a clock with a driving-wheel of any kind in A.D. 1088, still less that this could throw any light on what was done in A.D. 108. Had he known more of the generally accepted history of clockwork, it would not have been left to us to write the present monograph.

Ch'ao Yeh Ch'ien Tsai [743], by CHANG TSU [544], eighth century A.D. Quoted in *T'ai-P'ing Kuang Chi* [744] (Chi Ch'iao sect.), ch. 226 (item 5), p. 3*b*, ed. LI FANG [545], A.D. 981

In the Ju-I reign-period (A.D. 692) an artisan [944] from Hai-chow [859] was presented to the empress Wu Tsê T'ien [952].[1] He made a 'Wheel for Reporting the Twelve (Double-) Hours' [269]. When the wheel came to the exact south position [270], the Wu [271][2] door opened, and a jack with a horse's head appeared.[3] The wheel revolved round the four directions (as time passed) without the slightest mistake [272].[4]

Here then we may have the single ancestor of the jack-wheels which Su Sung piled up layer upon layer.

In the same year we find a reference to a Korean king receiving from a Buddhist monk a demonstrational armillary sphere set in a box-like casing.[5] The king was Hyoso Wang (Hsiao Chao Wang [1003]) of Silla, and the monk was named Tao-Chêng [679]. It is said that the sphere was constructed according to the system of Li Shun-Fêng [502], with whom Tao-Chêng had almost certainly studied during his previous decades of residence in T'ang China. The earth model at the centre seems to have been spherical or hemispherical, while 'day and night were produced by the appearance and disappearance of the sun and moon'. There is no definite statement that the sphere was made to rotate automatically, though the implication is considerably strengthened by the subsequent studies on the 'nine roads of the moon'[6] by Tao-Chêng's student monks, who are said to have added thirty-six holes. As will appear from what we have to say later on (p. 106) this can hardly mean anything else than the plugging in of small models of the sun, moon, and planets, along the ecliptic, to follow their motions in relation to that of the stars.[7] Pending further information about Tao-Chêng's apparatus, however, the honour of having constructed the last in the series of power-driven astronomical instruments ('proto-clocks' as we might call them) before the invention of the escapement early in the eighth century, goes to a remarkable Chinese technician Kêng Hsün, about A.D. 590.

[1] Wu Hou [953], personal name Wu Chao [548] (A.D. 625–705) might be described as China's Queen Elizabeth I. In 684, after the death of the emperor Kao Tsung, she displaced his successor and ruled the empire alone till 705. See Fitzgerald (1).

[2] The point of this is that the cyclical signs were used as azimuth points as well as names of hours. South thus corresponded to the noon double-hour and the horse.

[3] As it should, according to the animal cycle.

[4] The artisan was a man of skill; he also made braziers mounted in Cardan suspensions.

[5] *Samguk Sagi* [821] (Historical Record of the Three Kingdoms, Silla, Pakche and Koguryŏ), *c.* A.D. 1150, ch. 8, p. 4. The statement here is very brief, and we depend upon Rufus (1), p. 14, who did not indicate his exact Korean sources. The question (like Korean astronomy in general) calls for much further investigation.

[6] On this see Needham (1), vol. 3, p. 393.

[7] This plugging in of models was also a feature of the Hellenistic anaphoric clocks (cf. p. 64 above).

6 THE ADVENTUROUS LIFE OF KÊNG HSÜN (SIXTH CENTURY)

Sui Shu [714], ch. 78, pp. 7*b* ff., by WEI CHÊNG [510] *et al.*, *c.* A.D. 636

Kêng Hsün [549] whose *tzu* was Tun-Hsin [550] was born at Tan-yang [860].[1] He had a great sense of humour and was good at argument, while his technical skill was matchless.[2] In the time of Hou Chu [954] (last ruler of the Ch'en dynasty, ended A.D. 587)[3] he was a client [945][4] of the governor of Ling-Nan, Wang Yung [551]. After the governor died, Kêng Hsün did not return home but joined the tribes-people of Yüeh (in the south) and won their hearts. When the Li tribe rebelled in the Hui commandery, he put himself at their head, but was overcome and captured by the general Wang Shih-Chi [552].[5] Meriting death, he explained some of his ingenious ideas, so the general pardoned him and admitted him among his domestic slaves.

After a long time Kêng Hsün met his old friend Kao Chih-Pao [553] whose knowledge of the heavens had brought him to the position of Astronomer-Royal, and from him (Kêng) Hsün received instruction in astronomy and mathematics. (Kêng) Hsün then conceived the idea of making an armillary sphere [154][6] which should be turned not by human hands but by the power of (falling) water [273]. When it had been made he set it up in a closed room and asked (Kao) Chih-Pao to stand outside and observe the time (as shown by the) heavens (i.e. star transits). (His instrument) agreed (with the heavens) like the two halves of a tally. (Wang) Shih-Chi, knowing of this, reported the matter to the emperor Kao Tsu [955] (reigned A.D. 581–604) who made (Kêng) Hsün a government slave and attached him to the Bureau of Astronomy and Calendar.

Afterwards he was given to Hsiu [554] the prince of Szechuan, whom he followed to I-chow, and who greatly trusted him. When the prince was impeached and punished (Kêng Hsün) was again in danger of death, but Ho Ch'ou [555][7] spoke to the emperor Kao Tsu[8] for him, saying,

[1] A town and lake south of Nanking.

[2] The *Pei Shih* [747] (History of the Northern Dynasties), ch. 89, p. 31*a*, accords with the *Sui Shu* version closely (both are seventh-century texts). It also says that Kêng Hsün was witty and talkative, fond of discussions and extraordinarily ingenious.

[3] Ch'en Shu-Pao [470] (A.D. 553–604). Not a noteworthy ruler.

[4] The institution of clientage, in which more or less unattached persons accepted the protection of powerful families, existed in China anciently as in Rome. See Wilbur (1), p. 185; Yang Lien-Shêng (1), p. 116; Swann (1), pp. 290, 431, 445.

[5] Biography in *Sui Shu*, ch. 40, p. 10*a*.

[6] This passage is quoted in *Hsü Shih Shuo* [748], ch. 6, p. 11*a*, and *Yü Hai*, ch. 4, p. 26*a*. The term used for the instrument points to its having been a demonstrational armillary sphere, but it may have been a globe. Su Sung seems to say (p. 28 above) that it was the first celestial globe to be sunk in box-casing representing the terrestrial horizon, but evidence to be given presently will show that this cannot be true.

[7] Ho Ch'ou was an outstanding technologist, fl. A.D. 581–600. Besides being a military official he was Director of the Imperial Workshops. In charge of government textile production, he also controlled arsenals and built bridges. He must have known Li Ch'un [664], whose magnificent segmental arch bridges, built seven hundred years before their European parallels, still exist. Ho Ch'ou also revived the art of glass-making and played a prominent part in the development of porcelain.

[8] Personal name Yang Chien [556] (A.D. 540–604), first emperor of the Sui. Proclaimed this dynasty in 581. A forceful monarch, but not particularly interested in scientific and technical matters.

'Kêng Hsün's ingenuity is almost magical and your servant would deeply regret it if the court should suffer such a loss'. So the emperor made a special pardon for him. Later on (Kêng) Hsün made a 'stop-watch' clepsydra [274][1] and everyone said how marvellous it was.

When Yang Ti [956][2] came to the throne (Kêng Hsün) presented 'advisory inclining vessels' [275][3] and the emperor, being delighted with them, gave him his liberty as a freedman. A little over a year later he was appointed Superintendent of the Right Imperial Workshops. In the seventh year (of the Ta-Yeh reign-period, A.D. 611) the emperor led an expedition to the east, whereupon (Kêng) Hsün wrote to him warning him that he would not be able to defeat the Liao-tung people and that the expedition would be a failure. The emperor, furious, ordered his attendants to have him beheaded, but Ho Ch'ou pleaded desperately for him, so that he escaped. After the defeat at P'ing-yuan, the emperor thought that (Kêng) Hsün had been right after all, so he made him Assistant in the Bureau of Astronomy. Then came the rebellion of Yüwen Hua-Chi [558] and the assassination of the emperor (A.D. 618). After following Yüwen to Li-yang, (Kêng Hsün) said to his wife, 'I have studied the affairs of men below and investigated the patterns of heaven above. Yüwen is doomed. The House of Li[4] will conquer. I know whither I must go.' And he tried to get away, but Hua-Chi had him killed.

He wrote a book in one chapter entitled *Niao Ch'ing Chan* [745] (Predictions from the Flight of Birds), which had quite a good circulation.

From the above account, so laconically vivid a picture of a man's life (one only out of thousands in the dynastic histories), one can see that in early seventh-century China even a technician's career could be eventful enough. All we need remember from it, however, is the statement that his sphere was turned by water-power—there is no mention of any wheel. In other cases, as we trace the thread back through the centuries, we hear only that it was turned 'by a mechanism', even water being unmentioned. But as 'the unresting follows the unceasing', a water drive is to be surmised. Here, for example, is the account of the making of a demonstrational armillary sphere (or globe) by the great Taoist physician,

[1] We are satisfied that this is the correct meaning of this technical term. See below, p. 88.

[2] Personal name Yang Kuang [557] (A.D. 580–618), the second (and substantially the last) emperor of the Sui. Renowned in history for his vast expenditure on the enlargement and extension of the Grand Canal, an essential transport route. Though his court is accused of debauchery and extravagance, he encouraged classical literature and medicine before the turmoil of the later years of the reign. His particular significance to us is that he was famous, even before attaining the throne in 604, for his interest in mechanical toys—such as boatloads of puppets moved by concealed paddle-wheels, palace doors which opened of themselves, etc. Details will be found in Needham (1), vol. 4, pt. 1. Yang Ti collected quite a band of technicians like Kêng Hsün at his court.

[3] These mysteriously named objects were an ancient part of the ethical-didactic armamentarium of the Chinese emperors. They were hydrostatic trick vessels which lay on their side if empty, stood straight upright when half full, and fell over again if filled to the brim. The effect was easily obtained by built-in compartments. The oldest reference (*Hsün Tzu* [813], ch. 20, p. 1*a*), which would be of about 250 B.C., attributes their origin to the time of Confucius (sixth century B.C.), but they may safely be taken as of the Warring States period. The point of the idea was, of course, an ocular demonstration of the necessity of moderation in all things.

[4] The ruling family of the T'ang Dynasty.

alchemist and pharmaceutical naturalist T'ao Hung-Ching [520] (A.D. 452–536). We might date it about 520.[1]

T'ai-P'ing Yü Lan [749], ch. 2, p. 10*a*, ed. Li Fang [545], A.D. 983

The *Liang Shu* [750] (History of the Liang Dynasty) says[2] that T'ao Hung-Ching made a (demonstrational) armillary sphere [161] more than 3 ft. high, with the earth situated in the middle; the 'heavens' rotated and the 'earth' remained stationary—and it was all moved by a mechanism [276].[3] Everything agreed exactly with the (actual) heavens.

7 A SURVEY OF CHINESE CLEPSYDRA TECHNIQUE

But still we have not finished with Kêng Hsün. He occupies the position of an important forerunner to the first inventors of the mechanical escapement clock for quite another reason. In order to understand this we must engage in a brief study of the development of the clepsydra in China—a study which would in any case be indispensable since the motive power of the clocks of I-Hsing and Su Sung was water from a clepsydra dripping into the scoops of the wheel.

The general course of clepsydra technique in China may be summarised as follows (see Fig. 38). The most archaic device was no doubt the outflow type,[4] received in very ancient times from the culture centres of the Fertile Crescent.[5] This seems to have continued in occasional or parallel use down to quite late times.[6] But from the beginning of the Han onwards (i.e. from *c.* 200 B.C.) the inflow type, with an indicator-rod borne on a float, came into general, even universal, use. At first there was only a single reservoir but it was very soon understood that the falling pressure-head in this greatly slowed the time-keeping

[1] Thus contemporary with the Gaza clock (see p. 97, n. 2).

[2] In fact it does not (ch. 51, p. 17*a*) say more than the first few words. The real source is the *Nan Shih* [751] (History of the Southern Dynasties) which has the whole statement (ch. 76, p. 11*a*).

[3] Not only that, he wrote a book about it, the *T'ien I Shuo Yao* [752] (Essential Details of Astronomical Instruments), which is listed in *Sui Shu*, ch. 34, p. 15*a*. One version of this had diagrams with (planetary?) tracks and mnemonic rhymes.

[4] The inverse variant of this, the floating bowl with a hole in its bottom, which takes a definite time to sink, was also known in China, and occasionally used. For example, the T'ang monk Hui-Yuan [526] arranged a series of lotuses to sink one after another during the 12 double-hours (*T'ang Yü Lin* [754], ch. 5, p. 31*b*).

[5] Older commentators believed that there were mentions of the clepsydra in the *Shih Ching* [753] (Book of Odes), dating from the seventh century B.C. and earlier, but this is doubtful. Nevertheless there is nothing historically improbable in such references, if they were to be philologically substantiated. Contacts between Babylonia and China must have occurred long before. Shiratori (1) has drawn attention to what is probably our earliest reference to the clepsydra in China. The *Shih Chi* [834] (ch. 64, p. 1*b*), relating the life of a general and politician of the sixth century B.C., Ssuma Jang-Chü [693], who served Prince Ching of the State of Ch'i, says that while waiting for a rendezvous with another leader, Chuang Chia [694], Ssuma 'set up a sundial and started the water-clock dripping'. As Ching reigned from 546 to 488 B.C. this is a reference of venerable antiquity, contemporary with Confucius, and there seems no reason for rejecting it.

[6] Cf. Su Sung's list of clepsydras (p. 25 above) and again on p. 90 below.

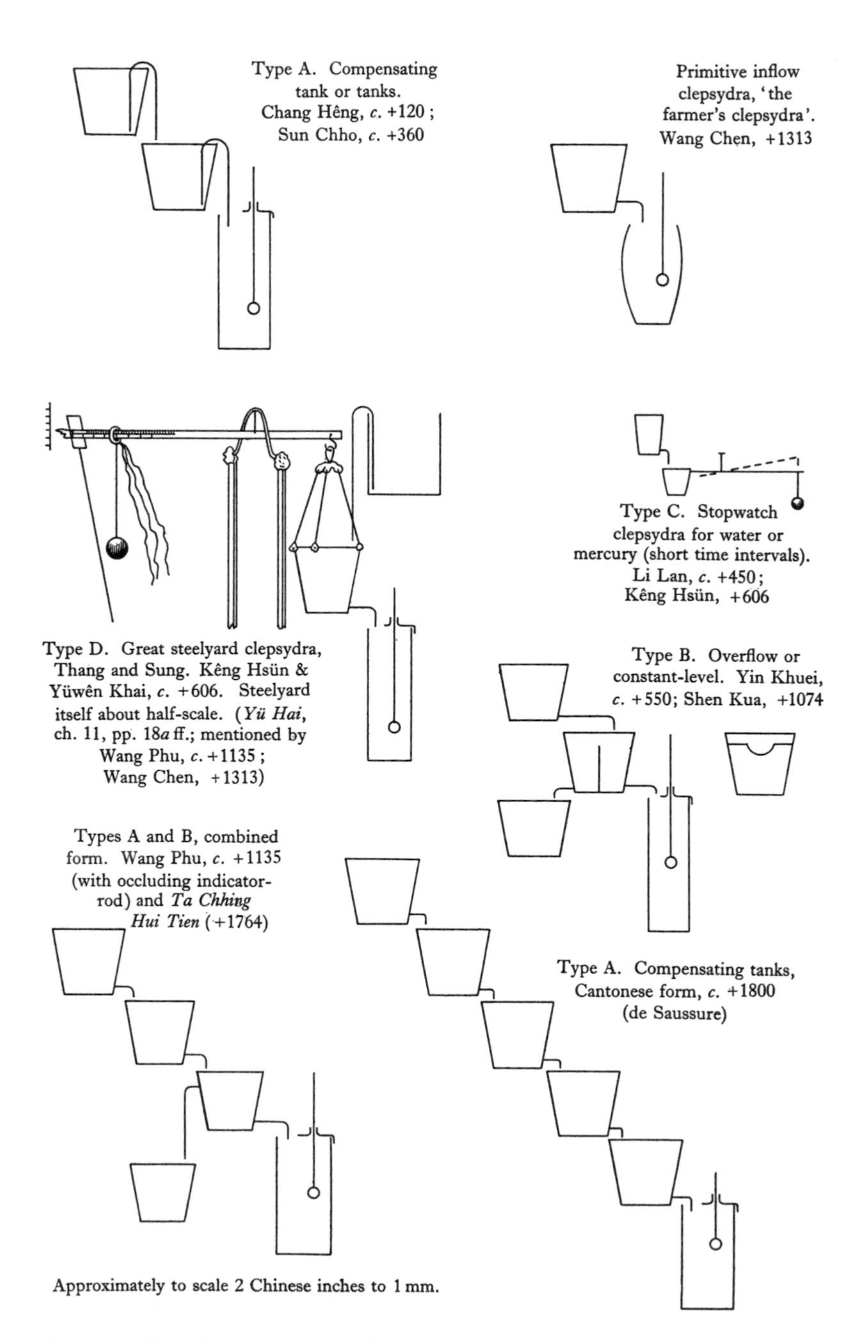

Fig. 38. The principal types of clepsydras used in ancient and medieval China (from Needham (1), vol. 3, p. 316).

as the reservoir emptied. Throughout the centuries two principal methods were used to avoid this difficulty; on the one hand the extremely simple and ingenious plan of interpolating one or more compensating tanks between the reservoir and the inflow receiver,[1] on the other hand, the insertion of an overflow or constant-level tank in the series.[2] The popularity of these methods varied from time to time.[3] The first type (which might be called polyvascular) was an admirable means of cumulative regulation, for at each successive stage the retardation of flow due to diminishing pressure-head is more and more fully compensated. The level of an *n*th tank in a series (counting the initial reservoir), for instance, falls according to an expression something like $1 - \frac{1}{n!}t^n$. As many as five tanks above the receiver are known to have been used.[4] The obscurity of some of the fragmentary descriptions which have come down to us led Maspero (1)[5] to believe that the overflow tanks in the second type were in certain periods (e.g. in the Liang and Sung) provided with some kind of ball-valves, even incorporating the use of mercury, but we are convinced that this was a mistake.[6] Where mercury entered in was in connection with clepsydras which involved weighing with balances.

[1] Chang Hêng had at least one compensating tank in his series in the early second century A.D. There was also at least one in the clepsydra for which Sun Ch'o [523] wrote an epigrammatic inscription in A.D. 360 or so (*T'ai-P'ing Yü Lan*, ch. 2, p. 13*a*; Maspero (1), p. 194). Sun Ch'o is otherwise chiefly known as a Confucian-Buddhist-Taoist syncretist.

[2] The earliest description of this which we have is from the pen of Yin K'uei [534], fl. *c.* A.D. 540 (see Maspero (1), p. 193). There is then a quite elaborate account by Shen Kua [475] in A.D. 1074 (Maspero, p. 188). Maspero also translated the definitive description of the overflow clepsydra in its final form from the *Ta Ch'ing Hui Tien* [755], ch. 81, p. 2*b*; this refers to the mid-eighteenth century. It is interesting to note that the earliest form of the overflow system (sixth to thirteenth centuries) embodied a tank with a partition out of which a semi-circular piece was cut. In later times there was simply an overflow pipe fixed in at the desired level.

[3] Whichever method was used there were alternative ways of delivering the water, either by pipes fitted into the base of the vessels, or else by siphons. The latter seem to be already present in Chang Hêng's second-century clepsydra. They had the advantage of permitting very delicate regulation of the pressure-head.

[4] Cf. an illustration of a page from a Cantonese almanac of 1831 figured by de Saussure (2).

[5] Pp. 190, 200 ff., 203, etc.

[6] The confusion arose partly because it was apparently the custom among Chinese medieval clepsydra technologists to refer to the delivering ducts or pipes, made of jade or other hard material, with carefully calibrated lumen, as 'weights' (*ch'üan* [277]). Shen Kua says this in so many words, in the passage already referred to (*Sung Shih*, ch. 48, p. 15*b*)—'This is called the "weight", because it delivers (lit. weighs) a greater or lesser (flow of water).' A good translation would perhaps be 'regulator'. On the use of gem stones by the Greeks for this same purpose see Drachmann (2), p. 18. The terminological confusion between 'weighing' and regulating goes back to Yin K'uei in the sixth century A.D., for in the fragments preserved in *Ch'u Hsüeh Chi* [756], ch. 25, p. 2*b*, he refers to the delivery pipe as *hêng ch'ü* [278], the 'weighing-canal', when its function as a regulating-canal is meant. The fact that there really were also clepsydras of various kinds embodying steelyards has naturally made confusion worse confounded. But we are confident that weighted ball-valves may be excluded.

These have hitherto been somewhat overlooked. They included at least two kinds, one in which the typical Chinese steelyard (the balance of unequal arms)[1] was applied to the inflow receiver itself, and another in which it weighed the amount of water in the lowest compensating tank. The former type naturally dispensed with the float and indicator-rod, and was usually made small and portable.[2] Here mercury was sometimes used,[3] with reservoir, delivery pipe, and receiver, all made of a chemically resistant material, jade. Such instruments,[4] which depended on the approach of the steelyard arm to the horizontal,[5] were well adapted for the measurement of small time intervals, as by astronomers in studying eclipses, or by others for the timing of races, and we may reasonably call them 'stop-watch' clepsydras.[6] The weighing of water in the compensating tank demanded a larger apparatus, which was used for public and palace clocks throughout the T'ang and Sung periods. It made possible the seasonal adjustment of the pressure-head in the compensating tank by having standard positions for the counter-weight graduated on the beam; and hence controlled the rate of flow for different lengths of day and night. It also of course avoided the necessity for an overflow tank, and warned the attendants when the clepsydra needed refilling.[7]

[1] For a discussion of the history of this, cf. Needham (1), vol. 4, pt. 1.

[2] There are references to special carriages for transporting them (*hsing lou ch'ê* [279]), as in *Shih Wu Chi Yuan* [759], ch. 2, p. 47*a*. Most of these point to the Sui period as the time when portable steelyard clepsydras became popular.

[3] Particularly suitable for rapid timing of short intervals.

[4] They seem to begin with the Taoist Li Lan [546] of the Northern Wei dynasty (*c.* A.D. 450). They are described also by Wang P'u [547], *c.* 1135, in his *Kuan Shu K'o Lou T'u* [757] (Illustrated Treatise on Standard Clepsydra Technique), a fragment of which is reproduced in the *Liu Ching T'u* [758] (Illustrations of Things mentioned in the Six Classics) by Yang Chia [559] *c.* 1155. Cf. p. 91.

[5] This procedure may be not without relevance to the history of pointer-readings in general.

[6] The technical term by which they can be recognised is *ma shang pên ch'ih* [280], 'rapidly pouring' clepsydras.

[7] Clepsydras combined in various ways with steelyard balances were also well known in the Arabic culture-area. Khanikov (1) has translated parts of a book by Abū'l-Fatḥ al-Manṣūr al-Khāzinī entitled 'The Book of the Balance of Wisdom' (*al-Kitāb Mīzān al-Ḥikma*) and written in A.D. 1122. Al-Khāzinī seems to have used an outflow vessel on the end of his steelyard (Khanikov's translation, pp. 17, 24, 105). Cf. Winter (1).

Particularly interesting in this connection are the accounts in the *Chiu T'ang Shu* (ch. 198, p. 16*a*) and the *Hsin T'ang Shu* (ch. 221 B, p. 10*b*) of the great steelyard clepsydra on one of the gates of the city of Antioch. In the entry on the Byzantine Empire, the former says (we quote the longer of the two accounts): 'On the upper floor of the second gate there is hung a golden steelyard with twelve golden balls suspended from it to show the twelve (double-) hours of the day. When each (double-) hour arrives, one of the golden balls drops beside a lifesize golden human figure, making a clanging sound to indicate the divisions of the day without the slightest mistake.' This account, written about A.D. 945, and its parallel a century later, apply to affairs of the seventh to the ninth centuries. All these things, therefore, are long after the time of Kêng Hsün (late sixth century). Hirth (1), pp. 53, 57, 213, who translated these and other passages, noted that the Chinese recognised this device clearly as a clepsydra since it was classified under this head in the encyclopaedia *Yuan Chien Lei Han* [762] (a conflation of T'ang encyclopaedias made in the eighteenth century), ch. 369, p. 9*a*.

We have already seen that Kêng Hsün made a stop-watch clepsydra—presumably a prototype for reproduction by the artisans. But the next quotations demonstrate that it was really he, in conjunction with his more famous contemporary Yüwên K'ai [560],[1] who produced the standard type of balancing compensator tank just described. Immediately adjacent to them, therefore, we place the detailed description of this type which is contained in the *Yü Hai*, but which has not so far been studied. It will show us in passing what the principal types of clepsydra were in the T'ang and Sung. And it will also show how the balancing compensator clepsydra contributed to the basic invention of I-Hsing and Su Sung.

Lou K'o Fa [763] (Clepsydra Technique), by LI LAN [546], *c.* A.D. 450. Fragments preserved in *Ch'u Hsüeh Chi* encyclopaedia [756], ch. 25, p. 2*b*, by HSÜ CHIEN [561], A.D. 700

Water is placed in a vessel whence it issues by a bronze siphon [73] in shape like a curved hook. Thus the water within is conducted to the silver dragon's mouth which delivers it to the balance vessel (*ch'üan ch'i* [281]). One *shêng*[2] of water dripping out weighs[3] one *chin*,[4] and the time which has elapsed is one quarter (-hour) [21].

[This is the small steelyard clepsydra.]

(It is by means of) jade vessels, jade pipes, and liquid pearls [282] (that the) rapid stop-watch [280] portable clepsydra [279] (is constructed). 'Liquid pearls' is only another name for mercury [283].

[This is the small portable steelyard clepsydra using mercury.][5]

Sui Shu, ch. 19, pp. 27*b* ff., by WEI CHÊNG *et al.*, *c.* A.D. 656. Copied in *Yü Hai* (Encyclopaedia of WANG YING-LIN, A.D. 1267), ch. 11, p. 12*a*

At the beginning of the Ta-Yeh reign-period (A.D. 606) Kêng Hsün [549] made 'advisory inclining vessels' in the ancient style, for filling with clepsydra water, and presented them to the emperor (Yang Ti) who was very pleased. The emperor asked him and Yüwên K'ai [560] to follow the methods developed by the Northern Wei Taoist Li Lan [546]. The old Taoist technique used a small steelyard clepsydra [204, 287], with a vessel which weighed the water—thus it was portable. Accordingly (Kêng Hsün and Yüwên K'ai) set up (sun-) shadow

[1] Yüwên K'ai (A.D. 555–612), was engineer, architect, and Minister of Works under the Sui dynasty for thirty years. He built very large wagons (probably sailing carriages) for the emperor Sui Yang Ti, carried out irrigation works, and superintended the construction of the Sui Grand Canal. He also wrote on the Ming T'ang (the cosmological temple of the ancient Chinese emperors) and made a wooden model of it.

[2] *Shêng* [284] is usually translated pint. Swann (1) gives the Han *shêng* as 0·36 U.S. dry pint.

[3] The words used, *ch'êng chung* [285], imply the steelyard.

[4] *Chin* [286], the catty or pound. At this period it corresponded to 0·22 kilo, or about half a pound.

[5] Our interpretations differ from that of Maspero (1), pp. 192 ff.

measuring (gnomons by which to) graduate the scales of indicator-rods (or steelyard beams), and square vessels to hold the water above. These time-keeping devices were established below the drum (-tower) in the front of the Ch'ien Yang Hall [957] (of the Palace) at the eastern capital (Loyang). They also made portable stop-watch clepsydras [274, 279] for telling the time. These two instruments, the sun-computing dial [288] (sundial), and the clepsydra [289], can measure (the phenomena of) Heaven and Earth, and therefore they are the basic means of calibrating the armillary sphere and the celestial globe. As they have undergone many changes during the centuries, it was necessary to discuss them in some detail here.

Yü Hai (Encyclopaedia of Wang Ying-Lin, A.D. 1267), ch. 11, pp. 18*a* ff. Greatly abridged in *Sung Shih*, ch. 76, pp. 3*b* ff.,[1] by Toktaga & Ouyang Hsüan, *c.* A.D. 1345

The *Kuo Shih Chih* [746][2] says that ((after the time of the invention of the clepsydra by Huang Ti,[3] who got the idea from watching dripping water, the Three Dynasties[4] appointed officials [931] to take charge of it. In later generations people made either outflow clepsydras (*hsia lou* [289]), or inflow float clepsydras (*fou lou* [290]), or (water-) wheel clepsydras (*lun lou* [291]), or clepsydras with steelyards and balance weights (*ch'üan hêng* [292])—various different kinds.[5]

[1] Those parts of this text which are found in the *Sung Shih* are enclosed between double brackets.

[2] For a long time we were not able to identify this source, but a note of L. Carrington Goodrich in his new edition of Carter (1), p. 66, directed us to the interesting discussion of it by Pelliot (2). Pelliot gives good reasons (pp. 44 ff.) for thinking that the *Kuo Shih Chih* was probably the bibliographical section of a book now long lost, the *Liang Ch'ao Kuo Shih* [819] (A History of Two Courts), completed in A.D. 1082 by Wang Kuei [675]. This work dealt in detail with the reigns of Jen Tsung [946] and Ying Tsung [1001], i.e. the period between A.D. 1023 and 1067. Alternatively it may have derived from a similar book, the *Liang Ch'ao Shih*, finished and presented in A.D. 1016 by Wang Tan [676]. This dealt with the time of the emperors T'ai Tsu [1002] and T'ai Tsung [948], i.e. 960–97. It will be seen that the paragraph on clepsydras and water-driven clocks which the *Yü Hai* quotes from the *Kuo Shih Chih* gives just the kind of summary usual in those covering notes with which Chinese bibliographers customarily prefaced or concluded their classified book lists.

[3] Legendary emperor.

[4] This is the customary phrase for the Hsia dynasty (legendary), the Shang (Yin) kingdom, and the Chou dynasty. According to the traditional dating, the first of these occupied the first half of the second millennium B.C., the second lasted from 1520 to 1030 B.C., and the third from then onwards until the unification of the empire under the Ch'in dynasty in 221 B.C. The Chou emperors were more like bronze-age proto-feudal 'High Kings' than emperors in the later sense. After the beginning of the iron age they became less important than the leader (hegemon) of the feudal princes, who changed from time to time according to the power of the feudal States.

[5] It is interesting to compare this list with others also dating from the Sung. We do not know the date of it, but it may well belong to the eleventh century. Another list of that time (A.D. 1090) has already been given (p. 25 above), namely the types of clepsydras which Su Sung used to check his mechanical clock. These were: (*a*) inflow float and indicator-rod clepsydra (*fou chien* [203]); (*b*) steelyard clepsydra (*ch'êng* [204, 287], assuming the correctness of our emendation there made, which the following list confirms); (*c*) sinking indicator-rod, i.e. outflow clepsydra (*ch'en chien* [205]); (*d*) 'unresting' or 'continuous' clepsydra (*pu hsi* [206]).

The *Hsiao Hsüeh Kan Chu* (A.D. 1299, but probably compiled at least some twenty years earlier), ch. 1, p. 32*b*, gives exactly the same list of four clepsydra types in the same order. Wang Ying-Lin adds: 'These were the water-clocks used by Su Sung at the beginning of the Yuan-Yu reign-period (1087).' It is fairly evident that the list in our present text is also the same, the order being slightly modified to run: (*c*), (*a*), (*d*), (*b*). The only obscurity concerns the last type, (*d*). The first thought which might occur to the mind is

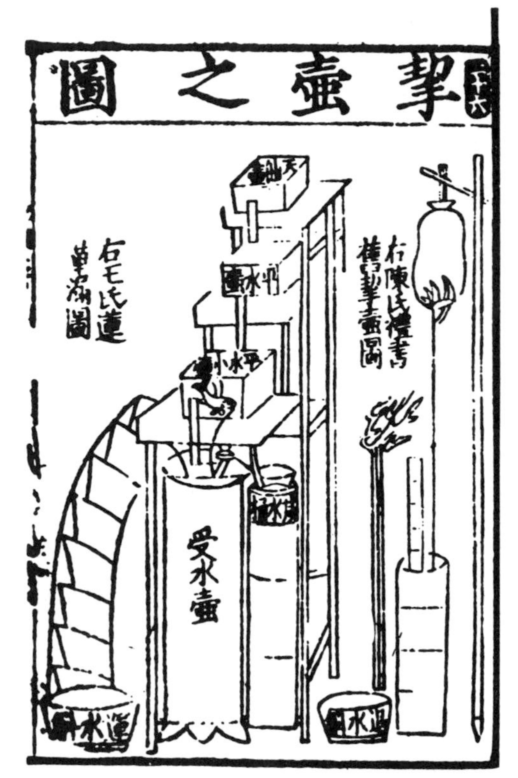

Fig. 39. The oldest printed illustration of clepsydras in any culture. From a Sung edition of the *Liu Ching T'u* of Yang Chia (*c.* A.D. 1155), enlarged by Mao Pang-Han (*c.* A.D. 1170). On the right, a drawing of the archaic inflow type of camp clepsydra described in the *Chou Li*, with a torch beside it to ward off the frost. On the left, 'Mr Wang's lotus clepsydra', i.e. the design of Wang P'u (*c.* 1135), which includes an overflow tank and an occlusive indicator-rod. Cf. p. 40, n. 3 above.

that it was the same thing as the mechanical clocks of I-Hsing and Su Sung. But then the latter would not have used it for testing his mechanical escapement clock. The probability therefore is that it was some derivative of the scoop-wheel and trip-lug device which dated from the time of Chang Hêng (second century A.D.), see below, p. 109.

We do not know of any description of this. Nor is there any illustration of one of these water-wheel clepsydras, unless we were to recognise one in the accompanying figure (Fig. 39). It is entitled 'Picture of Clepsydras', and derives from an early (Sung) printed book in the Cambridge University Library collection of microfilms. The book is the *Liu Ching T'u* [758] (Illustrations of Things mentioned in the Six Classics), an encyclopaedia compiled by Yang Chia [559] about A.D. 1155 and enlarged by Mao Pang-Han [665] about 1170. On the right there is a drawing supposed to represent the camp clepsydra described in the *Chou Li* [764] (Record of the Rites of Chou), an early Han book, *c.* 200 B.C. It is the most archaic inflow type, and has a torch beside it to ward off the frost. In the centre there is the regular medieval type with an overflow tank. But to the left of it is something which seems to be a wheel, though more probably it is a bad drawing of a flight of steps giving access to the top reservoir for filling. The accompanying text comes from an earlier book, the *Kuan Shu K'o Lou T'u* [757] of Wang P'u [547], whose floruit was in the neighbourhood of A.D. 1135. Unfortunately it throws no light on water-wheel clepsydras, being but a fragment devoted partly to the small steelyard pattern, and partly to a variant of the central design in which the rising indicator-rod finally cut off the supply of water. In any case, the picture is of great interest as the oldest printed illustration of a clepsydra in any culture.

In our present (the Sung) dynasty)) offices like the Astronomical Bureau have Ch'ieh Hu Chêng [958] (Clepsydra Superintendents) who take charge of the time-keeping (lit. hours and quarters) instruments. One was placed just inside the gate ((in the eastern gallery of the Wên Tê Hall [937] (of the Palace))), a drum-tower and bell-tower being set up in the hall court to left and right.

The method of this clepsydra was to have a water-balance [293] with a wooden beam as a steelyard [294], above which there was an inscription carved, saying *T'ien Ho* [295] (The Celestial River).[1] ⟨The *Chih* says that it was the water that was made to act as a balance.[2]⟩ The (lowest vessel) was sufficiently high and wide to hold indicator-rods [296], of which there were four, each made of wood, 3·5 ft. long, marked with the hours and quarters and night-watches,[3] and put in the (lowest vessel of the) Celestial River.[4] They were used alternately by day and night. Then at the right-hand end of the steelyard there was a bronze ring [297] attached by a hook. And there was made an inverted lotus of bronze [298] from which hung down three chains of bronze [299] attached to the bronze vessel [300] (i.e. the compensating tank above the inflow receiver).

There was a large lacquered (wood) box called the Water Box [301] having inside it a bronze tank [302] with a bronze siphon [73] at one corner leading the water over into the vessel. To the left-hand end of the steelyard there was attached a large bronze ring [297] with a thick bronze chain [299] hanging down from it and carrying a bronze weight [277] in the form of an elephant [303] (or some image).[5] There was also an iron rod 5 ft. long,[6] inside an iron lotus-shaped guide [304], bent at its upper end so as to form a square ring [305], and this was called the Cock Rod [306].[7] Whenever (the passage of) time (lit. hours and quarters) made (it necessary to) change (the setting),[8] those in charge of the time-keeping put the end [65][9] of the steelyard into the square ring, and with thick plaited cords laid hold of the large bronze ring above the weight drawing it back or forth (along the steelyard) [307]. And the balance [287, 308] was supported upon two large posts of wood like those which have stag's heads on the top and are used for supporting the transverse beams of bells [309], and they were painted with multicoloured designs and golden dragons. Between them there was an iron arch [310] with a large iron hook and ring attached to it from which the balance was suspended.

[1] We suspect that this sentence originally said that graduated scales were inscribed on the beam, but later this was misunderstood. The Celestial River would be a literary reference to the whole water-course from reservoir to receiver. One may remember that in Su Sung's system it was the technical term for the flume at the top of the water-circulation system (cf. p. 42 above and Fig. 20).

[2] This uncertainty was due to a corruption in the text, the characters for wood and water being easily mistaken for one another if badly written.

[3] Each one would have corresponded to 3 double-hours and 25 quarters (*k'o*). It is permissible to suppose that the four were for the periods midnight–dawn, dawn–midday, midday–dusk and dusk–midnight. The length as given was standard for clepsydra indicator-rods.

[4] This proves that it was the compensating tank that was weighed.

[5] We are not sure whether to take the word *hsiang* literally here.

[6] Emending from the 50 ft. given in the text.

[7] Cock-crow? Cock-spur? No clue.

[8] I.e. when the indicator-rods had to be changed seasonally or the tank's counter-weight moved in position on the beam.

[9] The technical term here used is also one of those found in Su Sung's descriptions of levers and linkwork.

The whole system was very clever and ingenious, but we do not know who originated it, though it has been used for a long time, right through the T'ang and Wu Tai (Five Dynasties) periods (i.e. from A.D. *c.* 620 to 910, and from 910 to 960).[1]

Whenever there was addition or diminution (of the water in the hanging bronze vessel) of the clepsydra (the beam) moved accordingly and gave an indication to the keepers at the gate. And the guards of the Imperial Palace Gate and the Locked Gate every day, one quarter of an hour after the middle of the *mao* (double-) hour (i.e. *c.* 6.15 a.m.), opened the locks of the Palace Gates. Each (double-) hour contained 8 quarters and 20 min. At each quarter a drum was beaten, and after eight drum-beats the hour-tablet [311][2] was hung up by the officials in charge. ⟨There were seven of these, made of ivory and with gilded characters carved on them.⟩ The remaining 20 min. (were signalised by) a cock-crow (trumpet?), and after this was finished fifteen drum-beats were given to indicate the middle of the (double-) hour.

⟨⟨Another version says that ((in the third year of the Ching-Yu reign-period (A.D. 1036) the clepsydra was again investigated, and it was found that the water was flowing sometimes too slow and sometimes too fast, so the suggestion of one of the officials was adopted, and a compensating tank [43] was added (to the system),[3] together with two siphons [73], and twenty-one indicator-rods.[4] But almost half of the hours and quarters overlapped with each other,[5] so at the beginning of the Huang-Yu reign-period (A.D. 1049), the emperor ordered Shu

[1] This is the passage which points directly at Kêng Hsün and Yüwên K'ai as having been the most important originators of this method about 610. Shiratori (1), p. 314, has drawn attention to a very relevant statement by the Arab traveller Sulaimān al-Tājir, who was in China in 851. His *'Akhbār al-Ṣīn wa'l-Hind* (Information on China and India) has been many times translated. Speaking of the gates and drum-towers of Chinese cities, Sulaimān says: 'Each city has four gates, at each of which there are five trumpets, which are sounded by the Chinese at certain hours of the day and night. There are also in each city ten drums, which they beat at the same time, the better to show publicly their loyalty to the emperor, as well as to give knowledge of the hours of the day and the night; and they are also furnished with sun-dials and clocks with weights.' So at least we may english the first translation, made into French by Eusebius Renaudot in 1718 (p. 25), with its startling final words. There was no difficulty about the word *'alāmat*, 'placards' or 'sign-boards' for sundials, but 'weights', *wazn*, is distinctly in the text, and subsequent translators were very uncertain what to do about it. Reinaud and Ferrand simply said 'instruments with weights for counting the hours', while the most recent translator, Sauvaget (1), still more cautious (p. 15), wrote 'they also have apparatus for marking and measuring the hours'. One cannot too much approve the hesitation of these arabists in ascribing weight-driven mechanical clocks to ninth-century China, but we can now see much more clearly, and one can hardly doubt that Sulaimān was referring to the balancing steelyard clepsydras of Kêng Hsün and Yüwên K'ai. His witness is particularly interesting because it refers to provincial cities and not to the capital; this type of clepsydra therefore must have been quite widespread.

[2] Here is one of the origins of the tablets carried by some of Su Sung's jacks. Evidently they had once been large objects hung up by real people.

[3] Presumably this was inserted between the reservoir and the hanging compensating tank.

[4] We cannot explain this number. Perhaps it was intended to say twelve per day, or twenty-four per year. In any case it would suggest that the understanding of how to regulate the clepsydra by adjusting the position of the weight on the steelyard beam had been lost.

[5] The *Sung Shih* text enlarges on this, inserting the following passage: 'Now (at the proper time) in the four seasons, sunrise should be reported at a quarter of an hour past the middle of the *mao* (double-) hour. But at the middle of each (double-) hour one quarter (of the next half) had already been reported, so that when the eighth quarter came, the beginning of the next (double-) hour had already been reported. Thus two (double-) hours overlapped with each other almost by half.' Cf. Maspero (1), pp. 210 ff.

I-Chien [474], Yü Yuan [515], and Chou Ts'ung [516], to make an additional compensating tank, and to improve the evenness of the water flow, dividing the day and night into 100 quarters....[1]))⟩⟩

The upshot of all this information is that the work of Li Lan in the fifth century and of Kêng Hsün at the end of the sixth was one of the direct antecedents of the invention of the mechanical escapement clock by I-Hsing and its development by Chang Ssu-Hsün and Su Sung. From the weighing of a water-receiving vessel it was not such a far cry to the weighing of one which both received it and delivered it, and that in turn pointed the way to the controlled retention of successive water-receiving and delivering scoops on a wheel by a weigh-bridge. The wheel itself (if we are right in our surmise about the system employed by Chang Hêng) would have derived from the non-chronometric motions of a trip-lug turning from time to time as the scoops on the wheel filled. It is also worth noting that Kêng Hsün had the idea of filling his hydrostatic trick vessels slowly with clepsydra water—again a suggestion that something could be done with it other than letting it raise a float or run to waste.

Before leaving the subject of clepsydras reference may be made to the arrangements for constant flow in the clock of Su Sung (p. 40 above). It is very curious that he seems to have had only a single compensating tank and not an overflow tank, though the system for this had been clearly described by Shen Kua only twenty years before. Perhaps Su Sung really had one and the description dropped out of the text later on, or perhaps his 'water-level arrow' was one of Wang P'u's occlusive floats.

8 THE BIRTH OF THE CELESTIAL GLOBE (FIFTH CENTURY)

We can now resume our historical investigation, pressing backwards in time to the ultimate origins of the mechanical clock. The following quotations have to do with the work of Ch'ien Lo-Chih (*c.* A.D. 435–40) of the Liu Sung dynasty, and Ko Hêng (*fl. c.* A.D. 250) of the State of Wu in the Three Kingdoms period. They are particularly precious because they give the link with the greatest pioneer of this train of scientists and technicians, namely Chang Hêng, who worked in the first decades of the second century. It was the recovery of the remains of his old instruments in the early fifth century, when the north-western capital was recovered from the Turkic and Hunnish houses by a Chinese dynasty once again

[1] The *Sung Shih* continues after this at some length about indicator-rods, and lengths of day and night varying with the seasons. It throws no more light on the large steelyard clepsydra.

powerful, that stimulated Ch'ien to construct new astronomical apparatus—and his imperial patron to support him. Here we are distinctly told that at least one of the new instruments was rotated by water from a clepsydra. We hear also, and in the same connection, of an earlier device of the same kind, due to Ko Hêng, who worked in the south-east somewhere during the third century, in the tradition of Chang Hêng but without access to the remains of his instruments. The trail thus leads back to him from Kêng Hsün through T'ao Hung-Ching and Ch'ien Lo-Chih. Lastly, the passages are particularly interesting because they speak with crystal clarity of the existence of the demonstrational armillary sphere as well as that used for observations. It is striking that the idea of having a small model earth at the centre of the rings occurred so early in China, since in Europe it appears only rather late—about the end of the fifteenth century.[1]

Sung Shu [765], ch. 23, p. 8*b*, by SHEN YO [562], *c.* A.D. 500. Parallel text in *Sui Shu*, ch. 19, pp. 17*b* ff., by WEI CHÊNG *et al.*, *c.* A.D. 656

The armillary sphere [150] made by (Chang) Hêng [480] (A.D. 78–139) was handed down through the (San Kuo) Wei (A.D. 221–80) and the (Western) Chin (A.D. 265–317) periods. But after the defeat of China (at the end of the Western Chin dynasty, A.D. 317) at the hands of the (northern Turkic and Hunnish) barbarians, even the later instruments of (Lu) Chi [500] (fl. A.D. 220–45) and (Wang) Fan [478] (A.D. 219–57) were all lost (because the Chinese empire withdrew to Nanking south of the Yangtse, and of course the instruments of Chang Hêng were also left behind in Ch'ang-an [858], i.e. modern Sian in Shensi).

But in the fourteenth year of the I-Hsi reign-period (A.D. 418) when An Ti [959] of the (Eastern) Chin (dynasty) was still reigning,[2] Kao Tsu (=Wu Ti [960], first emperor of the Liu Sung dynasty),[3] (drove north again), captured Ch'ang-an [858] (the ancient capital) and recovered the old instruments of (Chang) Hêng.[4] Although the forms were still recognisable, the marks of graduation had all gone, and nothing was left of the representations of stars, sun,

[1] Cf. Price (2), from which paper we reproduce a demonstrational armillary sphere of this kind (Fig. 40), made by Carolus Platus at Rome in 1588. An early example is that in the German National Museum at Nürnberg, made by Johann Wagner in 1540.

[2] An Ti, personal name Ssuma Tê [563] (A.D. 382–418). Tenth emperor of the Eastern Chin dynasty. Incapable ruler, finally put to death by one of his generals, Liu Yü [564].

[3] Wu Ti, personal name Liu Yü [564] (A.D. 356–422). First emperor of the Liu Sung dynasty, formerly a general of Chin.

[4] This passage is found also in *Yü Hai*, ch. 4, p. 15*b*. A rather fuller version of the recovery of the instruments was given in the *I-Hsi Ch'i Chü Chu* [766] (Daily Records of the I-Hsi reign-period, A.D. 405–18), now lost, but quoted in *T'ai-P'ing Yü Lan*, ch. 2, p. 10*a* (cf. Maspero (1), p. 304). It mentions only an armillary sphere and a gnomon shadow template. The date of this source, however, is not much earlier than the *Sung Shu*.

Another parallel passage is in *Nan Shih* [751], ch. 1, p. 19*b*, but this is a century later. It gives the year of the fall of the capital as 413, and lists, besides the two instruments just mentioned, a hodometer, a south-pointing carriage, and the Great Seal of Ch'in Shih Huang Ti.

moon and planets.[1] In the thirteenth year of the Yuan-Chia reign-period (A.D. 436) the emperor[2] ordered Ch'ien Lo-Chih [666], the Astronomer-Royal, to re-make and cast a (demonstrational) armillary sphere [150]. Its diameter was slightly less than 6·08 ft. and its circumference slightly less than 18·26 ft. The earth was fixed in the centre of the heavens [312][3] and there were the two paths of the ecliptic and the equator, the two celestial poles south and north, with the twenty-eight lunar mansions (*hsiu*) [196] depicted round about, as also the Great Bear and the pole-star. Each degree corresponded to 0·5 inch. The sun, moon and five planets were strung along the ecliptic. A clepsydra was set up and the whole apparatus made to rotate by its water [313].[4] The transits of stars[5] at dawn and dusk (shown on the instrument) all agreed exactly with the actual movements of the heavens [314].[6]

In the seventeenth year (A.D. 440) a small astronomical instrument [156] (i.e. a globe) was also made, of diameter 2·2 ft. and circumference 6·6 ft., with (two)[7] tenths of an inch to a degree.[8] The twenty-eight lunar mansions (*hsiu*) [196] were fixed on, and pearls of three colours, white, black and yellow, represented the stars of the three schools of astronomers.[9] The sun, moon and five planets were again attached to the ecliptic.

Sui Shu, ch. 19, pp. 17*b* ff., by Wei Chêng *et al.*, *c.* A.D. 656. Partly quoted in *Yü Hai* (Encyclopaedia of Wang Ying-Lin, *c.* A.D. 1267), ch. 4, p. 18*b*

Ch'ien Lo-Chih, following the old theories, explained them by means of a (demonstrational armillary) sphere [150] and a (celestial) globe [152] made of bronze, taking 0·5 in. to the degree, with a diameter of 6·0825 ft. and a circumference 18·2625 ft.[10] The earth was (shown) fixed

[1] If a large number of stars really were represented, then the instrument was a solid celestial globe, but the text does not positively assert this, and there are grounds for thinking that it was a demonstrational armillary sphere (cf. pp. 98 ff.).

[2] Wên Ti [961], personal name Liu I-Lung [667] (A.D. 407–53). Third emperor of the Liu Sung dynasty. His reign was prosperous. Diligent and economical, he promoted learning and science, improved the laws and maintained peace. But this did not prevent him from being murdered by one of his sons.

[3] As noted above, this is a remarkable statement in view of the late development of the system in Europe. Presumably the model earth of these early Chinese demonstrational armillary spheres was supported upon a pin in the polar axis. It would be extremely interesting to know whether it was represented as a ball or a disc or a square plate. Any one of these methods would have had support from one or other of the ancient Chinese cosmologists (cf. Needham (1), vol. 3, pp. 210 ff.).

[4] The parallel passage in *Sui Shu*, ch. 19, p. 17*b*, is in the same words, but omits only the mention of the water-power drive. But this is present again in the *K'ai-Yuan Chan Ching*, ch. 1, p. 23*b*, and *Yü Hai*, ch. 4, p. 18*b*. Archives must have been available in the Southern Ch'i, and also probably in the early T'ang, which attested it.

[5] Perhaps the translation should be planets.

[6] The word used here, *ying*, is interesting. It was a technical term of Chinese natural philosophy going back at least to the Han, with the meaning of 'resonance'. This was of great importance in the world-view of organic naturalism. On the whole question see Needham (1), vol. 2, pp. 282 ff., 304.

[7] The brackets indicate a small lacuna in the text here.

[8] The dimensions indicate a value of π very close to 3. This is rather curious since at that time much more sophisticated values were available (see Needham (1), vol. 3, pp. 100 ff.).

[9] See n. 2 on p. 22 above.

[10] The term *shao* [315] here is one of the set of technical terms which Chinese mathematicians and astronomers used to indicate fractions; it means 'a quarter'. These terms could be used not only as ordinary fractions of whole numbers, but also, as here, in formulations equivalent to decimal places. Thus one-eighth could be written in a form equivalent to $0{\cdot}1\frac{1}{4}$ instead of 0·125. Here, of course, $18{\cdot}2625 = 365\frac{1}{4} \times 0{\cdot}05$, and 6·0825 is derived by taking π as 3.

PLATE VI

Fig. 40. A demonstrational armillary sphere of the European Renaissance, made by Carolus Platus at Rome in A.D. 1588. Though characteristic of the sixteenth century in the West, many instruments of this kind had been made by the Chinese from the first to the fifth century.

immovably within the sphere of the heavens. Besides a meridian circle through the south and north poles, there were two rings one for the equator and one for the ecliptic, on which were marked the divisions of the twenty-eight *hsiu* (lunar mansions); and the positions of the Great Bear and the Pole Star (were also shown). The sun and moon and five planets were indicated on the ecliptic. There was an axle (in the polar axis) permitting rotation like the real heavens, and so exhibiting 'the hours of the day and night dictated by the culmination of the constellations'.[1] Everything corresponded with the heavens like the two halves of a tally.

Towards the end of the Liang dynasty (A.D. 557)[2] this instrument was placed in front of the Wên Tê Hall [937] (of the Palace). As for its system of construction, you might regard it as an armillary sphere [150], but it had no sighting-tube inside it; or you might regard it as a celestial globe [152], but it had no earth horizon outside it. Thus it combined the two systems to make a third type of instrument. In function it was like a celestial globe, but it did not distort the pattern of the universe for it represented the earth occupying its proper position within the concavity of the heavens.

In the San Kuo (Three Kingdoms) period (A.D. 222–80) in the State of Wu there was also Ko Hêng [565][3] (fl. A.D. 250) who was a perfect master of astronomical learning and was capable of making ingenious apparatus. He altered the astronomical instrument [156] in such a way as to show the earth fixed at the centre of the heavens, and these were made to move round by a mechanism [276] while the earth remained stationary.[4] (This demonstrated) the correspondence of the (shadows on the) graduated sundial with the motions (of the heavens) above [316]. This it was that Ch'ien Lo-Chih imitated.

In the seventeenth year of the Yuan-Chia reign-period (A.D. 440), he (Ch'ien) also made a small astronomical instrument [156] taking 0·2 in. to the degree, with diameter 2·2 ft. and circumference 6·6 ft. The twenty-eight *hsiu* (lunar mansions) and all the constellations both north and south of the equator were indicated by pearls of three colours, white, green and yellow, according to the three schools of astronomers.[5] The sun, moon, and five planets were

[1] The quotation comes from the Han apocryphal treatise *Shang Shu Wei K'ao Ling Yao* [728], in *Ku Wei Shu* [767], ch. 2, p. 2*a*.

[2] A sidelong glance at contemporary events at the other end of the Old World reminds us that rather earlier in this dynasty (*c*. A.D. 510) there had been constructed at Gaza in the Byzantine dominions an important striking water-clock which we know from the description of Procopius, *Ekphrasis Horologiou*, edited and translated by Diels (2). This apparatus, which had neither dial-faces nor any astronomical components, but which told the time by jack-work, dropping balls, and lights illuminated in turn, is particularly important because it links the Alexandrian mechanical clepsydra traditions with those of the later Arabic world. Cf. p. 67 and Fig. 34.

[3] One wonders whether Ko Hêng was a relative of his older contemporary Ko Hsüan [566] (fl. 238–55), the Taoist and a key figure in the development of the Taoist church, and of Ko Hsüan's nephew Ko Hung [496] (A.D. 280–360), who under the pseudonym of Pao P'u Tzu [567] became the most celebrated alchemist in Chinese history. Ko Hung was a scientific thinker of remarkable talent.

[4] A parallel passage from the *Chin Yang Ch'un Ch'iu* [768], by Sun Shêng [568], is quoted in 983 by the *T'ai-P'ing Yü Lan*, ch. 2, p. 10*a*. So also, much earlier, in the commentary (*c*. A.D. 425) of P'ei Sung-Chih [677] in the *San Kuo Chih*, ch. 63, p. 5*b* [820], at the end of the biography of the mathematician and statesman Chao Ta [678].

[5] Cf. pp. 22 and 96 above and p. 98 below.

attached to the ecliptic, and the rotation of the heavens demonstrated with the earth (-horizon) across the middle.[1]

This sphere and globe, made in the Yuan-Chia reign-period, were both transported to Ch'ang-an (modern Sian) in the ninth year of the K'ai-Huang reign-period (A.D. 589) after the conquest of the Ch'en dynasty (by the Sui), and then at the beginning of the Ta-Yeh reign-period (A.D. 605) they were moved to the astronomical observatory [962] at the Eastern Capital (i.e. Loyang), (of the Sui dynasty).

Arriving now at the term of our historical inquiry, we have to consider what little is known of the work of that rather enigmatic initiator of powered astronomical instruments, Chang Hêng. As a preliminary to this, however, we must finally clear up the question of the demonstrational armillary sphere, for it is probable that one of Chang Hêng's instruments was of this kind, and not a solid celestial globe. As has already been suggested (p. 23 above), it seems that in the early centuries there was a gradual shift of terminology, the words *hun hsiang* [152], which had originally meant the demonstrational armillary sphere (in which the model earth replaced the sighting-tube), coming to mean the solid celestial globe. In the following passage, taken from ch. 2 of the *Hsin I Hsiang Fa Yao*, Su Sung quotes from the third-century astronomer Wang Fan [478] to show that it was the latter who first removed the model earth from the demonstrational armillary sphere and replaced it by sinking the sphere in a flat-topped box to symbolise the terrestrial horizon. Evidence has already been given that the demonstrational instrument of I-Hsing in the eighth century was one of this type. It is probably best to take as true celestial globes only those instruments of which it is said that they depicted the stars of all the constellations according to the colours of the three schools of astronomers. On this criterion, it was Ch'ien Lo-Chih who was the first to make a solid celestial globe. Wang Fan's essay, the *Hun T'ien Hsiang Shuo*, which in view of the foregoing discussion we had better translate as 'Discourse on Instruments which are Models of the Celestial Sphere', was preserved in three of the dynastic histories (the *Chin Shu*, the (Liu) *Sung Shu*, and the *Sui Shu*), as also in the *T'ai-P'ing Yü Lan* encyclopaedia, and the *K'ai-Yuan Chan Ching* astronomical compendium, whence it has been reconstructed as far as possible in the great fragment collections.

[1] It seems clear from what follows that the globe still treasured at the end of the Liang dynasty (*c.* A.D. 555) was one of those made by Ch'ien Lo-Chih. Cf. p. 18 above. It also seems clear that what he made to rotate by water-power was the demonstrational armillary sphere and not the globe. This is a rather important point in connection with the nature of the drive, since the texts distinctly say that the former was cast (of bronze) and that the latter was made of wood.

Once the demonstrational armillary sphere had been sunk in the flat-topped box, it was quite a natural improvement to make a solid ball with a star-map on it. But this did not happen in the third century.[1] Soon after the time of Wang Fan we have a further reference to the demonstrational armillary sphere in the essay of Liu Chih [569] (fl. A.D. 274) entitled 'Discourse on the Heavens' (*Lun T'ien* [769]). 'The (demonstrational) armillary sphere [150]', he says,

images forth the round body of the heavens, and holds within itself the square earth. Like a wheel continuously rotating, it has two hubs, one of which, the (north) polar star, called the 'pole of motion' [317] we can see, while the other is at the south, underneath the earth and invisible, wherefore the men of old did not name it.... The pole being in the middle, there is a red ring representing the equator a little more than 91° from the poles. And there is a yellow ring for the ecliptic, for studying the motions of the sun and its varying distances from the equator at solstices and the intersection of its path with the equator at the equinoxes. Both rings have graduations....[2]

Hsin I Hsiang Fa Yao, ch. 2, pp. 2*a*, *b*, by Su Sung, A.D. 1092

The celestial globe [152] has not traditionally been kept in the Bureau of Astronomy and Calendar [917]. Ours is based on what is described in the (astronomical chapter of the) *Sui Shu*, with some modifications....

According to this chapter, the celestial globe [161] has the *chi* [164] (i.e. the movable rings), but not the sighting-tube, *hêng* [163]. At the end of the Liang dynasty there was one made of wood in the Secret Treasury, as round as a ball, several arm-spans in circumference....[3]

What we have now made is the same apparatus, simplifying the methods of Liang Ling-Tsan and Chang Ssu-Hsün regarding the sun and moon and five planets, which are tied around the 365 degrees (of the ecliptic) and rotate following the celestial movements.

Wang Fan said: 'In the (demonstrational) armillary sphere method [318] the earth should be placed at the centre of the heavens [319]. But as this is not very convenient, we may look at the matter in the opposite sense, and simply represent the earth by a (flat-topped) box outside. The perspicacious will see that there is no real difference—it depends on the point of view. Although the appearance may look unusual and strange, it is in perfect agreement with the principles involved [320]. This is indeed an ingenious thing.'[4] Thus the plan of having the earth sphere (or horizon) [9] represented outside the celestial globe (or demonstrational armillary sphere) [152] originated with Wang Fan.

[1] As already suggested, the limiting factor was perhaps the preparation of standard star-maps based on the *Hsing Ching* [722]. Chang Hêng had made something of this kind, but the manuscript was soon lost, and it fell to Ch'en Cho [499], who had been Astronomer-Royal of the Wu kingdom, to compile the first standard maps in A.D. 310. We are told explicitly that Ch'ien Lo-Chih's celestial globe or globes of about a century later were based on these. See *Sui Shu*, ch. 19, pp. 2*a* ff.; *Ch'ou Jen Chuan* [770], ch. 5.

[2] *CSHK* (Chin sect.), ch. 39, pp. 5*a* ff.

[3] Cf. pp. 18 and 97 above. There is a discrepancy about the size of this, if it was really made by Ch'ien Lo-Chih. But perhaps he constructed more than one.

[4] This quoted passage is identical verbatim, except for a few words, with that in the fully reconstructed essay *CSHK* (San Kuo sect.), ch. 72, p. 5*b*.

9 THE GRAND ANCESTOR OF ALL CLOCK DRIVES: CHANG HÊNG (SECOND CENTURY)

Now we can come to Chang Hêng. There is nothing at all enigmatic about him as a person. Born in A.D. 78 under the Later Han dynasty, he lived till about 142, attaining the greatest distinction as mathematician, astronomer, calendar expert, and geographer. Many biographies of him have been written,[1] besides the ample study in the *Ch'ou Jen Chuan*, ch. 3 (cf. Fig. 41). Unfortunately all the books he wrote have perished, but we have titles such as *Suan Wang Lun* [771] (perhaps: Mathematics of the Co-ordinate Network), and *Fei Niao T'u* [772] (perhaps: Diagrams of Flying-Bird Calendrical Methods). He was well known for his attainment of a more accurate value of π, which he gave as equivalent to $\sqrt{10}$; as we know from the third-century commentary on the *Chiu Chang Suan Shu* [774]. Particularly well attested and described is the seismograph which he set up at the capital in A.D. 132. This gave warnings of the directions and severity of earthquakes, and the method he adopted persisted for many centuries afterwards before it finally died out (see Needham (1), vol. 3, p. 627). The ingenuity of this device is so remarkable that one need not feel much difficulty regarding Chang Hêng's achievement of applying a power-drive to an astronomical instrument, though it is tantalising to have so little information as to how he did it. The very early period of his work makes it necessary to scrutinise rather carefully the reports of this 'clock drive'. This must be sufficient apology for any repetitiousness in what follows. Readers not philologically inclined may like to turn directly to p. 105.

The oldest description of what Chang Hêng did is contained in the *Chin Shu* (ch. 11, p. 3*b*) [775] which quotes some writing of Ko Hung [496], long since lost, but which must certainly date from soon after A.D. 310. We have already read the first part of the passage, since it was quoted by Su Sung in his Memorial (cf. p. 23 above). Ko Hung continues:

> Chang Hêng made his bronze armillary sphere and placed it in a secret chamber, where it rotated by the force of flowing water. Then the order having been given for the doors to be closed, the observer in charge of it would call out to the watcher on the observatory platform, saying the sphere showed that such and such a star was just rising, or another star just culminating, or yet another star just setting. Everything was found to correspond with the phenomena like (the two halves of) a tally. Thus Tshui Tzu-Yü [672] wrote the following inscription

[1] As by Chang Yü-Chê (1, 2) in English; and by Sun Wên-Ch'ing (1, 2), Yen Yü (1), and most recently Li Kuang-Pi & Lai Chia-Tu (1), and Lai Chia-Tu (1), in Chinese.

Fig. 41. Chang Hêng (A.D. 78–142), Astronomer-Royal in the Later Han dynasty, eminent mathematician and engineer, the first in Chinese history to build a machine rotating a celestial globe or demonstrational armillary sphere by water-power. An impression by a modern artist, Chiang Chao-Ho, made for a 1955 series of Chinese postage-stamps commemorating ancient and medieval discoverers and inventors (courtesy of Dr Kuo Mo-Jo).

on the stele of Chang Hêng: 'His mathematical calculations exhausted (the riddles of) the heavens and the earth. His inventions were comparable even to those of the Author of Change. The excellence of his talent and the splendour of his art were one with those of the gods.' And indeed, this was demonstrated by the armillary sphere and the seismographic apparatus constructed by him.

At first sight, most of the other mentions of Chang Hêng's second-century power-drive come from sources as late as the T'ang period (seventh century). Of these we have (*a*) the *Sui Shu* (History of the Sui Dynasty) written by Wei Chêng and others about A.D. 656, (*b*) a fragment contained in the *Ch'u Hsüeh Chi* encyclopaedia written by Hsü Chien in 700, (*c*) another fragment contained in the commentary of Li Shan [570] (fl. 660) on the 'Inscription for a New Clepsydra' (of 507) in the *Wên Hsüan* [773] anthology, (*d*) an important passage in the *Chin Shu* [775] (History of the Chin Dynasty), written by Fang Hsüan-Ling [571] in 635. Of course most of these sources were based on actual archives extant at the time. Nor is this situation particularly bad when compared with European parallels, for most of what we know of the views of the pre-Socratic philosophers (seventh to fifth centuries B.C.) seems to come from quotations in patristic writers as late as the fourth century A.D. Besides the above, we have one useful later source, the *Sung Shih*, written in the fourteenth century.

The *Sui Shu* mentions Chang Hêng several times in its astronomical section (ch. 19). On p. 2*a* it describes the writing of his book *Ling Hsien* [742] (The Spiritual Constitution of the Universe) and some details of the star catalogue to which it formed the introduction. More immediately relevant is the passage on p. 7*b*.

Chang P'ing-Tzu (Chang Hêng) [it says] made a bronze armillary sphere, (placing it) in a closed room [321] and applying the water of a clepsydra (lit. dripping water) to rotate it [190]. What it showed agreed with (the phenomena of) the heavens like the (two halves of) a tally deed [322].

Still more explicit is the passage on pp. 14*b*, 15*a*.

In the reign of Huan Ti, in the seventh year of the Yen-Hsi reign-period (A.D. 164),[1] the Astronomer-Royal Chang Hêng again (cast) a bronze instrument [160] on the scale of 0·4 in.

[1] This is undoubtedly a mistake, for it is agreed on all hands that Chang Hêng died about A.D. 142. As suggested above, the mistake may be due to a confusion between the Yen-Hsi and Yen-Kuang (122–5) reign-periods. Huan Ti's [963] personal name was Liu Chih [572] (133–67); he was the tenth emperor of the Later Han. If the former reign-period is correct, the emperor would have been An Ti [959], personal name Liu Yu [573] (94–125), the sixth emperor. Without doubt the chief period of Chang Hêng's scientific activity and highest rank was the reign of Shun Ti [964], personal name Liu Pao [574] (116–45), who reigned from 126 to 144. Of all these emperors much the same may be said: though not bad young men personally, they were incapable of defending the empire adequately against recurring natural calamities and the perennial raids of the Huns.

to the degree, its circumference being 14·61 ft. It was placed in a closed chamber [321] and rotated by the water of a clepsydra (lit. dripping water) [190]. One observer watched it behind closed doors and called out to another observer who was looking at the heavens on the observatory platform, saying when such and such a star should be rising, or making its transit, or setting, and everything corresponded like the two halves of a tally.

This is quite a circumstantial description. The significance of it will appear in a moment. First a few words on the two clepsydra fragments.

Both were preserved in the *Ch'u Hsüeh Chi* encyclopaedia of about A.D. 700 (ch. 25, pp. 2*a*, 3*a*). One says that the (inflow) clepsydra consists of bronze vessels containing water set at different levels and delivering water to each other by means of siphons. The receiver vessels are two in number, one for day and one for night, each having an indicator-rod on a float. The other says that on the covers of each inflow receiver there should be small statuettes (a policeman by day and an immortal by night) guiding the indicator-rods with their left hands and pointing to the graduations on them with their right hands. This second fragment was also preserved in the commentary by Li Shan [570] (fl. A.D. 660) on the *Hsin K'o-Lou Ming* [776] (Inscription for a New Clepsydra), written in 507 by Lu Ch'ui [575], and incorporated a few decades afterwards by Hsiao T'ung [576], prince of the Liang, in his literary anthology, the *Wên Hsüan* [773] of 530 (ch. 56, p. 13*b*). These fragments might at first sight seem to have nothing to do with the case, but Ma Kuo-Han [577], in his great collection of fragments (1853), the *Yü Han Shan Fang Chi I Shu* [777], gave them (ch. 76, p. 68*a*) under the surprising joint rubric 'Lou Shui Chuan Hun T'ien I Chih' [779], i.e. 'Apparatus for Rotating an Armillary Sphere by means of Water from a Clepsydra'. So also did Yen K'o-Chün [578] in his collection of 1836, the *Ch'üan Shang-Ku San-Tai Ch'in Han San-Kuo Liu Ch'ao Wên* [778] (Hou Han sect.), ch. 55, p. 9*a*. Both had long been preceded by Wang Ying-Lin in A.D. 1267 (*Yü Hai*, ch. 4, p. 9*b*, ch. 11, p. 7*a*). The significance of this caption will appear in a moment.

Let us now turn to the account in the *Chin Shu* (History of the Chin Dynasty), due to Fang Hsüan-Ling in A.D. 635. In ch. 11, p. 5*a*, we read:

Then, at the time of the emperor Shun Ti (A.D. 126–44), Chang Hêng constructed a celestial globe (or demonstrational armillary sphere) [152], which included the inner and outer circles [323], the south and north celestial poles, the ecliptic and the equator, the twenty-four fortnightly periods, the stars within (i.e. north of) and beyond (i.e. south of) the twenty-eight *hsiu* (equatorial lunar mansions), and the paths of the sun, moon and five planets. The instrument was rotated by the water of a clepsydra (lit. dripping water) [190] and was placed inside a (closed) chamber above a hall [324]. The transits, risings and settings of the heavenly bodies (shown on the instrument in the chamber) corresponded with (lit. resonated with) [1] those in the (actual) heavens [325] following the trip-lug and the turning of the auspicious wheel [326].[2]

Here then the other accounts are confirmed, but with the addition of the very intriguing last half-sentence. Our translation of *kuan-li* [327] as trip-lug is temerarious and requires explanation. It must be some technical term but lexicological investigations have thrown no particular light on it. That *kuan* is the word used from six to ten centuries later for the stopping devices of the mechanical escapement we already well know, but we cannot carry back this precise meaning to such an early date. Phrases somewhat similar have been met with elsewhere in mechanical connections, but their orthography slightly differs. For example, *kuan-li* [328] is the term used by Chou Ch'ü-Fei [579], in his *Ling Wai Tai Ta* [780] (Questions about what is beyond the Passes) [3] of A.D. 1178, for a fish-shaped movable stopper of silver placed inside a bamboo tube through which the tribespeople of the south-west drank wine in their ceremonies. If the sucking was either too fast or too slow the valves formed by the septae were occluded. The idea resembled therefore that of the 'inclining advisory vessels' already referred to (p. 84). Another form of *kuan-li* [329] was a term used for the interlocking pivots of the rings in the Cardan suspensions of paper lanterns used at festivals. It occurs in an eighteenth-century continuation of the *Hsi Hu Chih* [781] (Topography of the West Lake at Hangchow). If we analyse the semantics of the two words separately, we know first that *kuan* [330], a word in very frequent use, and in this

[1] On the philosophical significance of the term *ying* here used, see n. 6 on p. 96 above.

[2] The passage in the text continues, concluding the paragraph, as follows: 'Moreover, there grew on the steps (of the hall) the calendar-plant *ming-chieh* [334] which put forth leaves in accordance with the waxing of the moon, and shed them as it waned.' This is presumably a legendary embellishment. A study of the texts concerning the *ming-chieh* in the encyclopaedia *T'ai-P'ing Yü Lan*, ch. 873, p. 7*b*, shows that none of them is earlier than the Ch'an-Wei apocryphal treatises of the Later Han period. The only possible exception is a mention in the *Shang Shu Ta Chuan* [783] of Fu Shêng [580] of the Ch'in and Early Han, but that only mentions the auspicious plant and says nothing about its calendrical properties. The legend, if such it is, may remind us of the atmosphere of magic and credulity in which the work of men like Chang Hêng had to be carried out. A parallel legend, mentioned in the *Lun Hêng* [784] of Wang Ch'ung [581], about A.D. 83, and destructively criticised by him, spoke of a sensitive 'indicator-plant', the *ch'ü-yi* [335], which grew in the imperial palace courtyards, and pointed its branches at persons with traitorous intentions (ch. 17, p. 4*a*; tr. Forke (1), vol. 2, p. 320).

Using a different punctuation of the text, Liu Hsien-Chou (3), p. 4, takes the *kuan-li* and the auspicious wheel to refer also to the *ming-chieh* calendar-plant, and therefore regards the plant as a piece of jack-work. Perhaps it was; there seems no way of deciding. On the main issue of the nature of the *kuan-li* Dr Liu agrees with our interpretation.

[3] Ch. 10, pp. 14*a*, *b*.

doublet always written in the same way, has the meaning of closing and shutting (originally a bar across a door, or to put one in such a position), hence a barrier or mountain-pass, hence communication of any kind through any narrow way, hence to penetrate, to connect, to insert something in. *Li*, in its simplest form [331], has the sense of to wait, to arrive, to come, to fix, to push against, hence derivatively offensive or violent opposition. It is used of a soaring kite, forced sharply up and down by the wind. With the 'hand' radical [332], the word means strumming (the strings of an instrument) or poking in a consecutive and discontinuous manner, hence to insert the pegs or frets in a pear-shaped *p'i-p'a* [333] lute (cf. *P'ei Wên Yün Fu* [782], ch. 67B, pp. 2743.2 ff., 2781.3). Taking all these considerations into account, it seems justifiable to propose, at least, the translation of *kuan-li* as trip-lug in the present context. For it will be remembered (pp. 81, 94) that its function was to push a geared ring or wheel on by one tooth each time that sufficient weight of water had accumulated in one of the drive-wheel scoops to force the shaft round against the resistance of the astronomical instrument. If our reconstruction of how Chang Hêng's mechanism worked is wrong there is no obvious solution to the *kuan-li* problem.

Also of great interest is the appearance of the word *lun* (wheel) in this text. It is the only passage concerning Chang Hêng's device which mentions this. 'Auspicious' may well be the correct word for it, but it may be permissible to suggest that *jui* should be emended to *tuan* [336] so that we should read 'upright wheel', which would make particularly good sense. In any case, in the light of all else that we know about this piece of the history of engineering, we can hardly accept the counsel of despair adopted by other translators, who have written 'according to the turning wheels of calamity and good fortune'. Evidence presented elsewhere (Needham (1), vol. 4, pt. 1) shows that in Chang Hêng's time the vertical noria was almost certainly known, though the most common type of water-mill was probably that with a horizontal water-wheel. Moreover the operation of trip-hammers for grain-pounding by lugs on a water-powered shaft was one of the most characteristic and widespread machine designs in the second century A.D. in China.

Before leaving these T'ang records there is one more point to be made. They cannot have been projections back to Chang Hêng's time of considerations inspired by I-Hsing's advanced clockwork of A.D. 725, for every one of them was written in the previous century. Their writers could have known no more than the tradition of Kêng Hsün and his predecessors.

It may, however, be interesting to look briefly at the later tradition concerning Chang Hêng. There is a passage about him in the *Sung Shih*. It is interesting for several reasons. It accepts the tradition of his power-drive. It recognises the greatness of his invention of having a demonstrational armillary sphere in the closed room as well as the observational one on the platform outside. It says that this was equivalent to the invention of Li Shun-Fêng in the seventh century when he constructed the first armillary sphere with three shells of rings. And it shows that the meaning of the expression *hun hsiang* [152] for the demonstrational armillary sphere was still understood in the fourteenth century, for otherwise there would have been no sense in saying that the two forms were combined by Li Shun-Fêng.

Sung Shih, ch. 48, p. 3*b*, by TOKTAGA & OUYANG HSÜAN, *c*. A.D. 1345

As for the Component of the Six Cardinal Points (the outer fixed rings), and the Component of the Three Arrangers of Time (the intermediate shell of rings), and the Component of the Four Displacements (the inner sighting-tube carriage equatorially mounted); these three, which form the three layers, originated from Li Shun-Fêng [502] (fl. A.D. 627–80). It was I-Hsing [477] who added (temporarily) the ecliptically mounted sighting-tube. (This was an innovation) like that which Chang Hêng made when he constructed a (demonstrational) armillary sphere [152] based on the old methods of Lohsia Hung [512] (fl. 140–104 B.C.) and Kêng Shou-Ch'ang [582] (fl. 75–49 B.C.) (for the observational armillary sphere). He set it up in a closed room and made it rotate by the water of a clepsydra (lit. dripping water) [190] in order to correspond with the readings of stellar motion in degrees obtained from the (observational) armillary sphere [162]. Thus the (demonstrational) armillary sphere [152] was always something essentially different (from the observational armillary sphere). In the T'ang, Li Shun-Fêng and Liang Ling-Tsan [481] followed the plan of it, and began to combine (lit. use) it with the (observational) armillary sphere (in the three-shell type of instrument).

The duty which now immediately presents itself is to search for any contemporary or near-contemporary texts which would give evidence concerning Chang Hêng. At first sight none appears. A false start was made by Maspero ((1), p. 329) in attributing the important *Sui Shu* passage (ch. 19, pp. 14*b*, 15*a*) on Chang Hêng quoted above (p. 101) to Yü Hsi [583] (fl. 307–38) and not to the seventh-century writers of the *Sui Shu*. Yü Hsi's *An T'ien Lun* [785] (Discourse on the Conformation of the Heavens) is indeed quoted on p. 14*b*, but the citation seems to end naturally after only a couple of lines. This was certainly the opinion of Ma Kuo-Han (*Yü Han Shan Fang Chi I Shu*, ch. 77, p. 4*b*), who evidently would have attributed our passage to the *Sui Shu* itself, and we suppose that Ma should take precedence of Maspero. Then unfortunately there is nothing about the rotated instrument in the biography of Chang Hêng in the *Hou Han Shu* [786]

(History of the Later Han Dynasty), written by Fan Yeh [584] in A.D. 450 (ch. 89, pp. 1*a* ff.). There is mention, of course, of the *Ling Hsien* book, the commentary to which quotes a statement by Ts'ai Yung [585] (A.D. 133–92) about the three cosmological theories, saying that only the Hun T'ien (celestial sphere) system is near the mark, and that the bronze instruments in the observatory [337] all follow this theory.

More important is the passage in one of the Later Han apocryphal books, the *Shang Shu Wei K'ao Ling Yao* [728] (Apocryphal Treatise on the Historical Classic; the Investigation of the Mysterious Brightnesses). It is a text which may well be contemporary with Chang Hêng. The reader is already acquainted with this, for Su Sung quoted it in his memorial (p. 5*a*; cf. pp. 25, 26 above) from the *Sui Shu* (ch. 19, p. 13*b*). Let us repeat it.

If the (demonstrational armillary) sphere indicates a meridian transit when the star (in question) has not yet made it (the sun's apparent position being correctly indicated), this is called 'hurrying' (*chi* [210]). When 'hurrying' occurs, the sun oversteps its degrees, and the moon does not attain the *hsiu* (mansion) in which it should be. If a star makes its meridian transit when the (demonstrational armillary) sphere has not yet reached that point (the sun's apparent position being correctly indicated), this is called 'dawdling' (*shu* [211]). When 'dawdling' occurs, the sun does not reach the degree which it ought to have reached, and the moon goes beyond its proper place into the next *hsiu*. But if the stars make their meridian transits at the same moment as the sphere, this is called 'harmony' (*t'iao* [212]). Then the wind and rain will come at their proper time, plants and herbs luxuriate, the five cereals give good harvest and all things flourish.

Clearly this passage does not prove whether the demonstrational armillary sphere was mechanised or not. What it does explain is the idea which lay behind Chang Hêng's apparently mysterious arrangement of having one sphere in a closed room and another on the observatory platform. The calendar-making astronomers of his time were very much concerned with all divergencies or discrepancies between the indicated positions of the stars on the one hand and those of the sun and moon on the other. If, as is rendered highly probable by several texts (cf. above, pp. 24, 72, 77, 96, 98), the system was to have small objects representing sun, moon and planets attached in some way to the sphere or globe, yet freely movable thereon (e.g. beads on threads), then the computer inside the room would adjust their positions in accordance with the observations of the observer outside. And whatever formulae of prediction were used could be tested by having the computer say what ought to be happening, whereupon the observer would correct him. We have just been told (pp. 21, 100, 102 above) that it was the computer within who spoke first. In later centuries all this was no longer understood, and at least one version of the text (*Ku Wei Shu*, ch. 2, p. 2*b*) gave it in exactly the opposite sense, making the 'hurrying' and 'dawdling' refer to the stars and not to the demonstrational (or rather computational) instrument. This completely misled Maspero ((1), p. 338)[1] who, knowing of the tradition of the power-drive, supposed that the whole arrangement was astrological rather than calendrical, and that prognostication was made from comparisons of the speed of rotation of the instrument with that of the heavens themselves. Any such idea should now be set aside.

But if this passage throws no light upon the question of mechanisation, we have at least two pieces of evidence that Chang Hêng himself stated that he accomplished this. Already in this story we have come across these, and now it is time that they received their due emphasis. In his memorial Su Sung said (p. 3*a*; p. 21 above): 'Thus it was that Chang Hêng in his *Hun T'ien* [720] (On the Celestial Sphere) said that (one instrument) should be set up in a closed room and rotated by water-power....' So also, much earlier, Wei Shu, writing about A.D. 750, said of I-Hsing's clock: 'It was made to turn automatically by the force of falling water acting on a wheel. Discussing it, people said that (what) Chang Hêng (had described in his) *Ling Hsien* [742] (On the Spiritual Constitution of the Universe) could have been no better' (p. 80 above). It therefore seems probable that the claim to mechanised rotation really is Chang Hêng's own, and does not rest only on what the historians said in the T'ang. But the passages concerned are not in the extant fragments of either of these books today. The question thus arises of the date to which they were conserved in a more or less complete state. To some extent this is ascertainable from the bibliographies of the dynastic histories. Thus both books are listed, with one chapter each, down to the end of the T'ang period (*Sui Shu*, ch. 34, p. 15*a*; *Chiu T'ang Shu*, ch. 47, p. 5*b*; *Hsin T'ang Shu*, ch. 59, p. 12*b*). They were thus certainly available to Wei Shu, and it is not unreasonable to suppose that one of them might have lasted until Su Sung's time, after which it would have disappeared, like so many other books, during the retirement of the dynasty to the south after the fall of the capital to the Chin Tartars.

It is in the light of these considerations that we can clear up the problem of the caption used by Ma Kuo-Han. If we look at his chief source, the *Ch'u Hsüeh Chi*

[1] We are at a loss to understand how it came about that Maspero gave a reference only to the *Sui Shu* but in his translation followed the text of the *Ku Wei Shu*.

encyclopaedia compiled about A.D. 700 by Hsü Chien, we find the following (ch. 25, p. 3*a*):[1]

> The *Chin Ch'i Chü Chu* [787] (Daily Records of the Chin Dynasty)[2] says 'In the time of Hsiao Wu Ti[3] in the twelfth year of the T'ai-Yuan reign-period (A.D. 384) an official memorialised stating that when the Ch'u Palace [965] was first built there was no clepsydra. So they carefully studied the bronze clepsydra of the Yung An Palace [966] and established a post for an official to be in charge of it.' Now Chang Hêng, in his *Lou Shui Chuan Hun T'ien I Chih* [779] (Apparatus for Rotating an Armillary Sphere by Clepsydra Water) says 'The figure of an immortal is cast in bronze and placed on the left (receiver) vessel, while the figure of a policeman in gold is placed on the right one'.

Another part of this quotation appears under the same ascription on the preceding page (2*b*).[4] It thus becomes clear that these seven words were actually the title of a book, or more likely part of a book, namely Chang Hêng's *Hun I* [718], to which in all probability it formed a kind of appendix. This was before Hsü Chien as he wrote. One can only regret that the imagination of the later generations was so caught by the figures of the immortal and the policeman that they neglected to record what would now be much more precious to us, namely the way in which the demonstrational armillary sphere was rotated by their maker.

Whether the instrument which Chang Hêng had in the closed room was made of bronze or wood remains uncertain. In any case it is extremely probable that it was a demonstrational armillary sphere and not a solid celestial globe. Apart from what has been said above (pp. 23, 98) about these two types of instrument, we can reach the same conclusion in another way. As Maspero ((1), pp. 339 ff.) has well explained, Chang Hêng was much concerned with the establishment of the correspondence of the *hsiu* extensions on the equator with those on the ecliptic. These lunar mansions were, as we know, twenty-eight unequal spans of the equatorial circumference and, in the absence of adequate spherical trigonometry, Chang Hêng and his contemporaries had recourse to three-dimensional graphic methods. In his *Hun I* [718] or *Hun I T'u Chu* [721] (the title often varies), he said that 'as the bronze armillary sphere cannot be used in bad weather one should make a little (demonstrational) armillary sphere (*hsiao hun* [338])', and then, taking

[1] Copied in *Yü Hai*, ch. 11, p. 7*a*.

[2] This was compiled by Liu Tao-Hui [586]. *Sui Shu*, ch. 33, p. 9*b*, tells us that it had 317 chapters. No doubt the *I-Hsi Ch'i Chü Chu* [766] which we have already come across (p. 95, n. 4 above), in seventeen chapters, was a part of it.

[3] Personal name Ssuma Yao [587] (A.D. 362–96). Ninth emperor of the Chin. A young man who succeeded in winning nearly twenty-five comparatively peaceful years in a very troublous period.

[4] Also, of course, in the commentary on Lu Ch'ui's [575] inscription in *Wên Hsüan*, ch. 56, p. 13*b*.

a flexible bamboo ruler, mark out the *hsiu* extensions upon the ecliptic.[1] If his instrument had been a solid celestial globe it would have been more natural to use a thread, just as Su Sung and others had later on for attaching the sun and moon markers. The fact that he did not do this suggests rather strongly that what he was working with was a sphere with rings and not a solid globe. The only other thing we know about it is that Wang Fan [478] thought it was too large.[2] He therefore made his own smaller than Chang Hêng's but larger than those which he supposed were older.

This is the end of the trail. If our interpretation is right, the history of mechanical clocks begins, in a sense, not only seven centuries before the first escapement machines of the European fourteenth century, but even twice that lapse of time. And the link between the mechanical clock and the clepsydra turns out to be much closer than hitherto suspected. For the story seems to take its origin in the attempt to make water do something more interesting than just dripping, and so the art of the millwrights, to whom Chang Hêng most naturally turned, was involved from the very outset. In the China of the second century A.D. their craftsmanship was doubtless primitive, but during the previous hundred years the water-powered trip-hammer (*shui-tui* [339]) had come into widespread use (Fig. 42). Although the most characteristic form of mill-wheel in China in later centuries was the recumbent or horizontally mounted type, the vertical 'Vitruvian' type always persisted, and for the trip-hammer was the more suitable of the two. Moreover, metallurgical blowing-engines powered by water had become common in China during the first century A.D. If the Chinese vertical water-wheel really derived from the noria, the invention had been made in any case a good while before the time of Chang Hêng, so that the chief originality in what he did was to

[1] *Hou Han Shu*, ch. 13, p. 20*b* (commentary); *K'ai-Yuan Chan Ching*, ch. 1, p. 2*a*. *CSHK* (Hou Han sect.), ch. 55, p. 8*a*; *YHSF*, ch. 76, p. 66*b*.

[2] His criticism of the sizes of the earlier instruments and description of his own often occurs, and has generally (we think wrongly) been interpreted as referring to solid celestial globes. See *Sung Shu*, ch. 23, pp. 5*b* ff. (the oldest reference, *c.* A.D. 500); *Chin Shu*, ch. 11, p. 7*a*; *K'ai-Yuan Chan Ching*, ch. 1, pp. 16*a*, *b*; *CSHK* (San Kuo (Wu) sect.), ch. 72, p. 5*b*. Wang Fan (fl. A.D. 250) says that the original (demonstrational) armillary sphere was on a scale of 0·2 in. to the degree, and that Chang Hêng enlarged it taking 0·4 in. to the degree. 'I, Wang Fan, consider that the early sphere was too small, so that the stars were over-crowded and impossible to see clearly, while that of Chang Hêng was too large and difficult to turn round. I have therefore redesigned the instrument, taking 0·3 in. to the degree.' The phraseology does give the impression of a solid celestial globe, but Wang Fan may have been referring only to the *hsiu* determinant stars along the equatorial band, and he certainly says nothing about the stars of the three schools of astronomers, which were all over the visible heavens. A further point is that if Chang Hêng's instrument was so difficult to turn, it can hardly have been the one to which he applied water-power. But again, Wang Fan knew it rather more than a century later, when considerable corrosion might have taken place. Perhaps more probably Chang Hêng made several demonstrational armillary spheres.

Fig. 42. A battery of water-powered trip-hammers for cereal pounding (from the *T'ien Kung K'ai Wu* of 1637); note the trip-lugs on the shaft depressing the levers in turn. These machines were developed in the Han period.

PLATE VIII

Fig. 43. A hodometer of the Han dynasty (a rubbing of a stone cut following the *Chin Shih So*, ch. 1, p. 133, originally from the Hsiao T'ang Shan tomb-shrines of A.D. 128). The figures at the top automatically beat upon the drum counting the distance travelled (courtesy of Mr E. Bredsdorff).

arrange for a constant drip into scoops rather than a powerful flow and fall on to paddles. The trip-lug on his shaft merely corresponded to those which worked the grain-pounding trip-hammers all around him. But it was original, too, to harness it instead to a toothed ring or gear-wheel (no doubt with leaf-teeth) controlled by a ratchet (a device certainly known in the Han),[1] and so to make it rotate tooth by tooth whatever astronomical apparatus he wished to connect with it. There seems nothing at all in this arrangement which would have been beyond the powers of Han technicians.[2] Indeed the trip-lug, which constituted a pinion of one, has a distinctly Alexandrian air. One may compare it with the peg on the axle of Heron's taximeter (cf. Brunet & Mieli (1), pp. 514 ff., from *Dioptra*, ch. 34) in the previous century,[3] or with that on the axle of his organ-blowing windmill (cf. Woodcroft (1), p. 108, from *Pneumatica*, ch. 77). What connections there could have been between the engineers of Alexandria and Han China remains of course a completely unsolved question (cf. Needham (1), vol. 1, pp. 231 ff.).

Presumably Chang Hêng's simple machine was at rest during each period when water was slowly accumulating in one of the scoops. As soon as enough had collected, its weight overcame the resistance of the toothed wheel and armillary

[1] This is not the place to enlarge upon the knowledge of gear-wheels in China of the Han period (second century B.C. to second century A.D.); for a full discussion of the subject see Needham (1), vol. 4, pt. 1. But one may mention some of the most important of the finds of such objects from tombs dating from as early as the Warring States period (fourth century B.C.). Wang Chen-To (2) twenty years ago described and figured an earthenware mould for bronze toothed wheels marked with a pair of characters in Han script. This wheel is a ratchet with sixteen slanting teeth. More recently Ch'ang Wên-Chai (1) reported on five actual bronze ratchet wheels taken from a group of tombs of the third or fourth century B.C. at Yung-chi in Shansi. Most of these have forty ratchet teeth each, but one remarkable little wheel is apparently a pinion of five. The former have large square shaft-holes, but that of the latter is small and round. In all probability the ratchet wheels were parts of crossbow arming apparatus. Still more remarkable are the pair of gear-wheels taken from a tomb of the early first century A.D. near the Han capital of Ch'ang-an (modern Sian) in Shensi province, and illustrated by Liu Hsien-Chou (3). These enmesh closely with the aid of chevron teeth. One wheel has a square shaft-hole, the other a round one. We mention these details in order to show how extremely familiar gearing of various kinds must have been to Chang Hêng. The contemporaneity with Alexandria hardly needs emphasising, and there is no obvious reason for assigning priority to either end of the Old World.

[2] It may be worth remembering that less than a hundred years after the death of Chang Hêng a very elaborate puppet theatre was constructed by the engineer Ma Chün [588] (*San Kuo Chih*, ch. 29, p. 9*a*; *CSHK* (Chin sect.), ch. 50, p. 11*a*). The description states that the motive power was a water-wheel, probably horizontally mounted. Trip-lugs must have been used in abundance.

[3] It should be remembered that the hodometer or taximeter (Fig. 43) was also developed in Han China (for a full discussion see Needham (1), vol. 4, pt. 1). Liu Hsien-Chou (3) has pointed out that the Chinese form of it was particularly relevant to the history of clockwork, for apart from the use of pinions of one or three, the distances were sounded off by the beating of jack figures on drums.

Another piece of jack-work which should not be forgotten was the figure set up on the 'south-pointing carriage', a vehicle with gearing which made the figure point to the south whatever path might be traced out by the carriage. This originated in the Later Han about the time of Chang Hêng. Several reconstructions of it have been put forward, including a simple application of differential gearing; one of the most recent will be found in Liu Hsien-Chou (2), pp. 15, 220. For a full account of the matter see Needham (1), vol. 4, pt. 1.

sphere, and the trip-lug turned it round by one tooth, then coming to rest against the next. Although we are told optimistically that 'everything agreed with the heavens like the two halves of a tally' we are bound to assume that the chronometric properties of the device were extremely poor. So much of it depended upon play and resistance, the exact size of each scoop, the nature of the bearings of the polar axis, and similar factors. Nevertheless it is difficult to exclude Chang Hêng's device entirely from the horological definition of von Bertele (1). 'The fundamental solution', he wrote, 'of the problem of securing steady motion by intersecting the progress' of a powered machine 'into intervals of equal duration, must be considered the work of a brain of genius.' This object does not seem to have been attained in any of the designs of the Alexandrians, and if Chang Hêng did not attain it, he seems at any rate to stand at the threshold of the line of development which led to its attainment.

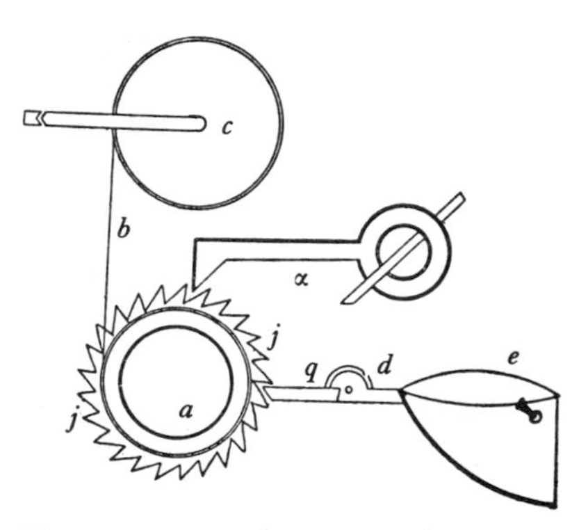

Fig. 44. A mechanism analogous to Chang Hêng's trip-lug for rotating a disc or globe by water-power, from al-Jazarī's treatise on striking clepsydras (A.D. 1206). The bucket tips when full, activating a ratchet which pushes round a gear-wheel by one tooth each time (Wiedemann & Hauser).

It is curious to notice that arrangements having distinct resemblances to his persisted through many later centuries. We do not speak of the work of his successors (Ch'ien Lo-Chih, T'ao Hung-Ching, Ko Hêng, Kêng Hsün and others down to the eighth century, and probably Tsêng Nan-Chung in the twelfth) but of men in quite other parts of the world. One of the mechanisms described by Ismā'īl ibn al-Razzāz al-Jazarī in his treatise on striking clepsydras of A.D. 1206, as interpreted by Wiedemann & Hauser (1), p. 147, has a tipping bucket attached to a hinged ratchet which pushes round a gear-wheel by one tooth each time that the bucket fills with water and comes to the emptying point (Fig. 44). This gear-wheel is connected by a cord with what seems to be the plate of an anaphoric clock. And then again, after four more centuries, the same arrangement is found in the book of Isaac de Caus on mechanical contrivances in 1644 (Fig. 45).

But none of these instruments involved that great invention, the mechanical escapement, which is what von Bertele's dictum really implies. As we show in the present monograph, the first of these was the weigh-bridge and stopping-lever system for a water-wheel, first devised by I-Hsing and Liang Ling-Tsan in the

PLATE IX

Fig. 45. Dial water-clock depicted by Isaac de Caus in 1644.

second or third decade of the eighth century A.D., developed by Chang Ssu-Hsün in the tenth and brought to fullest perfection by Su Sung and Han Kung-Lien in the eleventh. Though continuing in all probability after this time until the coming of the Jesuits, it does not seem to have undergone many further improvements. Whether it came to Europe exactly in that form, or whether only as a rumour, and if transmitted, just when, are questions discussed elsewhere (p. 196). The most important point about it is that it introduced a truly chronometric principle into the mechanical as opposed to the clepsydric part of the clock. At first this was not great, for it is obvious that in the Chinese astronomical escapement clocks the major part of the time-keeping was done by the constancy of the flow of water. The mechanism could only affect this in so far as changing the weight on the weigh-bridge would permit the scoop to fall before it was quite full. This is why we can see in it a missing link between the clepsydra and the purely mechanical clock. For after the weight-drive had been adopted in fourteenth-century Europe, the verge and foliot escapement took over the great part of the time-keeping duty. This was still not completely embodied in the escapement since any considerable change in the weight hanging from the drum would affect the fastness or slowness of the clock.[1] Not until the introduction of the pendulum in the seventeenth century was a truly isochronous mechanism attained. That it had taken nearly two millennia is not surprising when one considers the leisurely growth of human technology as a whole before the Renaissance.

[1] It can be deduced from elementary mechanical considerations that the periodic time of a verge-and-foliot escapement is given by

$$T=4\left(1+\frac{\epsilon}{\delta}\right)\sqrt{\frac{L(L+J)\delta}{JM}},$$

where L = the moment of inertia of the foliot,
J = the moment of inertia of the rest of the mechanism,
M = the moment of the restoring couple,
$\pm\delta$ = the angle of displacement of the pallets,
ϵ = the angle of free motion of the pallets.

For our purposes

$$T \propto \sqrt{\frac{1}{M}}.$$

In practice the going error in verge watches is found to be of the order of ± 15 min. per day, which corresponds to a variation of $\pm 1\%$. If we assume similar errors for all verge-and-foliot escapements, we may deduce that the moment of the restoring couple must vary by $\pm 2\%$ due to friction, or that the movement of the weights and the foliot are accurate to within about $\frac{1}{2}\%$. Cf. von Bertele (1).

VII

THE DEVELOPMENT OF CLOCKWORK AFTER SU SUNG

We can now see that the idea of a powered time-keeping machine was a secondary development. What came first was the idea of applying a power-drive to a demonstrational armillary sphere in imitation of the actual heavens, thus meeting a need of the calendrical computers. Let us therefore return to our starting-point and prepare to follow the events which took place after the completion of the monumental clock of Su Sung. We may take that wintry day towards the end of A.D. 1088 when the workmen were busy in the imperial palace at K'ai-fêng erecting and assembling the full-scale wooden prototype of this masterpiece of Han Kung-Lien and his colleagues. From there onwards let us follow the course of history and see what we can find between that time and the arrival of Matteo Ricci in Peking in 1600—a missionary indeed but just as well and truly the advance-guard of the scientific and technological movement of Europe.

The first passage which we shall present throws light on a number of points. First it shows that the order was duly given to cast all the bronze parts required for Su Sung's clock. It also shows that his senior astronomical colleagues were doubtful whether his mechanised armillary sphere would also be suitable for all the actual observations which were wanted. Another sphere, purely for observations, was therefore put in hand. The considerable restlessness just before Su Sung's time suggests that conditions in the empire, doubtless including rising pressure from the nomadic north, were causing the imperial astronomers to verify their data and computations. For one of the most important ways of pleasing Heaven was to have a well-ordered calendar, and the members of the Bureau concerned were undoubtedly looked upon by courtiers as well as common people as men invested with occult astrological powers.

Sung Shih, ch. 80 (Lü Li Chih), p. 25*a*, by TOKTAGA & OUYANG HSÜAN, *c.* A.D. 1345

In the seventh year of the Hsi-Ning reign-period (A.D. 1074) the Astronomer-Royal (Shen Kua) [475] presented a new armillary sphere and float clepsydras at the Ying-Yang Gate. The emperor[1] with his ministers examined them and asked many questions, to all of which Shen Kua replied in detail, giving the reasons why the improvements had been made. (These were the instruments which he himself had designed). All the astronomical officials, at the imperial order, assembled to test their correctness, which proved to be incomparable. So the emperor ordered them to be placed in the Astronomical Department [915] of the Han-Lin Academy,[2] and in the seventh month Shen Kua was honoured with the title of Right-hand Counsellor,[3] while Huangfu Yü [589], the Director of Astronomical Observations (Western Region), and others, were also given their due rewards.

Previously (Shen Kua) had presented to the emperor three (written) discussions, one on the armillary sphere, one on the float clepsydra, and one on gnomon shadows—details about these are recorded in the astronomical chapters of this *Sung Shih*. The emperor followed his advice and authorised the improvements in apparatus, technique and calendar which he suggested....

In the first month of the fifth year of the Yuan-Fêng reign-period (A.D. 1082), Wang An-Li [590],[4] a Han-Lin Academician, stated that the wooden models of the armillary sphere and float clepsydra newly presented by Ouyang Fa [591],[5] an armillary official, had the merits of up-to-date equipment, avoiding the errors of the old types. He (Wang) and others were of the opinion that those then in use by the Bureau of Astronomy were erroneous and ought to be discontinued. All the apparatus which had come down from the Chih-Tao reign-period (A.D. 995–7) and the Huang-Yu reign-period (A.D. 1049–53), whether spheres or gnomons, was inaccurate, and ought to be checked item by item. It was decided to do this.

In the fourth year of the Yuan-Yu reign-period (A.D. 1089), the Han-Lin Academician Hsü Chiang [592] and others referred to the imperial edict establishing a Bureau for Constructing the Yuan-Yu Armillary Sphere and Celestial Globe.[6] This body had received imperial authorisation to make a wooden model of a water-driven armillary clock, which should be thoroughly tested by observations of the celestial bodies, and if the results were satisfactory, cast full-scale in bronze. Now (said Hsü Chiang) it has proved to be in perfect agreement with the heavens, so the order has been given to have it cast, and named the 'Yuan-Yu Armillary Clock' [340]. 'Now what we used to call the armillary sphere', he went on, 'had a spherical outer form, with the (equatorial) constellation (-positions) in degree-marks distributed round it. Inside there were the concentric rings [193] and the sighting-tube [163] which were used for observations of the heavens. Thus two separate things are being made, the armillary sphere and the celestial globe. The former is (traditionally) for observing the true numbers of the degrees of heavenly

[1] This was Shen Tsung [951]; cf. n. 3 on p. 62 above.

[2] The chief of the secretariats for drafting imperial edicts and providing expert advisers for the emperor; cf. n. 3 on p. 16 above.

[3] Or Policy Monitor, or Adviser.

[4] Wang An-Li wrote on astronomy. His *T'ien Wên Shu* [788] (Treatise on Astronomy) appears in the dynastic bibliography (*Sung Shih*, ch. 206, p. 8*b*). It had sixteen chapters. We do not know when it was lost.

[5] Ouyang Fa wrote at least three astronomical books, all of which are mentioned in the bibliography (*Sung Shih*, ch. 206, p. 9*b*) but none of which survived. There was the *Hun I* [789] (On the Armillary Sphere) in twelve chapters, the *K'o Lou* [790] (On the Clepsydra) in five chapters, and the *Kuei Ying Fa Yao* [791] (Essentials of Sundial Technique) in one chapter.

[6] This was of course Su Sung's. It is mainly from this passage that we know that his clock really achieved its final form, complete with bronze sphere and globe. Strong contributory evidence, however, will shortly appear (pp. 116–17, p. 127).

motion. The latter is (traditionally) placed within a closed room, performing the celestial rotation by itself and checking with what is observed by means of the armillary sphere. Now if you combine these two instruments into one (by mechanising both of them), they will simply do the same thing, that is, both follow the heavenly rotation. So I propose that another armillary sphere (for purely observational purposes) should be constructed.'[1] It was decided to do this.

In the seventh year (A.D. 1092) the emperor[2] asked the Minister of Personnel, Su Sung, to write an inscription for (his own) armillary clock [340], and in the sixth month the apparatus was completed. All the high officials were ordered to watch its displays.

The next that we hear of the great clock of Su Sung comes from a passage in the *Ch'ü Wei Chiu Wên* [792] (Talks about Bygone Things beside the Winding Wei (River in Honan)) written by Chu Pien [593] about A.D. 1140. It reveals to us for the first time something of the acrimonious political background of Su Sung's times. As we have already seen (p. 5) he was rather closely connected with, though probably not an active member of, the conservative party within the ranks of the bureaucratic scholars. When their enemies the reformers came to power in 1094 there was talk of destroying the clock which Su Sung had built. It was an old idea, of course, that a new reign-period or a new ministry must renew and change everything, just as every new emperor was expected to promulgate a new calendar. In one way this politico-philosophical conception contributed to keeping the astronomers busy, but in so far as it meant the abandonment of solid gains in instrumental technique it was quite at variance with the slow growth of practical scientific knowledge. In this particular case, as we shall see, the great astronomical clock remained untouched, ticking over the quarters inexorably until the impending fall of the capital to the Chin Tartars. But it was evidently regarded as one of the achievements of the conservatives, and the reformers looked upon it with much dislike.

Ch'ü Wei Chiu Wên, ch. 8, p. 10*a*, by Chu Pien, *c.* A.D. 1140

It was in the third month of the fourth year of the Yuan-Yu reign-period (A.D. 1089) that the bronze armillary sphere of Su Tzu-Jung (i.e. Su Sung) was completed....[3]

[1] The mechanising of the armillary sphere outside on the platform was a distinctive idea of Su Sung's. Hsü Chiang's argument runs somewhat counter to Su Sung's implication (p. 17) that it would be convenient for observers to have a kind of 'coarse adjustment' for their spheres. Perhaps Hsü Chiang was just being conservative. But his objection seems to have prevented the plan from ever being carried out again. However, Su Sung has more than seven hundred years' priority for the invention of the 'clock drive'.

[2] The young Chê Tsung [950]. See n. 1 on p. 16 above.

[3] Note that this statement confirms the evidence given in the previous text that the bronze parts of Su Sung's clock were cast in this year. Cf. p. 127.

In his earlier years Su Sung had had model armillary spheres of small size in his home,[1] and so had gradually come to understand the principles of them, but regretted that his mathematical knowledge was insufficient for embarking on the construction of a full-scale one. Later on, after he had been ennobled as a Ta Tsung Po by the Imperial Family, he found a good collaborator, whose name I have forgotten, among the First-Class Secretaries.[2] Su Sung referred his problems to this mathematician, who made the computations and checked up the results with ancient records; finally they unravelled all the secrets to their satisfaction. Several years' work was needed before the instruments were finished, and one was so large that a man could get inside and stand in it, as if it were a huge lantern. Holes were made in it according to the positions of the stars, and they all followed round as the instrument was slowly rotated by a water-wheel. Thus the transit stars appearing in the morning and the evening exactly at the right moment could all be seen through the holes.[3] The astronomers and calendrical pundits all flocked to watch its operation, and were quite astonished at it, for nothing like it had previously been achieved. Then Su Sung made diagrams of its form and system of construction, and included them in a book which he presented to the emperor, who ordered that it should be preserved in the Imperial Library.

But at the beginning of the Shao-Shêng reign-period (A.D. 1094) Ts'ai Pien [594] (Minister of State)[4] suggested that it (i.e. the armillary clock of Su Sung) ought to be destroyed as something which belonged to the previous (Yuan-Yu) reign-period. At that time Ch'ao Mei-Shu [668] (i.e. Ch'ao Tuan-Yen [595])[5] was Assistant Director of the Imperial Library, and as he greatly admired the accuracy and beautiful construction of Su Sung's instruments, he struggled to argue against Ts'ai Pien, but at first his efforts proved unsuccessful. However, he sought the help of Lin Tzu-Chung [596] who talked to Chang Tun [597] (Prime Minister),[6] and thus the

[1] This remark is of particular interest in view of what was said above about the 'governmental' or 'official' quality of medieval Chinese astronomy, and the prohibition of its study by unauthorised persons (p. 6). Presently we shall find another instance of astronomical instruments kept by a private family (p. 128). It seems clear that the 'official' aspect must not be pressed too far, and that the study of astronomy and calendrical science was quite possible in scholarly families connected with the bureaucracy.

[2] Han Kung-Lien [457] of course. See p. 19 above.

[3] All this is very odd. Writing as he was in the south some fifteen years or so after the fall of the capital, Chu Pien's memory must have been playing him tricks. We know very well the dimensions of Su Sung's two mechanised instruments. The diameter of the rings of the outer nest (the Liu Ho I, cf. p. 28) of the armillary sphere was 7·77 ft. (*Hsin I Hsiang Fa Yao*, ch. 1, p. 11*a*). The diameter of the ring of the innermost nest (the *ssu yu i* [5], cf. p. 28) was 6·0 ft., and the sighting-tube was 5·7 ft. long (ch. 1, pp. 18*a*, *b*). On the other hand, the diameter of the celestial globe itself was 4·565 ft. (ch. 2, p. 4*b*) and that of the rings surrounding it 5·47 ft. (ch. 2, p. 3*b*). Remembering that the foot of the Sung period was (unlike that of earlier periods) very close to our own, it is clear that a man could have stood up easily in the sphere but not in the globe. Holes corresponding to stars make no sense in relation to the sphere and, while they would have been possible on the globe, we can find no evidence in Su Sung's second chapter that this was the way in which the stars were marked. Nor is there any hint that it would have been possible to get inside the globe. Nor was the globe on an external platform so that stars could have been seen from inside it. As Chu Pien died in 1154 he would have been extremely young when Su Sung's clock was set up. He must have been writing from hearsay accounts.

[4] A.D. 1054–1112. A leading member of the reforming party and son-in-law to Wang An-Shih himself. Mild in manner but a ruthless politician.

[5] Born A.D. 1035. Descendant of a long line of high officials and writers.

[6] A.D. 1031–1101. Originally a friend of the poet Su Tung-P'o [466], but later of divergent political views and one of the leading reformers.

destruction of the clock was averted. However, after Ts'ai Ching [598][1] and his brother came into power nobody dared to say anything to prevent Su Sung's machinery being torn down. How shameful!

I SUNG DYNASTY POLITICS AND COMPETING CLOCK-MAKERS

Historians and sinologists have had much to say about the last years of the Northern Sung, but we doubt whether any of them have noted that this half of the dynasty expired in a blaze of horological exuberance. The following text throws further light on the political aspects of clock-making in the early twelfth century, but we shall postpone our comments on these until it has been read. Here we shall refer only to the more technical questions. The first paragraph gives the official form of the reformers' opposition to the work of Su Sung. It sounds very like an 'investigating committee' of Congress, and was probably justified by various alarming natural phenomena as well as the ominous activities of the Chin Tartars, but in fact it did not come to much. The subsequent paragraphs give new specifications for astronomical clockwork. Who was the man who proposed them? Wang Fu [599] is best known as an archaeologist,[2] for his book, the *Po Ku T'u Lu* or *Hsüan-Ho Po Ku T'u Lu* [793] (Illustrated Record of Ancient Objects), was finished in A.D. 1111. It was the catalogue of the archaeological museum of bronzes, stone pieces, inscriptions, etc. instituted by the emperor Hui Tsung, who had come to the throne just ten years before. We shall have more to say about him presently. It is quite clear that he disposed of the talents of certain more obscure Taoist technologists, and that the instruments which they proposed to construct were water-driven mechanical escapement clocks. What is said about the gearing is particularly interesting, for the great differences in the speeds of rotation of some of the wheels suggests a more complicated arrangement than anything in Su Sung's design. For example, it was evidently intended to show automatically the successive phases of the moon.

[1] A.D. 1046–1126. Elder brother of Ts'ai Pien. It was in 1101 that the two brothers obtained control of the administration. This was the year of the death of Su Sung, who must have suffered great uncertainty as to the fate of his clock.

[2] He was also a great deal more, for he made an adventurous (and somewhat unscrupulous) official career in the last decades of the Northern Sung. His first post was that of a Secretary for Textual Revision (Chiao Shu Lang [967]) in the Academy. After Ts'ai Ching came back into power (presumably in A.D. 1101) he became Chien I Ta Fu [968], one of the two Advisory Censors, and was later promoted to T'ê Chin Shao Tsai [969], minister to one of the princes. Towards the end of Hui Tsung's reign he advocated the appeasement of the Chin Tartars. Collecting six million strings of cash by tax-farming, he bought five or six empty cities on the borders and represented to the emperor that they had been recaptured from the Chin. Finally he was executed under Ch'in Tsung in 1126 during the disturbances which led to the fall of the capital.

There follows a historical disquisition quite in character for an archaeologist not very well qualified in astronomical science. It is more confusing than helpful, but will be explained in the annotations. Lastly comes the rather grandiose proposal to construct several of these complex astronomical clocks, a proposal which never got beyond the constructional stages, if indeed it went so far. For the capital was doomed, and we do not know what happened to Wang Fu's group of Taoist horologists.

Sung Shih (Lü Li Chih), ch. 80, pp. 25*b* ff., by TOKTAGA & OUYANG HSÜAN, *c.* A.D. 1354

In the tenth month of the first year of the Shao-Shêng reign-period (A.D. 1094) the secretariat of the Ministry of Rites was ordered to study the plans of the Bureau for the Construction of Armillary Spheres and Globes, and to call a conference of all the astronomical officials with a view to testing the (extant) pieces of apparatus, both old and new, selecting the most accurate for use, and finally to report back to the emperor.[1]

In the seventh month of the sixth year of the Hsüan-Ho reign-period (A.D. 1124) Wang Fu [599], who was minister to one of the princes, said: 'In the first year of the Ch'ung-Ning reign-period (A.D. 1102) I chanced to meet a wandering unworldly scholar [970] at the capital, who told me his family name was Wang [600] and gave me a Taoist book [341][2] which discussed the construction of astronomical instruments [193, 163] in detail. So afterwards I asked the emperor[3] to order the Supply Department [971] to make some models to test what it said.[4] This they did in the space of two months. The instrument [162] is round like a ball,

[1] This investigating committee was thus set up only a few months after the final completion of the clock of Su Sung, with its bronze instruments (cf. p. 11). It is not difficult to ascertain the kind of things which were worrying the government at the time, and which the reformers would have used as a pretext for this committee. According to the annals (*Sung Shih*, ch. 18, pp. 3*a* ff.): 'In the ninth month of the first year of the Shao-Shêng reign-period, Liu Chêng [601] was despatched to Hopei to rescue the starving population in the flooded districts.... On a *kêng-shen* day Venus appeared in the daytime.... On a *ting-mao* day the emperor ordered further relief works for the refugees in Hopei province and to the east and west of the capital.... On a *wu-ch'en* day, a shooting star appeared in the Purple Palace (North Polar Region)....' All this besides the Chin Tartars.

[2] This is yet a third book on astronomical clocks. That there was another by Juan T'ai-Fa [462] has already been mentioned (p. 11), though its date is not known. Here the date is known but not the author, and Mr Wang's book does not recognisably appear in the Sung dynastic bibliography.

[3] This was now Hui Tsung [972], personal name Chao Chi [603] (A.D. 1082–1135), eighth emperor of the Sung. Learned, accomplished, fond of Taoism, and favourable to the reform party, but incapable of organising effective resistance to the Chin Tartars. After having abdicated in favour of his son Chao Huan [604] (Ch'in Tsung [973]) (1100–60), both were taken prisoner early in 1127 by the Chin Tartars and lived the rest of their lives in northern captivity.

[4] The biography of Wang Fu mentions (*Sung Shih*, ch. 470, p. 6*a*) that at one point he requested that this Bureau should be re-established, with himself as director of it. Possibly this was preparatory to the construction of prototype models of the clocks which he had in mind. Besides his archaeological work, Wang Fu was also a geographer and cartographer. He seems to have revised and enlarged an earlier work of Wang Tsêng [602], the *Chiu Yü T'u Chih* [794] (Illustrated Account of the Nine Regions); cf. *Sung Shih*, ch. 204, p. 15*b*.

graduated in $365\frac{1}{4}$ degrees, and shows the south and north poles, the K'un-lun Mountains [861],[1] the ecliptic and equator, the twenty-four fortnightly periods [199], the seventy-two five-day periods [200], the sixty-four hexagrams [233], the cyclical signs, ten *kan* [229] and twelve *chih* [230], the 100 quarters of the day and night, the twenty-eight lunar mansions [196], the three walled regions [342] of the heavens,[2] and the stars of the whole heavenly round. The sun and moon (are shown) following the ecliptic. Each day the heavens rotate once to the left, while the sun makes one degree to the right. At the winter solstice point its path is 24° to the south of (lit. outside) the equator; at the summer solstice the same distance to the north of (lit. inside) it. At the spring and autumn equinoxes the ecliptic path crosses the equator, sunrise occurring at the cyclical (azimuth) sign *mao* [343] and sunset at the sign *yu* [344]. (While the heavens rotate once to the left) the moon makes thirteen degrees and a fraction (to the right). Starting bright in the west, its shape is (seen on the instrument) first like a hook, then only the lower half of it is visible in the west, then at the mid-month it becomes full and round, after that the lower half starts to disappear at the west, then only half can be seen in the east, finally at the end of the month it is quite hidden. The planets, too, can be seen (on the instrument) rising, approaching their transits, and setting north or south of the equator (lit. to left or right), sometimes tarrying, sometimes hurrying—everything is in perfect agreement with the phenomena of the heavens without a hair's-breadth of error.[3]

A jade balancing mechanism [163] (escapement) is erected behind (lit. outside) a curtain, holding and resisting [345] the main scoops [346]. Water pours down, rotating the wheel [347]. Lower there is a cogwheel [348] with forty-three (teeth). There are also hooks, pins

[1] A sign of the influence of Buddhist cosmology (Mt Meru) on the Taoists.

[2] Chinese uranography recognised among the most important of its constellations three 'walled regions', or patterns of stars surrounded by chains of stars like the walls of a palace or city. The first of these regions was the Tzu Wei Yuan [88] or Tzu Kung [88] (Purple Palace). It enclosed the north polar region embracing numerous stars in our constellations of Ursa minor, Camelopardus, Cepheus and Draco; and corresponded to the imperial palace on earth (cf. *Chin Shu*, ch. 11, p. 8*a*), just as the pole-star itself corresponded to the emperor. The second region was the T'ai Wei Yuan [349], which covered the greater part of our constellations of Leo, Coma Berenices and Virgo. This, says the *Chin Shu* (ch. 11, p. 9*a*), 'is the court of the Son of Heaven, (containing) the thrones of the five emperors and the mansions of the twelve feudal princes. By the boundaries are the nine ministers. Here is the balance which governs fairness and justice. Here is the heavenly court where legal matters are attended to, disputes settled, and virtues preached and promoted. It is also the place where the *hsiu* (lunar mansions) accept the tallies of their commissions, and the spirits receive instructions, make their reports and settle doubtful matters.' This description shows the celestial bureaucracy at work. The third region was the T'ien Shih Yuan [350] (Market-Place), which occupied most of Serpentis Caput and Cauda, Ophiuchus, Hercules and Aquila. The stars of this formation, says the *Chin Shu* (ch. 11, p. 11*a*), 'govern steelyards and balances and also assemblies of people. Another name for them is the Celestial Banner, for they govern affairs connected with (public) executions. When they are bright and numerous they predict a year of plenty. When Mars is found among them, it means that a disloyal official will be put to death. When a comet is seen leaving the group, it is a sign that the places of markets will be shifted or that the capital will be changed. When a guest-star (a nova) appears in its midst the omen means that soldiers will be mustered in great numbers. When it disappears, some noble person will die.' It may be worth while to have these glimpses of medieval Chinese State astrology. We thank Mr Ho Ping-Yü for permission to use his draft translation of this chapter.

[3] The concluding phrase makes it clear that the instrument is being described and not the actual phenomena in the heavens. The details about the stars also indicate that at least part of the equipment must have taken the form of a solid celestial globe.

and interlocking rods one holding another [351]. Each (wheel) moves the next without reliance on any human force. The fastest wheel turns round each day through 2928 teeth [219],[1] the slowest only moves by one tooth in every five days. Such a great difference is there between the speed of the wheels, yet all of them depend on one single driving mechanism. In precision the machine can be compared with Nature itself (lit. the Maker of All Things [352]).[2] As for the rest it is much the same as the apparatus made (long ago) by I-Hsing. But that old design employed mainly bronze and iron, which corroded and rusted so that the machine ceased to be able to move automatically. The modern plan substitutes hard wood for these parts, as beautiful as jade.[3] The old system had the sun and moon fastened to two rings (lit. wheels) fitted outside (the globe or sphere), and these two wheels sometimes obscured the graduation marks of the star positions, especially when one looked upwards at them. But according to the new system the sun and moon are both attached to the ecliptic, moving like ants on a millstone.[4] According to the old system the body of the moon was always round, so that one could not distinguish the phases. Now it is turned by the mechanism in such a way that it sometimes looks round, sometimes crescent-shaped; sometimes dark and sometimes visible; all in agreement with the phenomena of the heavens. Moreover, on the old (i.e. the T'ang) designs there were bells and drums sounding only at the (double-) hours and the quarters. The lengths of days and nights, times of dawn and dusk, the variations of the night-watches and their fractions, none of these could be distinguished. Now, however, the God of Longevity jack [353] stands and points (raising its arm) at the hours and quarters indicated by the twelve (double-) hour wheel as it turns round. Besides all this, there is a candle dragon [354][5] sitting above a bronze lotus, and

[1] We note that this number of teeth is 366×8. This suggests that there was one tooth for each degree, and that the wheel rotated eight times in each day and night.

[2] This is a phrase taken (perhaps rather significantly here) from the old Taoist classics such as *Chuang Tzu* of the fourth century B.C. As their organic naturalism was pantheistic if not atheistic, the references are always poetical and even somewhat mocking.

[3] Even in the later clocks of Europe wood was found a convenient material for construction. Farmhouse clocks made of wood for cheapness were turned out in great numbers from the seventeenth century onwards, South Germany and New England being among the centres of such productions. On another level there were the wooden clocks made by the Harrison brothers in the eighteenth century, one of which, dated 1715, is still running in the South Kensington Science Museum. With the Chinese traditions of very large clocks with water-wheels, where there must have been considerable splashing, the use of a light and non-corroding material was especially suitable.

[4] The analogy of ants walking on a millstone contrary to its revolution is of course very old. The *locus classicus* is in the *Lun Hêng* [784] (Discourses Weighed in the Balance) by Wang Ch'ung [581], *c.* A.D. 83 (ch. 32, tr. Forke (1), vol. 1, pp. 266 ff.). It is found again in Ko Hung's [496] *Pao P'u Tzu* [567] (*ap. T'ai-P'ing Yü Lan*, ch. 762, p. 8*a*, ch. 947, p. 5*b*), *c.* A.D. 340, and often afterwards. Just the same analogy was made by Vitruvius (IX, i, 15) a century before Wang Ch'ung, but derivation is unlikely at that time.

[5] We do not know what sort of a thing this was. But the spitting of 'pearls' into a bronze lotus so as to make a noise is exactly what was done in the striking water-clocks of the Byzantine and Arabic culture-areas. The description of Antioch in the *Chiu T'ang Shu*, ch. 198, pp. 16*a*ff., includes an account of the system (cf. Hirth (1), pp. 53, 213). This would have been written just before A.D. 945, and was repeated in the *Hsin T'ang Shu* and the *Sung Shih* (cf. Needham (1), vol. 1, pp. 193, 203). We have already suggested (p. 75) that embassies from Byzantium (e.g. A.D. 643, 667, 719, 720) might have supplied stimuli for the competitive inspiration of I-Hsing (A.D. 725). The dropping of balls automatically into bowls for striking the hours remained very characteristic of the Arab horologists, as we may see, for instance, in the treatise of al-Jazarī (Wiedemann & Hauser); cf. Fig. 34 above. But that was in A.D. 1206, just a century after the time of Wang Fu.

when the middle of every (double-) hour comes it spits forth a pearl which falls into the lotus, making a noise; and all goes round automatically—this is quite beyond what I-Hsing did.

This (astronomical clock), which shows the whole body of the heavens and the appearances therein may be called a *hsüan chi* [162].[1] The use of scoops and water (and the escapement mechanism) may be called the 'jade traverse' (*yü hêng* [163]).[2] Among the ancient people some meant by *chi hêng* [193, 163] the armillary sphere [154];[3] others said that whatever had the *chi* [193] but not the *hêng* [163] was the celestial globe [161].[4] Some said that the sighting-tube [132] of the armillary sphere [150] was the *hêng* [163].[5] All of them were wrong. Others had no idea at all what the *chi hêng* instrument was. Only Chêng K'ang-Chêng [605] (i.e. Chêng Hsüan) [606] (A.D. 127–200) realised that what turns round is *chi* [193], and what holds upright is *hêng* [163].[6] This statement, checked by our modern knowledge, seems very near the truth.

[There follows a short digression on the reflection of the sun's light from the moon.]

I (Wang Fu) therefore propose that the officials should be ordered to set up a bureau for the construction (of this kind of astronomical clock) according to the models. Auspicious places should be selected, either at the Ming T'ang [974] Temple,[7] or perhaps in a special clock-tower [117], where it can be consulted and compared with the heavenly phenomena. And three more should be built, one for the Imperial Treasury [975], one for the Bell-and-Drum Tower [976] and one on a portable stand to accompany the emperor on his travels. A book has been written about all this, for the information of posterity.'[8]

The emperor ordered that the matter of making and naming this '*Chi Hêng*' should be discussed. (Finally construction was ordered), with Wang Fu as director and Liang Shih-Ch'êng [607] as his deputy.

It is probably safe to conclude from the above that the actual construction of the new clocks was put in hand late in A.D. 1124 or in 1125. But the times were evil and the hour was waxing late. In the following year the emperor Hui Tsung, failing in the defence of the empire, abdicated in favour of his son. The capital itself (K'ai-fêng) was twice under siege by the Chin Tartars, and the second time,

[1] As already mentioned (p. 18 above), the *hsüan chi* is now considered to have been a kind of jade nocturnal, the 'circumpolar constellation template'. The *yü hêng*, which fitted into it, was a short sighting-tube made of jade, for observing the pole-star. On this double instrument, which is believed to go back to the beginning of the first millennium B.C., see Michel (1). The terms became traditionally applied in later writings to any sort of armillary sphere.

[2] Wang Fu here mixes up the escapement lever, also called *hêng*, with the ancient sighting-tube. Of course there is a time-keeping analogy, in veiled words, which it may have been his intention to use.

[3] This was not entirely wrong, for the circumpolar constellation template was held in the equatorial plane and so corresponded to the equator ring, but of course the sighting-tube of the armillary sphere was mounted so that it could be pointed to any part of the celestial sphere and not only to the pole-star.

[4] Or of course the demonstrational armillary sphere, not here specifically distinguished.

[5] There they were quite right.

[6] This statement is either the height of Wang Fu's confusion, or else a poetical (and far-fetched) analogy for the driving-wheel and the escapement. It is hard to tell whether he was genuinely at sea in this passage, or using some kind of concealed language, perhaps to seek the authority of the classics.

[7] See n. 1 on p. 89 above.

[8] Can this be a fourth book?

in September, it fell, and both emperors were taken away captive to Peking in the north. For a while the princes and the remains of the court wandered about behind the lines held tenuously by the Sung armies, first settling in one place and then in another, but in 1129 it was decided to choose Hangchow as the new capital and the move there was complete by 1133. Still for some time there were periodical retirements inland to Suchow and Nanking because of the naval strength of the Chin, which dominated the coasts of Chekiang, and it was not until 1139 that the Sung got the better of these forces and Hangchow became safe as the capital.[1] Its position and circumstances destined it to become one of the most glorious and wealthy cities of the world, the 'flower of cities all', as Marco Polo was to see it a hundred and forty years later. But more than this we cannot say here; we must return to the desolation of the old capital. One of the most curious things about the sieges of 1126 was the fact that the Chin Tartars, apparently envious of the technical culture of the Sung, exacted as tribute in the periodical armistices whole families of artisans and skilled workmen. The records in the *Sung Shih Chi Shih Pên Mo* [795] (ch. 56) tell us that the Chin people demanded from the city all sorts of craftsmen, including goldsmiths and silversmiths, blacksmiths, weavers and tailors, and even Taoist priests.[2] Some of the best men from the Bureau of Supply workshops may well have been lost in this way, and as we shall presently see, there are reasons for thinking that all the clock-making millwrights and maintenance engineers followed the Chin power and migrated to the north.

If we look over once again the memorial of Wang Fu we notice a very curious thing. Although his proposals were being made only thirty years after Su Sung's masterpiece had been set going in the imperial palace, although (as we know from evidence in texts shortly to be presented) this was still telling the time with a great grinding of gears every 10 min. or quarter of an hour, there is no mention of it whatever. One immediately suspects a political reason and further study makes this rather probable. In order to explain it, a little recapitulation of the political history of the times is necessary.

It will be remembered that the Ts'ai brothers first came to power in A.D. 1094 when the young emperor Chê Tsung was able to follow his own inclinations and dismiss the quarrelsome and divided conservatives.[3] It was immediately after

[1] A sketch of the whole story has been given by Ferguson (1).
[2] For this reference we are indebted to Dr Wu Shih-Ch'ang of Oxford.
[3] On the political history of the time, Williamson (1) and Ferguson (2) may be consulted.

this that the 'committee of investigation' of astronomical instruments was called, and that growing danger threatened Su Sung's clock. Its fate was still uncertain when its chief maker died in 1101. This was the year in which Hui Tsung succeeded to the throne, and throughout his reign he consistently supported the reformers, except between 1107 and 1112 when a particularly brilliant comet and the day-light appearance of the planet Venus caused a temporary restoration of the conservatives. Now one of the most striking features of the period was the alliance between the reformers and the Taoists, counterbalancing the strict Confucian orthodoxy of the conservatives. The full working out of the implications of this situation has never been attempted, but it would be extremely worth while, for the Taoists had been from the beginning opposed to feudal and feudal-bureaucratic society and were associated with all subversive movements throughout the centuries.[1] At the same time they were the curators and advancers of all kinds of proto-scientific and technological knowledge and practice—in pharmaceutical botany and mineralogy, in astronomy, alchemy, and various forms of engineering. The reforming party of the Sung were bureaucratic scholars who broke away from the typical Confucian ideas and were prepared to ally themselves with Taoist science and technology.[2] It was highly significant, for example, that Wang An-Shih [453], and again in 1104 Ts'ai Ching [598], included mathematics and medicine among the subjects which could be offered in the imperial examinations.

From the beginning of Hui Tsung's reign, therefore, welcome at court was forthcoming for Taoist adepts of all kinds. In A.D. 1104 Wei Han-Chin [608], a thaumaturgist from Chang Ssu-Hsün's province, Szechuan, was entrusted with the casting of nine urns of bronze or iron, in emulation of those legendary vessels which Yü the Great was supposed to have made, bearing upon their outer surfaces some kind of picture-maps of the provinces of the empire.[3] During the first decade of the century, Chu Mien [609], the son of a Hunanese pharmacist, was despatched on a tour of the country by Ts'ai Ching with authority to purchase or secure as tribute 'all kinds of valuable articles for the imperial palace. He forced the people to give up their paintings and writings, bronzes and jades, precious stones and ornaments, and every object which would help to adorn the palace or to gratify the luxurious taste of the court.' So, at least, is the way historians write about his activities, and there is no doubt that Chu Mien himself profited by them, but as

[1] The spread of printing and other causes had helped the rise of a new class of less aristocratic *literati*, more urban, less connected with the landowning gentry; cf. Kracke (2, 3).

[2] Cf. Needham (1), vol. 2, pp. 138, 155.

[3] *Sung Shih*, ch. 462, p. 10*b*.

historians have generally not been interested in the sciences, whether natural or humanistic, and as the history of the period was written by the conservatives and not the reformers, we may allow ourselves to read between the lines. We shall not be going too far astray therefore in supposing that the Taoists at court were extremely concerned with rare drugs, curious gems, and all kinds of natural productions. And this was where Wang Fu [599] came into the picture, for it was he who drew up the archaeological catalogue of the imperial museum in 1111.

In the following year two eminent Taoist adepts were presented at court, where they remained for some years close to the emperor. Wang Lao-Chih [610] was an intimate of Ts'ai Ching's also, and a friend of Wang Fu, for whom he predicted initial success but ultimate downfall.[1] After Wang Lao-Chih's death his position fell to another Taoist, Wang Tzu-Hsi [611], also known as a diviner, but of course skilled in many branches of technique as men were in days when magic and science were imperfectly separable. We are told that Wang Tzu-Hsi constructed a 'spherical image' (*yuan hsiang* [355]), possibly a celestial globe, which the emperor kept in a special pavilion. This suggests that perhaps we should recognise in him the 'Mr Wang', the unworldly scholar of 1102, who presented the book on clockwork to Wang Fu.[2] During this period the organised Taoist church received great imperial favour. In 1114 Ts'ai Ching proposed the construction of a new Ming T'ang (Cosmological Temple) with a square lake, and an imperial Taoist cathedral (Tao Kuan [977]);[3] the first part at least of the plan was carried out in 1115. A year later another Taoist came into prominence, Lin Ling-Su [612], whose specialities seem to have been rain-making and Taoist bibliography.[4] The Taoist patrology was incorporated into the imperial library, and two librarians specially appointed to look after it. Finally there was Liu Hun-K'ang [613] the geomancer.

But this entourage of virtuosi was not the kind of court which could have helped even a military emperor to organise the massive resistance needed against the overwhelming forces which the Chin were gathering. By 1123 all the northern provinces were lost to the Sung and the emperor, panic-stricken, abdicated in favour of his son. Then came the two sieges of K'ai-fêng in 1126 and the capture of both emperors. Hui Tsung died nine years later in exile in Kirin in

[1] *Sung Shih*, ch. 462, p. 11*b*.

[2] Cf. Wang Tzu-Hsi's biography in *Sung Shih*, ch. 462, p. 12*b*.

[3] *Sung Shih*, ch. 472, p. 6*b*.

[4] Unlike Wang Lao-Chih, both Wang Tzu-Hsi and Lin Ling-Su died in prison after losing the imperial favour. Wang Tzu-Hsi quarrelled with a powerful eunuch, which suggests covert Taoist–Buddhist antagonism.

the far north. The experimental co-operation of Taoists and reformers which might have profoundly affected Chinese society was ended. As for Su Sung's clock we shall soon see what happened to it. But before leaving this fascinating period it seems not unduly speculative to note the existence of two rival schools of horological artisans, one associated with the conservatives and one with the reformers. For Han Kung-Lien must have had his group of craftsmen just as Wang Fu (perhaps Wang Tzu-Hsi) had his. Presumably the political parties vied with one another in the planning and erection of monumental clocks. But that is indeed an unexpected conclusion in view of the usual belief that the clock was something absolutely new to the Chinese in 1600.

2 HOROLOGY IN THE SOUTH AFTER THE FALL OF THE CAPITAL (A.D. 1126)

In the following excerpts we shall trace what happened in the Sung in the south after 1126. Then we shall present some texts which reveal what happened in the north under the Chin.

Sung Shih, ch. 81, pp. 13*b* ff., by TOKTAGA & OUYANG HSÜAN, A.D. 1345. Cf. *Yü Hai*, ch. 4, pp. 47*b* ff. (Encyclopaedia of WANG YING-LIN, A.D. 1267)

In the second year of the Shao-Hsing reign-period (A.D. 1132)...it was suggested that an armillary sphere [150] should be made and Li Chi-Tsung [614] was ordered to make the necessary tests.[1]...In the first month of the third year (A.D. 1133) a wooden model of the armillary sphere was presented to the emperor[2] by Ting Shih-Jen [615] an official of the Bureau of Astronomy and Calendar (T'ai Shih Chü [917]), and Li Kung-Chin [616], and they were ordered to set it up in the Imperial Palace. They said that in the Eastern Capital (K'ai-fêng [862]) there had been four armillary spheres, adding 'the one in the Armillary Sphere and Clepsydra Laboratory [978] (of the Imperial Library) was called the Armillary

[1] Parallel statements to these occur in *Sung Shih*, ch. 48, pp. 18*a* ff.

[2] This was Kao Tsung [979], personal name Chao Kou [617] (A.D. 1107–87), the first emperor of the Southern Sung dynasty, and the tenth of the line. He succeeded to the throne after his father and brother had been carried into captivity by the Chin Tartars, and it was he who eventually fixed the new capital at Hangchow. His reign was marked by vacillation of policy towards the north, the conservatives and reformers having been replaced by a war party and a peace party, equally at odds with one another. But the fact that he reigned till 1162 and did not die till twenty-five years later tells its own story, for most of the last emperors of the Northern Sung died very young. In spite of the Chin pressure, or perhaps rather because it greatly slackened, a period of prosperity and culture supervened. Among the most illustrious of Chao Kou's subjects was the philosopher Chu Hsi [618], China's Thomas Aquinas, and the Neo-Confucian school was at the height of its creativeness. But much had been lost besides the clock-making artisans. Chinese literature of all kinds suffered severely during the retreat to the south, and of lost books it is often said—'failed to cross the River'.

Sphere of the Chih-Tao reign-period (A.D. 995–7) (Han Hsien-Fu's). The one in the Astronomical Bureau of the Han-Lin Academy [915, 916] was called the Armillary Sphere of the Huang-Yu reign-period (A.D. 1049–53) (Shu I-Chien's). The third was in the Astronomical Department of the Bureau of Astronomy and Calendar [915, 917] and was called the Armillary Sphere of the Hsi-Ning reign-period (A.D. 1068–77) (Shen Kua's). The fourth was in the Combined Platform [117] clock-tower observatory and was called the Armillary Sphere of the Yuan-Yu reign-period (A.D. 1086–93).[1] Each embodied over twenty thousand catties[2] of bronze.[3] Now if we halve this sum, we should use more than ten thousand catties. In fact for the Yuan-Yu one (A.D. 1086–93) we have an exact record from the time of construction showing the use of 8400 catties.'[4]

Sung Shih, ch. 81, pp. 15*a* ff. Cf. *Yü Hai*, ch. 4, p. 48*b*

In the fourteenth year of the Shao-Hsing reign-period (A.D. 1144) the Bureau of Astronomy and Calendar requested that an armillary sphere should be made.[5] Hsieh Chi [619], an Assistant Secretary in the Ministry of Works, said: 'I once inquired about the methods and systems of the armillary sphere [150] but the officials and students in the Astronomical Bureau had differences of opinion among themselves and we still lack information about how it should be made. Your humble servant's opinion is that we should first of all search for the system and method, sending out widely an appeal for scholars who are masters of calendrical, mathematical and astronomical knowledge, so that they can look into the rights and wrongs of the matter and see what is in agreement with the former systems.' Thus the son of Su Sung was called upon to search for any books which his father might have left behind in order to investigate the question. (This produced no results.)[6] Eventually the prime minister, Ch'in Kuei [620][7] said: 'None of the ministers of the court is able to understand this thing.' The emperor, Kao Tsung, said: 'Unfortunately the (necessary) documents are still lost. But I have already had a model made in the court.[8] Although it is small it can, however, be used. The shadow of the sun is measured

[1] This statement of Ting Shih-Jen and Li Kung-Chin appears again in Wang Ying-Lin's *Hsiao Hsüeh Kan Chu* encyclopaedia, ch. 1, p. 10*a*. He must have used the same archival source as the *Sung Shih* editors. The 'four famous astronomical instruments of the Northern Sung', as he calls them, are given the same locations in both texts. The conclusion that the capital (K'ai-fêng) had no less than four astronomical observatories is of some interest. Another parallel and identical statement is in Chou Mi's [540] *Ch'i Tung Yeh Yü* [739], ch. 15, p. 5*b* (A.D. 1290).

[2] At this time 1 catty was equivalent to *c.* 0·6 kilo.

[3] These statements are also found, in abbreviated form, in *Sung Shih*, ch. 48, p. 18*a*.

[4] Here again is evidence that Su Sung's instruments were actually cast in bronze. Cf. pp. 115, 116.

[5] Parallel statements are found in *Sung Shih*, ch. 48, p. 18*b*.

[6] The information about Su Sung's son is also found in *Sung Shih*, ch. 48, pp. 18*a*, *b*. His name was Su Hsi [621]. It is there implied that some of the father's books or writings were available, but that the son could not understand them. Probably the family had carried to the south only a dossier of manuscript notes, which, in the absence of Han Kung-Lien (p. 19) and others, could not be interpreted.

[7] A.D. 1090–1155; the famous leader of the peace party in the south, execrated in later generations because of the supposedly unpatriotic character of his policy. But the last word has not yet been said either on the reformers' party or the peace party, since history was written only by their enemies.

[8] In Chou Mi's *Ch'i Tung Yeh Yü* (Rustic Talks in Eastern Ch'i) of about A.D. 1290 we read (ch. 15, p. 5*b*): 'After the capital was moved to Lin-an (Hangchow) south of the Yangtze River, Yuan Chêng-Kung [622] (i.e. Yuan Wei-Chi [456], Su Sung's pupil, cf. p. 11 above), a Divisional Director of the Ministry of Works,

by day and the degrees and transits of the constellations measured by night....We should enlarge the dimensions of our instrument (to make a definitive one).' The emperor, therefore, authorised his prime minister, Ch'in Kuei, to select Shao O [623], a Palace steward, who was ingenious and intelligent, and to entrust him with the duty. The apparatus was completed after a year or two....[1]

Sung Shih, ch. 48, p. 19*b*

[After describing the armillary spheres of the Shao-Hsing reign-period, i.e. Kao Tsung's time, after the removal to Hangchow, the historian goes on to say:]

As for the water-power drive system [356] and the celestial globe, these were no longer available (for use with the armillary sphere). Later on (the great philosopher) Chu Hsi [618] (A.D. 1130–1200) had in his house an armillary sphere [150]. He tried hard to investigate the water-drive devices [357], but without any success. Although some writings of Su Sung were still in existence, they dealt principally with the details of the celestial globe and did not record constructional measurements, so it was very difficult to recover his system.

These texts show plainly that after the move to the south and to Hangchow the new emperor, interested though he was in reconstructing astronomical science at his capital, met with much difficulty in organising the equipment of a new observatory. Too much had been lost in the flight from K'ai-fêng. Even further out of reach was the aim of making another astronomical clock like Su Sung's, for even his own son could not produce adequate designs from the family papers. The disappearance of most of the skilled technicians must have been a heavy blow. This makes us realise one of the great obstacles to technological advance in ancient and medieval times—the rarity of trained artisans and engineers and the total lack of any continuity in the arrangements for training them. We should like to quote, in this connection, a fragment from one of the writings of Huan T'an [624] (40 B.C.–A.D. 25), that scholar of sceptical rationalist views who tried to measure temperature and humidity, and who was the first to write of the water-

presented to the emperor a wooden prototype of an armillary sphere. The emperor ordered the Ministry of Works to make one half the size of the prototype model, and 8400 catties of bronze were used, but it was not successful. (There seems to be a confusion here with the 8400 catties previously mentioned, p. 127.) In the seventh year of the Shao-Hsing reign-period (A.D. 1137) a small model was made (also by Yuan Wei-Chi?). In the fourteenth year of the same reign-period (A.D. 1144), the Palace Steward Shao O [623] was put in charge of making these instruments. Two were completed, one for the observatory [980] of the Bureau of Astronomy and Calendar [917], and another for the laboratory [981] attached to the Imperial Library [982]. Both were of excellent workmanship in refined bronze. Yet they were not half as good as those in the old capital.' This shows that the activity of Yuan Wei-Chi occupied the difficult years between the loss of K'ai-fêng in 1126 and the fixing of the capital at Hangchow ten years later. It was not surprising that he could do little; the necessary artisans were probably dispersed.

[1] Parallel statements in *Sung Shih*, ch. 48, p. 18*b*.

power trip-hammers becoming widespread in his time for pounding cereal grains. He was a great friend of the naturalist and philologist Yang Hsiung [625] (53 B.C.–A.D. 18) who, he said:

was devoted to astronomy and used to discuss it with the Taoists. He made an armillary sphere himself. An old artisan once said to him: 'When I was young I was able to make such things, following the method of divisions (graduations) without really understanding their meaning. But afterwards I understood more and more. Now I am 70 years old, and feel that I am only just beginning to understand it all, and yet soon I must die. I have a son also, who likes to learn how to make (these instruments); he will repeat the years of my experience, and some day I suppose that he in his turn will understand, but by that time he too will be ready to die!' How sad, and at the same time how comical, were his words![1]

The problem of the water-power clock mechanism occupied some of the best minds of the Southern Sung, as we see from the unexpected and extremely interesting information that the great philosopher Chu Hsi [618] himself gave much thought to it. The Neo-Confucian school, of which he was the most important exponent, systematised a world-view very much in accord with scientific thinking. It was a well-developed organic naturalism which interpreted the known universe in terms of matter on the one hand, and a principle of organic pattern manifesting itself at many different levels of organisation on the other.[2] It is characteristic that Chu Hsi has been regarded by western scholars as being at one and the same time China's Thomas Aquinas and her Herbert Spencer. The fact that Chu Hsi took an active interest in the redesigning of the power-drive for the clock, and also that he proved unable to achieve it, are both quite significant comments on Neo-Confucian philosophy. The fact that he had an armillary sphere in his college or private dwelling, i.e. in private hands, also indicates, like the instruments in Su Sung's family a hundred years before, that astronomical and calendrical studies were by no means entirely confined to official government circles. Meanwhile, in Chiangsi province, a lesser figure than Chu Hsi, but a man not at all negligible, Tsêng Nan-Chung [626], was also occupying himself with the problem of making striking clocks. As the next excerpt shows, he used water naturally enough, but evidently had to return to simpler methods than those used by Su Sung. Unfortunately it is not possible from the details given to say whether he made use of the simple float principle with which to work his jacks, or some kind of variation on the wheel of Chang Hêng. Tsêng Min-Chan [627] (i.e. Tsêng

[1] *CSHK* (Ch'ien Han sect.), ch. 15, p. 2*a*, from *Pei T'ang Shu Ch'ao* (A.D. 630) and *T'ai-P'ing Yü Lan* (A.D. 983); cit. also in *Yü Hai* (A.D. 1267), ch. 4, p. 29*a*.

[2] Cf. the account in Needham (1), vol. 2, pp. 455 ff., 565 ff.

Nan-Chung) was also responsible for the invention or popularisation of the double-pin gnomon equatorial sundial. There is a good deal about him in the *Tu Hsing Tsa Chih* [796] (Miscellaneous Records of the Lone Watcher) written about A.D. 1176 by his son or grandson Tsêng Min-Hsing [628]. We shall give first this excerpt and then pass on directly to the happenings in the north.

The last remark in the passage concerning Chu Hsi corresponds well with what we know from Su Sung's memorial itself (p. 27 above), namely, that the description of the celestial globe and its star-maps was originally a kind of appendix to his book. Only in later editions did it become the whole or part of ch. 2. We are almost bound to assume, therefore, that the main part of the book, together with the artisans and the clock itself, was taken away to the north by the Chin people. Yet the whole book appeared again in the south, in Chiangsu province, by A.D. 1172, when it was printed for the first time by Shih Yuan-Chih [464] (p. 12 above). The relevant activities of Chu Hsi in Fukien could hardly have been much earlier than 1170. Perhaps therefore it was his agitation which led to the recovery of the text, either because it had been mislaid in the south, or more probably because it was brought down from the north, for by this time there was plenty of intercourse between the two parts of the country.

Tu Hsing Tsa Chih [796], ch. 2, p. 11*a*, by TSÊNG MIN-HSING [628], *c.* A.D. 1176

The Yü-Chang [863] (i.e. Chiangsi province) water-clock and sundial were invented by Tsêng Nan-Chung [626] who had mastered astronomy when he was young. At the beginning of the Hsüan-Ho reign-period (A.D. 1119) he passed the examinations as *chin-shih* (third degree) and was given a post as magistrate at Nan-ch'ang [864] (in Chiangsi).[1] When the Auxiliary Academician of the Lung-T'u Pavilion [938], Mr Sun [629], became commander-in-chief of the army there (and got to know Mr Tsêng) he acquired great love and respect for him. So when the latter suggested the construction of sundials and water-clocks according to new methods, the general was delighted and authorised him to recruit artisans for the purpose.

Thus metal was cast into the shape of vessels and wooden rods were carved (with graduations). Behind the main vessel four basins and one tank were set up. The water in the main vessel (the inflow receiver) came from the basins (compensating tanks) and the water in the basins came from the tank (the reservoir). The water poured out through the open mouths of bronze dragon siphons [358]. Beside the indicator-rod stood two wooden figures, the left one in charge of the day quarter-hours [21] and the half double-hours [359] of the night. In front of it an iron strip was set up, on which it struck at each quarter-hour and each half double-hour. The right figure was in charge of the double-hours of the day [151] and the night-watches

[1] As nearly all of this province lies south of the Poyang Lake, it was more or less of a rear area in the defence against the Chin Tartars. And the end of the passage shows that Tsêng Nan-Chung was a Chiangsi man himself.

[360].[1] In front of it there was a bronze gong, on which it struck at each double-hour and each night-watch.[2] He (Tsêng Nan-Chung) also made two wooden dials [361] with diagrams on them. One was set upon a wooden support for reading (the hours by) the sun's shadow.[3] The other was rotated by water [362] to imitate the motion of the heavens [363].[4] This instrument was highly ingenious and the method very precise, so that it exceeded anything known in former times.[5]

(Tsêng) Nan-Chung often used to observe the phenomena in the heavens by night, and could predict the motions of the stars and constellations, saying that such and such a star (or planet) would pass such and such a degree [364] on such and such a night. One bitter winter, when it was extremely cold, he used to lie on his bed having removed some of the tiles from the roof, the better to watch the heavens. Once he fell asleep so that the frost came down on his body, and afterwards being invaded by cold he died. Alas! his knowledge was not handed down. Only the general constructional designs of his water-clocks and sundials were known to his son. Nowadays water-clocks of this kind are still being made in Chiang-hsiang [865][5] and some other districts.

(Tsêng) Nan-Chung's *ming* was Min-Chan. He was a native of Mu-pei [866] in the district of Lu-ling [867].[6]

3 THE FATE OF SU SUNG'S CLOCK IN THE NORTH AMONG THE CHIN TARTARS

Chin Shih [810], ch. 22, pp. 32*b* ff., by TOKTAGA & OUYANG HSÜAN, *c.* A.D. 1345

[This text says that in the Yuan-Yu reign-period Su Sung and Shen Kua both received orders to construct armillary apparatus. It then repeats all the information about Han Kung-Lien, etc. given in Su Sung's memorial.[7] It then continues:]

On the one hand there are the cross-struts [365] (in the polar-mounted declination-ring) supporting the sighting-tube by a pivot half-way along its length, so that it is freely movable

[1] On night-watches, see pp. 199 ff. below.

[2] Since there is no hint here of a water-wheel of any kind, the most probable supposition would be that we have here a striking clepsydra of the float type. It should be remembered that Tsêng Nan-Chung was working about A.D. 1135 or some seventy years before the description of al-Jazarī (1206) and thirty before that of Ridwan al-Sā'ātī (1168), for which see Wiedemann & Hauser. But there had been abundant contacts between the Arab and Chinese culture-areas during the previous three or four centuries (cf. Needham (1), vol. 1, pp. 214 ff.). The account of Tsêng's clepsydra with striking jacks, however, throws light on the transition in China from figures pointing at the indicator-rods to jacks actively announcing the passage of time.

[3] This undoubtedly refers to the equatorial or equinoctial sundial.

[4] This is very interesting as it seems to suggest once again an anaphoric clock of some kind. If so, the Hellenistic idea could easily have been transmitted through Arabic intermediation. Cf. pp. 64 ff. above.

A paraphrase of this passage occurs in the Ming book *Shang Yu Chi* [837] by Chang Hsi-Ming [698]. It is quoted in *Ch'ou Jen Chuan* [770], pt. 4, ch. 5.

[5] This must refer to northern Chiangsi along the course of the Yangtze River.

[6] Modern Chi-an city in southern Chiangsi.

[7] On p. 33*a* the word *lun* [367], wheel, is used for sets of armillary rings. This is an unusual (and very confusing) nomenclature; cf. pp. 68 ff. above.

up and down, and can point continually directly at the sun moving one degree eastwards while the heavens are moving a complete revolution westwards, as well as being capable of measuring the positions of the stars in all four directions. This arrangement follows more or less that introduced by Li Shun-Fêng [502], K'ung T'ing [507], Han Hsien-Fu [473] and Shu I-Chien [474].[1] On the other hand there are the component of the three arrangers of time[2] [4], and the diurnal motion gear-ring [10] rotated by water-power.[3] The arrangement of a water-power drive started with Chang Hêng [480] and was brought to perfection by Liang Ling-Tsan [481] and I-Hsing [477], then recovered by Chang Ssu-Hsün [479] in the T'ai-P'ing Hsing-Kuo reign-period (A.D. 976–83).[4] Han Kung-Lien [457] later changed the system somewhat by adding the diurnal motion gear-ring above the (vertical) transmission shaft [38], and the escapement mechanism [366] which moves the armillary sphere; this (the gear-ring) was his new contribution.[5]

[Then follows a rather elaborate account of the armillary clock mechanism of Su Sung and Han Kung-Lien, verbally very close to parts of the description in Su Sung's book itself, which perhaps suggests the conservation of the specifications only in the north. The writers of the *Chin Shih*, Toktaga [630] and Ouyang Hsüan [631], were of course working under the Mongol power in *c.* A.D. 1340, i.e. after the whole of China had been conquered by the northerners. The *Sung Shih* was also written by them, but omits the description of the clock mechanism, presumably because it was conserved in the Chin, but not in the Sung, archives.]

After the Chin (dynasty) had captured Pien (-ching) [868] (=K'ai-fêng, the Sung capital, in A.D. 1126) all the astronomical instruments were carried away in carts to Yen [869] (perhaps modern Peking or more generally the north-eastern regions). The celestial (gear-) wheel [18], the equatorial gear-ring [19], the time-keeping gear-wheel [368, 57], the celestial globe [369], the bells, drums and quarter-striking jacks [370, 22, 24], the upper reservoir [42], the scoops, sump and tanks [43–6, 48], etc. all broke or wore out after some years. Only the bronze armillary sphere remained in the observatory [224] of the (Chin) Bureau of Astronomy and Calendar [917]. But as Pien (-ching) and Yen were more than a thousand li away from one another, the (polar) altitude was quite different, so that it was necessary to alter the arrangement, setting the (south) polar pivot four degrees lower.[6]

[1] This is a group of makers of observational armillary spheres.

[2] The intermediate nest of the three nests of armillary rings in the fully developed Chinese armillary sphere.

[3] Characteristic of Su Sung's clock drive.

[4] Here the writer sketches the lineage of mechanisation without making any distinction between the escapement machines and those which preceded them.

[5] This confirms other claims, by Su Sung himself (p. 24 above), that this was a new invention in 1088. The present statement implies (*a*) that the observational armillary spheres had never previously been mechanised, and (*b*) that the demonstrational armillary spheres mechanised from Chang Hêng's time onwards must have been rotated by a shaft in the polar axis.

[6] Peking is actually just 5° lat. north of K'ai-fêng.

In the eighth month of the sixth year of the Ming-Ch'ang reign-period (A.D. 1195) there was a terrible storm with rain and wind, thunder and lightning. The armillary sphere, with its dragon columns, cloud-and-tortoise column, and water-level stand, was struck by lightning. The masonry of the observatory tower was also split asunder so that the sphere fell to the ground and was damaged. The emperor[1] ordered the officials to repair it, and it was replaced upon the tower. In the Chen-Yu reign-period (A.D. 1214–16) (the Chin court and people, pressed by the rising Mongol power) crossed the (Yellow) River (fleeing) southwards. It was proposed that the armillary sphere should be melted down to make things, but the emperor[2] did not have the heart to destroy it. On the other hand its bulk was so large it would have been difficult to transport it on a cart, so in the end it was left behind.

In the Hsing-Ting reign-period (A.D. 1217–21) the officials of the (Chin) observatory had no armillary sphere at all, and there was an insufficient number of observers, so they appealed to the emperor[3] asking that an armillary should be cast, and that more personnel should be appointed, in order that study and prediction could be pursued. The emperor Hsüan Tsung asked the opinion of the Minister of Ceremonies, Yang Yün-I [632], about it. He replied that there had been in force a prohibition of bronze (-casting)[4] and that neither public nor private supplies of copper would be enough for the purpose; moreover the coinage was unequal to the needs (of the country). Therefore the plan should be abandoned. But the emperor returned to the matter, so finally some more officials were appointed but no armillary sphere was made.

4 CLOCKWORK UNDER THE MONGOLS (THIRTEENTH AND FOURTEENTH CENTURIES)

The history of the Chin Dynasty of the Jurchen Tartars thus reveals to us something of the ultimate fate of the work of Su Sung. The Chin, in conquering the Sung capital, had obtained all that money (or rather military force) could buy, but they do not seem to have been able to do much with it. They had the instruments adjusted for the change of latitude, and evidently put the clockwork in operation for some years, but perhaps they did not treat the artisans they had captured very well for otherwise the parts could have been replaced when they wore out. Then just one hundred years after the making of the clock-tower, the armillary sphere was damaged in a frightful storm. It was like a prelude to the

[1] Chang Tsung [983], personal name Wanyen Ching [633] (succeeded to the Chin throne in A.D. 1189, died 1208), sixth emperor. Young, peace-loving, rather ineffective.

[2] Hsüan Tsung [984], personal name Wanyen Hsün [634] (succeeded to the Chin throne in A.D. 1213 by deposing an uncle, died 1224), eighth emperor. In 1215 the Mongols took Peking and the Chin court retreated south of the Yellow River, setting up at Ka'i-fêng. Wanyen Hsün's reign was mostly occupied by organising this withdrawal and fighting defensive actions. It is good to know that he refused to destroy Su Sung's armillary sphere, so that in time it reached the hands of the great astronomer Kuo Shou-Ching.

[3] Still Hsüan Tsung. Again we see his interest in astronomy.

[4] In times of shortage the metal was doubtless reserved for coinage and military uses only.

storm among men which was about to ensue, in the course of which the Mongol might, at first barbarous but quickly acquiring culture, unified the greater part of the Old World from Budapest to Tientsin. It was still in its barbarous phase when it liquidated the Chin State, forcing back into the Chinese homeland the former nomads who had adopted Chinese culture so fully, and perhaps ignorant that in due time it itself would bow to the same civilising influences. And so, when the Chin people crossed the Yellow River in a defeat which contrasted strangely with their conquests in the same crossing a century before, the sphere of Su Sung was left behind. Apparently it survived another half-century but its ultimate fate is unknown. By then the Mongols had educated themselves and in the fabulous realm of Khubilai Khan the Mongol emperor was served by Chinese scholars of the greatest distinction, such as the astronomer and engineer Kuo Shou-Ching (Fig. 46).

Yuan Shih [797], ch. 48, p. 1*b*, by SUNG LIEN [636] *et al.*, *c.* A.D. 1370

After the disturbances of the Ching-K'ang reign-period (A.D. 1126) the astronomical instruments (of the Sung) all fell to the Chin. Afterwards, when the Yuan (Mongol) dynasty arose and established its capital at Yen [869] (Peking) (in 1264), (the astronomical officials) began by using the apparatus left from the Chin. But the rings (of the armillary sphere)[1] had ceased to turn smoothly and could no longer be conveniently handled. Therefore, the Astronomer-Royal Kuo Shou-Ching [635] presented[2] (between A.D. 1264 and 1279) his inventions, the Simplified Instrument [371],[3] the Hemispherical Sundial [372][4] and many other instruments

[1] This presumably refers to Su Sung's.

[2] To the emperor, Khubilai [637] Khan, i.e. Shih Tsu [986] (A.D. 1214–94). Proclaimed emperor in 1260 at Shangtu during the invasion of China upon the death of his brother Mangu Khan who had been leading it. Received the submission of his younger brother Ariq-bügä in 1264. Resumed the conquest of China in 1262, gained Hsiang-yang after a celebrated siege in 1273, captured Hangchow in 1276. Capital fixed at Peking from 1264 onwards. The whole reign was one of great brilliance, though of quasi-foreign domination. If military expeditions to Japan, Burma, and Java were unsuccessful, vast internal projects were achieved, such as the completion of the Grand Canal and many other hydraulic works, some under the direction of Kuo Shou-Ching. The Yellow River sources were explored. Commercial activity reached great heights, as the story of Marco Polo shows, and many large enterprises in scholarship and learning were brought to successful conclusion. The astronomical observatory at Peking was fully reorganised under Kuo Shou-Ching.

[3] This was an astronomical instrument developed by Kuo Shou-Ching under the stimulus of the visit of Jamāl al-Dīn with an astronomical delegation from Hulagu Khan in Persia. It was essentially a dissected armillary sphere in which the rings were no longer concentric, and it constituted what may be called the invention of the equatorial mounting still to this day employed for telescopes. In all probability, it originated from the torquetum, a form of dissected armillary sphere which had been invented in Muslim Spain a couple of centuries earlier, but it dispensed with all the ecliptic components, and this was why it was called 'simplified'. See Needham (2), and (1), vol 3, pp. 370ff. This instrument, or an exact replica of it made by Huangfu Chung-Ho [638] in 1437, is still at Nanking.

[4] A scaphe. Small-sized ones of later date made in China still exist. The principle was the same as that of the large Indian *jai prakāś* instrument. Cf. Needham (1), vol. 3, p. 301.

Fig. 46. Kuo Shou-Ching (A.D. 1231–1316), Astronomer-Royal in the Yuan dynasty, Khubilai Khan's outstanding mathematician, astronomer and civil engineer, the inventor of the equatorial mounting for sighting-tubes still used for telescopes. A nineteenth-century Chinese artist's impression (after Favier, copied by Ch'en Ts'un-Kuei).

and gnomons, all elaborate and excellent.[1] His ingenious ideas and extraordinary knowledge were something quite beyond what the men of old had attained.

Yuan Shih, ch. 48, pp. 7*a* and *b*

The Illuminated Clock of the Ta Ming Hall [985]

The system of this illuminated clock [373] was as follows. It was 17 ft. high, and its framework was made of metal.[2] On the curved beam (of the framework) there were 'cloud pearls'[3] [374] in the middle, with (a thing shaped like) the sun to the left, and (a thing shaped like) the moon to the right. Underneath the 'cloud pearls' another 'pearl' hung down. At the two ends of the beam there were dragon heads which could open their mouths and roll their eyes.[4] This arrangement was connected with the regulation of the speed of the flowing water [375]. Above the middle of the beam, there were two dragons playing with pearls and following them up and down. This also was an indication of the evenness of the stream of water [376]. All these things were not just for (decoration).[5] As for the illuminated balls [377][6] they were made of gold and precious stones [378]. There were four levels one above another. The uppermost wheel [379] had four jacks in the shape of immortals (which) turned corresponding to the positions of the sun, moon and stars, having a leftward rotation each day. On the next level were the images of a dragon, a tiger, a bird and a tortoise,[7] each facing in the appropriate direction and jumping (in turn) at every quarter (-hour) to the sound of small bells rung from inside. The third (level) showed the hundred quarter (marks) and had twelve jacks each holding a (double-) hour tablet appearing to report at four doors whenever a (double-) hour began. Another figure also appeared at a door pointing its finger at the quarter-hour marks. The lowest (level) had a bell, a drum, a gong and a hand-bell each with its figure. At the first

[1] The *Yuan Shih*, ch. 15, p. 7*b*, mentions the completion of an armillary sphere in A.D. 1288, but the usual date of the equipment of the observatory is 1276–9. One of Kuo's spheres, or an exact replica of it made by Huangfu Chung-Ho [638] in 1437, is still at Nanking. Cf. Needham (1), vol. 3, pp. 367 ff.

[2] Or gilded metal, or even gold—the text is not precise.

[3] The association of pearls and dragons in China as an art motif is widespread and ancient. This has to do, however, with the fact that the dragon was the symbol for the eastern, or spring, palace (i.e. quarter) of the heavens, and that the first full moons of the year rose among constellations named as parts of a dragon body. It was not, until later times, connected with the conception of a dragon which eats the sun at eclipses, and from which we derive our term 'draconitic months' to designate the time between the passage of successive nodes upon the path of the moon.

[4] This decoration looks like some kind of amphisbaena (double-headed snake) such as is seen quite frequently among the motifs in the Buddhist 'Thousand Buddha Caves' (Ch'ien Fo Tung [870]) near Tunhuang in Kansu province.

[5] We do not understand the meaning of this, and have not been able to reconstruct the device.

[6] The whole description of this striking clock has a somewhat Arabic air. It seems strangely like the one figured in manuscripts of al-Jazarī's book at the beginning of the century (cf. Fig. 34 above). That also had four levels—from above downwards, the signs of the zodiac appearing and disappearing, lamps being lit, eagles or falcons dropping golden balls into cups, and lastly an orchestra of jack musicians. Part of the framework certainly has a 'curved beam', i.e. an arch. As early as 1262, before Khubilai had established himself at Peking, Kuo Shou-Ching had made for him a 'Precious Mountain Clock' (*Pao Shan Lou* [380]), presumably at Shangtu [871] (*Yuan Shih*, ch. 5, p. 2*a*).

[7] The symbolical animals of the four palaces or quarters of the heavens.

quarter, the bell was struck, at the second the drum was beaten, at the third the gong sounded and at the fourth the hand-bell was rung. All this took place at the middles as well as the beginnings of the (double-) hours. The mechanism was hidden inside the casing and was driven by water [410].[1]

From the foregoing texts it is safe to conclude that Kuo Shou-Ching made for Khubilai Khan at least two great clocks with striking jack-work depending on the principle of the water-wheel escapement. Though the *Yuan Shih* does not specifically say that the clock in the Ta Ming Hall was designed by him, its description occurs in the middle of a series of descriptions all devoted to his inventions, so that the conclusion is inescapable. The jacks seem to have been quite similar to those of Su Sung. Here then was the horological tradition of the earlier (Northern) Sung time persisting and manifesting itself in the north. Kuo Shou-Ching himself was one of the most outstanding scientific men in Chinese history. He prepared a new calendar; he directed the reconstruction of the Grand Canal and the making of a new extension of it, with lock-gates, to reach to a point near the capital; he developed spherical trigonometry largely from indigenous Chinese sources; he invented the equatorial mounting still in use, and he made, with forty-foot gnomons, what are perhaps still the most accurate measurements of the sun's solstitial shadows. Of more immediate interest to us here is the fact that he constructed, not only clocks with fanciful jacks, but also a celestial globe rotated by mechanical means exactly in the tradition of Chang Hêng and Su Sung. And the particular interest of this apparatus is that it was the first of the line to come to the notice of observers from Europe.[2] Let us first read the account of it in the History of the Yuan Dynasty.

[1] This final remark indicates fairly clearly that a clock with a water-wheel is being described, and not simply a clepsydra with float-worked mechanisms. Besides, a little before, there is definite mention of jack-wheels.

[2] It is well worth noting how few men it took to span all the centuries of clockwork drive mechanisms. Ricci saw the globe of Kuo Shou-Ching, but Kuo in the thirteenth century A.D. certainly knew the armillary sphere of Su Sung of 200 years earlier. There is reason to think that Su Sung and his collaborators in the eleventh century had access to the tenth-century designs of Chang Ssu-Hsün, if not those of the eighth-century workers (I-Hsing and Liang Ling-Tsan). They in their turn must have been acquainted with what Kêng Hsün had done in A.D. 590, if not indeed with the remains of the instruments of Ch'ien Lo-Chih which are known to have lasted at least as late as 605. And finally Ch'ien Lo-Chih's work in 436 was directly inspired by the recovery of the remains of Chang Hêng's apparatus. Thus from Chang Hêng to Matteo Ricci there were only six intermediate men (perhaps only four) spanning fifteen centuries.

For part of the tradition, that of the astronomical constants involved, Su Sung was but one stage removed from Chang Hêng, for (as we have seen, pp. 36, 78, 96 above) they persisted with a strange conservatism which suggests transmission of many points of detail.

Yuan Shih, ch. 48, p. 5*b*,[1] by SUNG LIEN *et al.*, *c.* A.D. 1370

The celestial globe [152] (of Kuo Shou-Ching) is a ball as round as a crossbow-bullet, 6 ft. in diameter, the degrees and minutes of the sphere being marked both on meridian and equator. The equator [381] is in the centre, equidistant from the two poles by a quarter of the whole circumference in each case. The ecliptic [382] is elevated above and depressed below the equator by 24 degrees and a small fraction in each case. Since the elevations and depressions of the moon's path [383] vary, a bamboo ring divided into degrees equally throughout is used to verify the intersections with the ecliptic, and moved from time to time accordingly. By observations first taken on the Simplified Instrument [371] (the Equatorial Torquetum), the right ascension (lit. position within which *hsiu*) and declination (lit. (north) polar distance) are ascertained. The bamboo ring then being placed upon the globe in accordance with these data, the proximity and angle of the moon's path with the ecliptic and equator becomes very obvious. Thus a check can be made with the positions predicted by computations. The globe is placed upon a square box, the south and north poles being below and above the surface by 40 degrees and a large fraction, half of the globe being visible and half concealed. Within the box there are hidden toothed wheels set in motion by machinery for turning the globe [384].

This is perfectly clear, though we might wish for more details of the mechanisation. Alas, today we cannot simply go and study the apparatus, as we can examine Kuo Shou-Ching's equatorial torquetum and his armillary sphere at Nanking, for at some later time it was melted down. However, there exist several Jesuit accounts of it, the first being that of Matteo Ricci (Fig. 47) himself. In the early spring of the year 1600, during his second visit to Nanking, he made friends with some of the members of the staff of the Bureau of Astronomy and Calendar, who came to visit him and to discuss with him the new scientific knowledge which he had brought from Renaissance Europe. We can hear his own words, taken from his journal (ed. d'Elia (1), vol. 2, pp. 56 ff.):

Occasion offered for the Father (Ricci) to go and see the emperor's mathematical instruments, which are set up on a high hill within the city (probably at that time the Pei Chi Ko [987], North Pole Pavilion) on an open level place surrounded by very beautiful buildings erected of old. Here some of the astronomers take their stand every night to observe whatever may appear in the heavens, whether meteoric fires or comets, and to report them in detail to the emperor. The instruments proved to be all cast of bronze, very carefully worked and gallantly ornamented, so large and elegant that the Father had seen none better in Europe. There they had stood firm against the weather for nearly 250 years,[2] and neither rain nor snow had spoiled them.

[1] This passage was translated long ago by Wylie (1), Sci. Sect. p. 12, but we have been obliged to depart from his version at some points.

[2] 350 would have been a nearer estimate.

There were four chief instruments (i.e. the celestial globe, the armillary sphere, a large gnomon, and the equatorial torquetum).[1] The first was a globe [385, 161] having all the parallels and meridians[2] marked out degree by degree, and rather large in size for three men with outstretched arms could hardly have encircled it. It was set into a great cube of pure bronze which served as a pedestal for it, and in this box there was a little door through which one could enter to manipulate the works.[3] There was however nothing engraved on the globe, neither stars nor terrestrial features. It therefore seemed to be an unfinished work, unless perhaps it had been left that way so that it might serve either as a celestial or a terrestrial globe.[4]

So far Matteo Ricci. That the markings were absent, or possibly very faint, may mean that the globe had suffered more from the weather of three and a half centuries than Ricci thought. In any case there can be little doubt that it was intended as a representation of the heavens and not of the earth. Unfortunately we learn nothing further about it from the biography of Kuo Shou-Ching (*Yuan Shih*, ch. 164, pp. 5*b* ff., 7*b*).

Nearly a century later the globe was described again by Louis Lecomte (A.D. 1655–1728), another of the Jesuit missionaries, who published a valuable book[5] on China and Chinese culture in general in 1696. In this we read (1697 ed., p. 65):

Upon this Platform the Chinese Astronomers had placed their Instruments, which tho' but few, yet took up the whole Room. But Father Verbiest, when he undertook the Survey and Management of the Mathematicks, having judged them very useless, perswaded the Emperor to pull 'em down, and put up new ones of his own contriving. These old Instruments were still in the Hall near the tower, buried in dust and oblivion. We saw them, but thro' a window close set with Iron Bars. They appeared to us large, well cast, and of a shape not much unlike

[1] It is generally considered that the armillary sphere and the equatorial torquetum of Kuo Shou-Ching are now conserved on the terrace of the National Observatory on the Purple Mountain at Nanking, where one of us (J.N.) has had the opportunity of seeing and studying them. But it is known that exact copies were made by Huangfu Chung-Ho [638] in A.D. 1437, so that the extant instruments may be his.

[2] This was slightly incorrect nomenclature. The parallels on Kuo Shou-Ching's globe were certainly declination circles parallel with the equator, and not parallels in the western sense aligned with the ecliptic and denoting celestial latitude.

[3] There is an interesting variation here in the version of Nicholas Trigault. Trigault (A.D. 1577–1628), another of the Jesuit missionaries, took Ricci's journal and descriptive notes after his death in 1610 and rewrote them to some extent, issuing them under his own name in 1615 at Augsburg under the famous title 'De Christiana Expeditione apud Sinas'. As translated from the Latin by Gallagher (1), p. 330, this phrase reads 'there was a small door, for entrance, to turn the sphere', suggesting that it had to be rotated by hand. But Ricci's words are alone authentic, and he uses the phrases 'per poter entrar dentro a maneggiarlo' which implies manipulation of machinery.

[4] It is true that a terrestrial globe was one of the instruments (or designs for instruments) which Jamāl al-Dīn had brought from Persia to Peking in 1267, some years before the equipment of the Peking observatory was finally put in hand (cf. Hartner (1)). But there is no record of its having been adopted by the Chinese astronomers.

[5] 'Memoirs and Observations made in a late Journey through the Empire of China' (Fr. Paris, 1696, Eng. tr. London, 1697).

PLATE XI

(*a*) (*b*)

Fig. 47. Matteo Ricci, S.J. (Li Ma-Tou, A.D. 1552–1610), founder of the Jesuit mission in China, and altogether remarkable both as a scientist and a sinologist. A nineteenth-century Chinese artist's impression, based on contemporary documents (after Favier).

Fig. 48. Ferdinand Verbiest, S.J. (Nan Huai-Jen, A.D. 1623–88), second Jesuit Director of the Bureau of Astronomy and Calendar, chiefly responsible for the re-equipping of the Peking Observatory in 1674. A nineteenth-century Chinese artist's impression, based on contemporary documents (after Favier).

PLATE XII

(*a*)

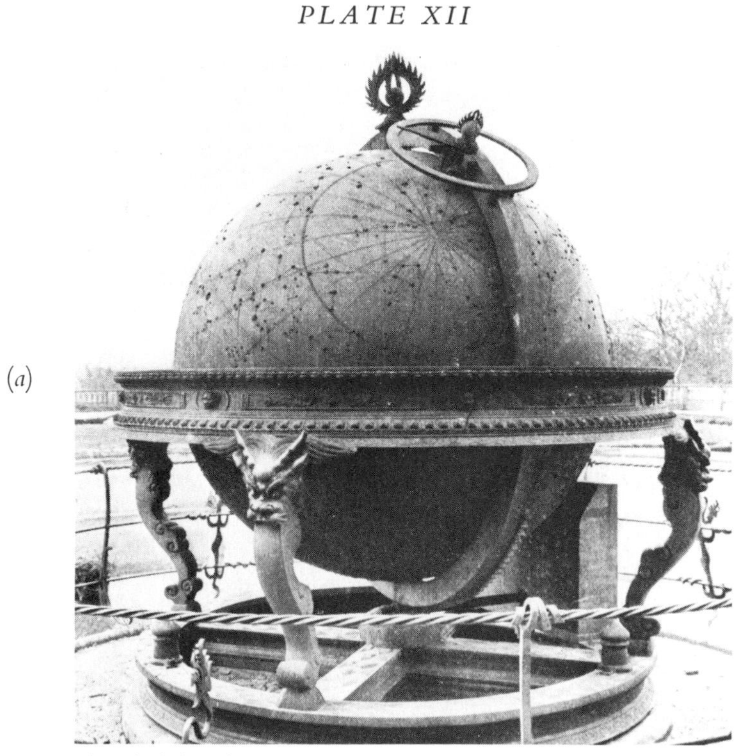

Fig. 49. The Jesuit celestial globe of 1674, still at the Peking Observatory; note the casing for the gearing below at the right (photograph R. Müller). The partial gear-ring to allow of variable polar altitude settings can be seen at the lower part of the meridian ring.

(*b*)

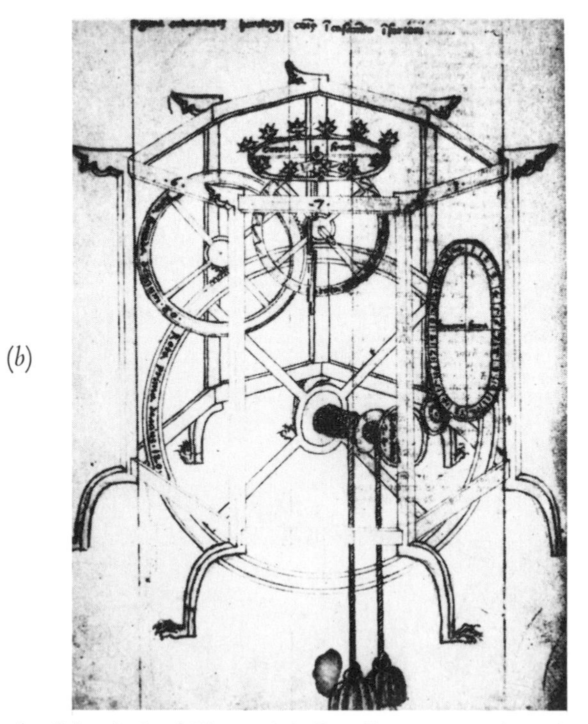

Fig. 50. Schematic sketch of the clock of Giovanni de Dondi, A.D. 1364, the earliest known depiction of a weight-driven mechanical escapement clock (after Lloyd).

our Astronomical Rings. But they had laid in a by-Court a Celestial Globe of about three Foot diameter; that we had a full view of; it was something enclining to an Oval divided with little Exactness, and the whole Work very coarse....

From this account one feels some doubt whether the globe which Lecomte saw was the same as that described by Ricci, for the size was so much smaller and the workmanship less impressive. Ferdinand Verbiest (A.D. 1623–88; Fig. 48) was another of the Jesuit missionaries, an eminent astronomer and mathematician, who was entrusted with the refitting of the Peking observatory in 1669. This resulted in the establishment of many beautiful instruments, made under Jesuit direction, on a tower of the eastern wall of the city of Peking, where they still are.[1] Several accounts of them were published in Chinese and one in Latin.[2] Verbiest himself appears not to have destroyed any of the old instruments of Yuan or Ming times, but later Jesuit directors of the Astronomical Bureau did, notably Bernard Kilian Stumpf (A.D. 1655–1720), who in 1715 (with imperial authorisation) melted some of them down to make quadrants (Pfister (1), p. 645). This was no doubt at a time of temporary bronze shortage, but the loss to history was incalculable, and the disappearance of Kuo Shou-Ching's celestial globe might well be traced to this period.

Again we are approaching the end of our journey, which will naturally centre on what the Jesuits found in China and the interest which was aroused in the up-to-date mechanical clocks of Renaissance Europe which they brought with them. But before speaking of this, we have to take leave of the millennial indigenous tradition of clocks driven by water-power. To do this we have to enter the private apartments of the imperial palace, where about the middle of the fourteenth century we find the last emperor of the Yuan dynasty, in extraordinary contrast to the hard-riding desert warriors who were his ancestors, busied like Louis XVI with making clocks himself. The fact that there were 'scoops which made the water circulate up and down' leaves no doubt that the imperial jack-work and time-announcing immortals were in the tradition of Su Sung and I-Hsing. But this apotheosis of the water-wheel escapement was almost exactly contemporary

[1] One of us (J.N.) has had the opportunity of visiting this historic observatory thrice (1946, 1952, 1958). The instruments are well preserved and still in their traditional positions in the open air. The observatory has become a public museum and the halls in the gardens below house an admirable exhibition on the history of astronomy and calendrical science. The Jesuit celestial globe (Fig. 49) repays close study. Its position in the lineage beginning with Chang Hêng renders it particularly venerable.

[2] This was Verbiest's 'Astronomia Europaea sub Imperatore Tartaro-Sinico Cám Hy (K'ang-Hsi) ex Umbra in Lucem Renovata' (1687).

(A.D. 1360) with the astronomical masterpiece of the de Dondi family in Europe,[1] made all of metal, powered by the falling weight, and using the verge-and-foliot (actually a crown-wheel) escapement. The moment was a turning-point in the history of the clock.

Yuan Shih, ch. 43, pp. 13*b*ff., by SUNG LIEN *et al.*, *c.* A.D. 1370. Parallel text in *Hsü Thung Chien Kang Mu* [798], ch. 27, pp. 11*a* ff.,[2] by SHANG LU [639] *et al.*, A.D. 1476.

The emperor (Shun Ti)[3] himself made (in his workshops) (A.D. 1354) a boat 120 ft. long and 20 ft. wide, manned by twenty-four figures of sailors all in cloth of gold, and holding punt-poles, which sailed about on the lake between the Front and the Back Palaces (having mechanical arrangements so that the) dragon's head and tail could move about, while it rolled its eyes, opened its mouth and waved its claws.[4]

He himself also constructed a Palace clock [386] 6 to 7 ft. high and half as wide. A wooden casing hid many scoops [387] which made the water circulate up and down [388] within. On the casing there was a 'Hall of the Three Sages of the Western Paradise', at the side of which there was a Jade Girl holding an indicating-rod for the hours and quarters. When the time arrived this figure rose up on a float and to left and right appeared two genii in golden armour, one with a bell and the other with a gong. At night these jacks struck the night-watches automatically without the slightest mistake. When the bells and the gongs sounded, lions and phoenixes on each side all danced and flew around. East and west of the casing there were 'Palaces of the Sun and Moon', in front of each of which stood six flying immortals [389]. Whenever the noon and midnight (double-) hours arrived, these figures went in procession two by two across a 'Bridge of Salvation' so as to reach the 'Hall of the Three Sages', but afterwards they withdrew and returned to their original positions. The ingenuity of all this was beyond belief and people said that surely nothing like it had ever been seen before.

The final blow to the indigenous tradition of water-powered clocks and jacks might well be dated about A.D. 1368 when the new Ming dynasty's forces captured Peking and ended the dominion of the Mongols. Hsiao Hsün [641], about twenty years later, in his *Ku Kung I Lu* [799], left a vivid account of the architecture and contents of the Yuan palaces which were destroyed by order of

[1] See the description by Lloyd (1) and Fig. 50.

[2] A paraphrase translation of the passage has been given by Wieger (1), p. 1735.

[3] Shun Ti [988], personal name Toghan Timur [640] (A.D. 1320–70), also canonised as Hui Tsung [989], tenth and last emperor of the Yuan (Mongol) dynasty. Though he reigned more than thirty years the dynasty was continuously decaying for social and economic reasons, in spite of the service of some forceful ministers. Fig. 51.

[4] An interesting parallel to this puppet boat, though on a much smaller scale, can be found among the mechanical toys described in the book of the Beni Musa brothers, who flourished at Baghdad in the ninth century (illustration in Feldhaus (1), p. 236). They also knew and used the system of a bucket filling and periodically tipping so as to rotate a scoop-wheel and make puppets move by trip-lugs on its shaft.

the first Ming emperor.[1] Hsiao Hsün was present as a Divisional Director of the Ministry of Works (Kung Pu Lang Chung [990]) and therefore had opportunity to see in detail the beauty of the buildings which had housed more than a thousand concubines, as well as the arrangements of the workshops in which so many ingenious toys had been made by the emperor and his workmen for their amusement. Hsiao Hsün describes (p. 7*b*) dragon-fountains with balls kept dancing on the jets, tiger robots, dragons spouting perfumed mist, and several dragon-headed boats full of mechanical figures. The destruction of all these things, understandable though it was as the act of a dynasty which was coming to power because it represented the resentment of the mass of the people at the economic exploitation to which they had been subjected, and could direct this against a quasi-alien race, was nevertheless very unfortunate. Much that could have been put to better use probably perished—the Ming, like another revolution later, 'had no use for' clock-makers. It would be an undue simplification to regard this nationalist uprising as anti-technological, for there is much evidence connecting its success with the first adequate tactical use of metal barrel cannon, then a new invention. No doubt the horological tradition had become smothered in its own jack-work, and was hopelessly identified with the 'conspicuous waste' of the Mongol court. But its death, if indeed it did quite die at that time,[2] was a circumstance of peculiar historical importance, for it meant that when the Jesuits arrived 250 years later there was extremely little to show them that mechanical clocks had ever been known in China. Moreover, for reasons not yet at all clear, there was in Ming times a general decline in most of the autochthonous traditions of physical science and technology,[3] so that there was no one who could explain Chinese mathematics, astronomy or other sciences to the Jesuit missionaries. This situation was naturally exploited by the Jesuits in several ways. To the Chinese they emphasised as much as they possibly could the superiority of the natural sciences of early seventeenth-century Europe because by the aid of it they wished to convince them of the superiority of European religion. To the Europeans they praised the ethical and

[1] Another description of their contents was given by Ch'üan Hêng [680] in his *Kêng Shen Wai Shih* [822]—'An Unofficial History (of the Emperor born in the) *kêng-shen* Year', i.e. (Toghan Timur). But although this has something to say (ch. 2, p. 8*a*) of the mechanical toys, including 'the dragon ship, built with great art and device, with a dragon which could move its tail and mane mechanically, and beat the water with its claws', there is no mention of the striking clock with its parade of puppets.

[2] As we shall see presently, it did not die, but was transformed into a minor industry which made clocks depending on the falling of sand instead of water; see pp. 156 ff. below.

[3] The ceramics industry was an exception. Moreover the decline did not affect the biological sciences, especially medicine and pharmaceutics, which in the Ming dynasty reached levels unprecedentedly high.

social philosophy of China as much as possible in order to raise the prestige of the Jesuit mission, designed to convert not savages, but a highly civilised people. The general upshot was that the indigenous scientific and technical tradition of China was very much underestimated in the West.[1]

5 ENTRY OF THE JESUITS WITH THEIR 'SELF-SOUNDING BELLS'

The horological history of the Jesuit period raises four distinguishable questions: (*a*) the story of the entry into China and the travels of the missionaries, in which clockwork played an unexpectedly prominent part, (*b*) the reception of the Jesuits at the imperial court and their subsequent role there as clock-makers as well as astronomers and mathematicians, (*c*) what the Jesuits had to say about the indigenous time-keeping apparatus of Ming China, and (*d*) what the Chinese had to say about the up-to-date clockwork which the Jesuits brought with them, and its relation to earlier inventions described in the Chinese historical literature. Let us consider these points in order.

It will readily be allowed that few historical events have been so rich in consequences as the decision taken by certain southern Chinese officials in 1583 to invite into China some of the Jesuit missionaries who were waiting in Macao. It was the first decisive step in the long process of unification of world science in Eastern Asia, and the better mutual understanding of the great cultures of China and Europe. The two men chiefly concerned were Ch'en Jui [642] (*c.* 1513–85 A.D.) who was for a short time Viceroy of the two Kuang provinces (Kuangtung and Kuangsi),[2] and Wang P'an [643] (*c.* 1539–1600 A.D.) who was Governor of the city of Chao-ch'ing in Kuangtung [872].[3] They were particularly interested in reports that the Jesuits had, or knew how to make, chiming clocks of modern type, i.e. constructed of metal with spring or weight drives and striking mechanisms. These became known, in the talk and correspondence of the time, as 'self-sounding bells' (*tzu ming chung* [214]), by a direct translation of the word 'clock' or 'cloche' (*glocke*). This is an important point, for an entirely new name suggested an entirely new thing. The mechanical clocks of the medieval period had been, as we know, extremely cumbrous and probably never very widespread; moreover no response had been forthcoming to Su Sung's appeal for a special name to distin-

[1] This impression has recently been relayed once more to the western reading public by Cronin (1) in his study (in many other ways admirable) of the life of Matteo Ricci.

[2] D'Elia (1), vol. 1, p. 164; Trigault (Gallagher tr.), p. 137.

[3] D'Elia (1), vol. 1, p. 176; Trigault (Gallagher tr.), p. 145.

Fig. 51. Toghan Timur (A.D. 1320–70), Shun Ti, last emperor of the Yuan dynasty, himself a horological engineer. A nineteenth-century Chinese artist's impression (after Favier).

guish them from non-mechanised astronomical instruments.[1] It was therefore not surprising that the majority of Chinese, even scholars in official positions, now got the impression that the principle of the mechanical clock was a new invention of dazzling ingenuity—almost magical in its precision—which European intelligence alone could have brought into being (cf. Fig. 52). Nor was it surprising that the missionaries (as men of the Renaissance) quite sincerely believed in this higher European intelligence, for they sought to commend the religion of the Europeans as something equally on a higher plane than any indigenous faith.

Both Ch'en Jui and Wang P'an obtained the clocks which they desired.[2] The latter had especially ordered an iron one from Macao, but eventually gave it back to the Jesuits for use in their house.[3] This residence in Chao-ch'ing, the first set up in China by the great pioneer Matteo Ricci (A.D. 1552–1610),[4] had a clock-face on the street,[5] with a public self-sounding bell, and this was one of the charges against him when the new Viceroy Liu Chieh-Chai [644] closed the mission-house in 1589.[6] Meanwhile, two years before, a magnificent spring clock with three bells chiming the half-hours and quarters had been sent out from Rome as a gift for the emperor,[7] and its peregrinations in the care of the Jesuits seeking to present it form a prominent part of the events of the following years. Ricci took it to Nanking in 1598,[8] and it made him popular there among the virtuosi on his way north to Peking for the first time.[9] Cases highly ornamented were made for it at

[1] Besides, by this time, the clock as a piece of purely time-keeping apparatus had become in both East and West entirely divorced from its original astronomical connections. As Price (1) has said: 'We must cease to regard the mechanical clock as a time-teller improved beyond all measure by the invention of an escapement, but must regard it instead as a type of astronomical machine which was known and improved throughout the middle ages.' The mechanical clock, as a pure time-teller, 'was a fallen angel from the world of astronomy'.

[2] D'Elia (1), vol. 1, p. 166; Trigault (Gallagher tr.), p. 138.

[3] D'Elia (1), vol. 1, pp. 201, 211; Trigault (Gallagher tr.), pp. 160, 168.

[4] A few words are necessary about this great and remarkable man. Born in Macerata in Italy, he profited by a good scientific education under Jesuit care, but proved also to be endowed with an extraordinary memory and an extraordinary gift for languages. Thus when he took charge of the Jesuit mission in China he was able to win the friendship and collaboration of some of the most enlightened Chinese scholars of his day, wrote numerous books in classical Chinese style, spoke Chinese with elegance and eloquence, laid the foundations of the high astronomical and mathematical competence of the mission, organised the translation of Euclid into Chinese, and was in every way a noble representative of most of what was best in European culture. If his conversions to Christianity were less numerous than he might have hoped, that lay to the count of basic cultural differences. The Chinese appreciated him more as a scholar and scientist. The reader is referred to a recent historical novel on him by Cronin (1), which gives a good idea of his life, though often seriously underestimating, as we have said, the cultural, and especially the scientific, achievements of the Chinese.

[5] D'Elia (1), vol. 1, pp. 252, 259; Trigault (Gallagher tr.), pp. 194, 201.

[6] D'Elia (1), vol. 1, p. 265; Trigault (Gallagher tr.), p. 206.

[7] D'Elia (1), vol. 1, p. 231; Trigault (Gallagher tr.), p. 180.

[8] D'Elia (1), vol. 2, p. 4; Trigault (Gallagher tr.), p. 296.

[9] D'Elia (1), vol. 2, pp. 39, 87; Trigault (Gallagher tr.), pp. 320, 348.

Fig. 52. A 'self-sounding bell', i.e. a striking clock of European style such as the Jesuits introduced to China from 1585 onwards (from the *Huang Ch'ao Li Ch'i T'u Shih*, ch. 3, p. 68*a*, A.D. 1759).

Nan-ch'ang and Nanking in 1600,[1] and the fact that the Wan-Li emperor asked why it had not arrived was the heaven-sent circumstance which got Ricci and his companions out of the clutches of the eunuch Ma T'ang [645] who had detained them while on the way to Peking the second time.[2] Later, in the following year, it was duly installed in the palace in a special tower made for it by the Ministry of

[1] D'Elia (1), vol. 2, p. 99; Trigault (Gallagher tr.), p. 355.
[2] D'Elia (1), vol. 2, p. 121; Trigault (Gallagher tr.), p. 369.

Works[1] and the Jesuits were entrusted with the regulation of it and the training of certain eunuchs in clock maintenance and repair.[2] This was in 1601, at the very beginning of the long period of residence of the Jesuits at the Chinese capital. But wherever they were in the country, clocks and clockwork were not far away. When Lazare Cattaneo (A.D. 1573–1606) got into trouble at Shaochow in Kuang-tung it was partly owing to the grudge of one of the officials, Tung T'ing-Ch'in [646], who had not been allowed to borrow a mechanical clock as long as he wished.[3] When Benedetto Goes (A.D. 1562–1607) passed through Kashgar in 1604, on his epic journey across Central Asia from Delhi to Suchow (in Kansu) which proved that China was the same as Cathay, he presented a portable iron watch to the local prince.[4] Finally, the very exceptional favour granted by the Wan-Li emperor to the Jesuits on the occasion of the death of Matteo Ricci, a gift of land for a splendid tomb, was partly in gratitude for a small portable clock of which the emperor was extremely fond.[5] To sum up the matter, it is quite clear that one of the reasons why the early Jesuit missionaries were so much welcomed by the Chinese was for their interest in clocks and clock-making, hardly less indeed than for their skill as mathematicians and astronomers.

By the second decade after the death of Ricci collaboration between Jesuit missionaries and Chinese scholars had become a well-established tradition. It is to the work of one such pair that we owe the earliest Chinese description of a verge-and-foliot escapement clock. And this instrument was one in which the two traditions of clock-making came together, for while a mechanism of Western type kept equal double-hour time in the front of the cabinet, a scoop-wheel system of the Chinese type told the night-watches at the back.

In Ricci's time the Jesuit order was capable of attracting for its overseas missions some of the best minds of Europe. It was a mobilisation of oecumenical idealism something like that which the League or the United Nations have now and then commanded in our own time. Johann Schreck (A.D. 1576–1630) was born a Swiss at Constance and while still very young became known throughout the realm of German culture for his brilliant attainments in medicine, natural philosophy and mathematics. Passing to Italy in the early years of the seventeenth century he gained further renown and was, with Galileo, one of the first half-dozen members

[1] D'Elia (1), vol. 2, p. 128; Trigault (Gallagher tr.), p. 374.
[2] D'Elia (1), vol. 2, pp. 126, 159, 313, 471; Trigault (Gallagher tr.), pp. 373, 392, 536.
[3] D'Elia (1), vol. 2, p. 382; Trigault (Gallagher tr.), p. 288.
[4] D'Elia (1), vol. 2, p. 416; Trigault (Gallagher tr.), p. 506.
[5] D'Elia (1), vol. 2, p. 579; Trigault (Gallagher tr.), p. 571.

of the Cesi Academy. He had already become a personal friend of Kepler in the north. This was the man who, as Têng Yü-Han [695], a dedicated exile,[1] found himself in 1626 working with a distinguished Chinese scholar, Wang Chêng [696],[2] on a book which should provide for the first time in Chinese dress the principles of Renaissance mechanics and an account of their applications by the engineers of Europe. This was the *Ch'i Ch'i T'u Shuo* [835] (Diagrams and Explanations of Wonderful Machines),[3] which appeared in the following year. At the same time Wang Chêng published a smaller companion book, the *Chu Ch'i T'u Shuo* [836] (Diagrams and Explanations of a number of Machines), mainly of his own invention or adaptation.[4] Here we find (pp. 12*b* ff.) the description of the clock. Wang called it a 'Wheel Clepsydra' (*lun hu* [449]).

It was a cabinet of two storeys about 2½ ft. in each dimension and 1½ ft. in depth (see Fig. 53). The gear train of the clockwork is seen in the middle compartment; it was supported on iron bars and pillars, the wheels being made of the purest iron,[5] and driven by a falling weight of lead. Above the gear-wheels the foliots are seen. Wang Chêng described them in the following words:

> Beside all this there is a cross-shaped (device) having two hitting teeth [92, 219] (i.e. pallets) to left and right. All the wheels (of the gear train), impelled by the motion transmitted through them, would move quite fast, if it were not for the hitting teeth (pallets) in the centre, which slow down the movements, seeming to be pushed to the left yet arresting on the right. This is a very delicate piece of machinery, and the very essence of the 'Wheel Clepsydra'. But though everything depends on it, it is extremely difficult to describe in writing, and can hardly be represented in a diagram.

Such is the earliest account of the verge-and-foliot escapement in Chinese.[6]

[1] Schreck latinised his name, punningly, as Johannes Terrentius, hence the English form John Terence, best not used. He had arrived in China in 1621 and for the last few years of his life was engaged (apart from his work in physics and engineering) in the beginning of the calendar-reform which the Jesuits were asked to undertake. On this see Needham (1), vol. 3, p. 444.

[2] A.D. 1571–1644; biography in Hummel (1), pp. 807 ff. The late Professor Fritz Jäger of Hamburg made himself an authority on Wang Chêng and all his scientific work. One of us (J.N.) has had the privilege of consulting the unpublished posthumous papers of Prof. Jäger; for this our thanks are due to Professor Wolfgang Franke, the present head of the Department of Chinese Studies at that university.

[3] On this book see Jäger (1). Identifications of the European sources will be found in Needham (1), vol. 4, pt. 1.

[4] Pfister (1), p. 157, is wrong in saying that this book was also under joint authorship with Schreck, and that it was concerned with 'les machines indigènes'. Wang Chêng was decisively the first 'modern' Chinese engineer, indeed a man of the Renaissance, though so far from its birthplace. This was well brought out by Liu Hsien-Chou (4), in an interesting article on him published during the last war.

[5] The number of teeth on the larger wheels visible in the picture are given as 60, 48 and 36 respectively.

[6] We are indebted to Dr Lu Gwei-Djen for her collaboration in the study of Wang Chêng's description.

(*a*)

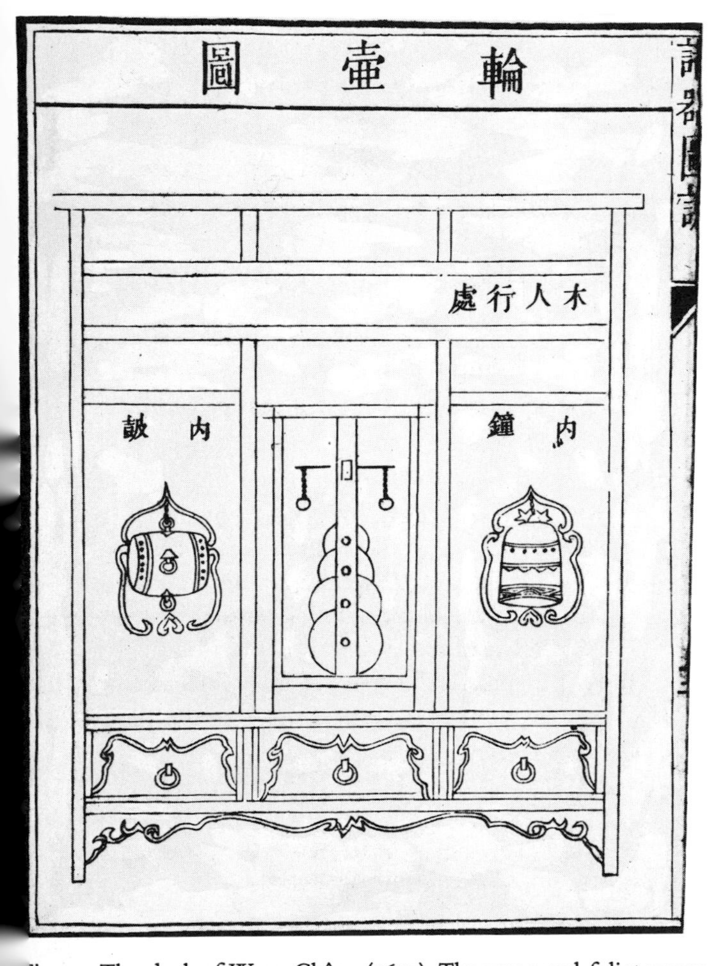

Fig. 53. The clock of Wang Chêng (1627). The verge-and-foliot escapement and the gear train can be seen in the centre compartment, with the drum on the left and the bell on the right. Above, right across, is the runway of the perambulating jack. Below are three drawers. See pp. 146–7 and 212.

(*b*)

Fig. 54. The K'ang-Hsi emperor, Elghe Taïfin or Shêng Tsu (r. 1662–1722). A nineteenth-century Chinese artist's impression, based on contemporary documents (after Favier).

Naturally no clock in this part of the world would have been complete without jack-work. In Wang Chêng's design it took the form of a perambulating figure with a pointer, driven by a belt or chain from the uppermost wheel of the clockwork, and indicating each of twelve tablets for the double-hours in turn. Nothing is said of a mechanism to return it to its initial position each day, but we learn that the figure operated catches which beat upon the drum in the left-hand compartment, while other mechanisms were so arranged as to sound both the drum and the bell in the right-hand compartment at other times.[1]

The most extraordinary thing about Wang Chêng's clock, however, is that it seems to have incorporated at the back of the cabinet a time-keeping mechanism of the traditional Chinese type. For inside, he says, 'there is a night-watches clepsydra (*kêng lou* [360, 401]), with two flumes (*ts'ao* [449*a*]) and two tubes (*t'ung* [449*b*]) filled with lead shot (*ch'ien tan* [449*c*]), and (the rest of the appropriate) machinery'. Obviously this can have nothing to do with clockwork of the Western type, but it does suggest that Wang adapted one of the wheel-clock designs of the Ming (see pp. 155 ff. below) to use small lead shot instead of sand. Presumably because it was so familiar he tells us little about it. One cannot say, therefore, whether it had an escapement of the ancient type bearing upon the buckets, or whether it worked simply by reduction gearing. Presumably also his reason for adopting the old bucket-wheel method for the unequal night-watches was that he thought it would be easier to alter from time to time during the year. This must have involved, one would suppose, a nice adjustment of the orifice through which the lead shot escaped, but Wang Chêng says nothing at all about these matters. Nevertheless, his clock is of considerable interest not only as the first Chinese conception which used the weight drive and the verge-and-foliot escapement, but also as incorporating in one and the same instrument the new clockwork from the West with the age-old Chinese 'water-wheel' time-measurer.

One at least was built at the capital, and kept time with great distinctness and precision. Those who saw it, says Wang Chêng, 'were not like the men of old, who seeing a sheep's carcase hung up, mistook it for a hungry horse!'

[1] The description in Wang Chêng's book, and the illustration, looking so unlike a modern clock, have not until now been fully recognised and identified. A close study of them, however, was made by the late Professor Fritz Jäger with the assistance of Dr Li Shu-Hua, and we have been happy to find that they also saw clearly the presence of a mechanical clock of Western type. They were inclined, however, to interpret the 'cross-shaped device' as an anchor escapement, which would have been quite impossible at the date in question. Moreover, not knowing what the present researches have revealed about the Chinese tradition of 'water-wheel clocks', they were unable to arrive at any explanation of the 'night-watches clepsydra' with its lead shot.

6 THE JESUITS AS MAKERS AND REPAIRERS OF CLOCKS 'IN NOMINE DEI'

As time went on, the four eunuchs trained by Ricci proved quite unable to cope with the horological requirements of the Chinese imperial court; indeed they were only the forerunners of a whole tribe of artisans, both European and Chinese, who succeeded one another during the following two centuries. After a while the Jesuits recruited professional clock-makers into their ranks and incorporated these pious workmen into the China Mission. The full history of this interesting alliance of technology and religion has never been written, but Pelliot (1) started a sketch of it,[1] and much more can be found by anyone willing to delve in Pfister's register (1). Gabriel de Magalhaens, who was born the year before Ricci died, was one of the earliest missionaries to answer this need; in about 1670 he made for the K'ang-Hsi emperor (Fig. 53) a chiming clock which played tunes while at each hour little guns were shot off.[2] In 1700 Thomas Pereira rebuilt the College chapel at Peking, giving it a clock with a carillon activated by a toothed drum.[3] A whole lifetime of clock-making was devoted to the court by the Swiss lay brother François Louis Stadlin, who was in attendance from 1707 onwards,[4] and by another lay brother, Jacques Brocard, from France.[5] The distinguished mathematician Pierre Jartoux was specially trained in clock-making and repairing before he came out in 1701.[6] In 1736 Valentin Chalier completed for the emperor

[1] Pelliot's article was a review of the book of Chapuis, Loup & de Saussure, *La Montre Chinoise*, which should also be consulted.

[2] Gabriel de Magalhaens (A.D. 1609–77) was a Portuguese. His clock is described in Verbiest's *Astronomia Europaea...* (1687), pp. 92 ff. Most of the section entitled 'Horolotechnia' concerns drum-operated carillons and the effect which they produced on the Peking populace. The K'ang-Hsi emperor himself came to visit the Jesuit turret-clock which played Chinese tunes each hour. Bernard-Maître (1) has transcribed for us a letter of Verbiest's, written on 11 May 1684: 'Rex venit ad aedes nostras, intravit omnia cubicula, vidit organum et 2 campanas majores horologii suspendendas in turre.' This was the time when the emperor was studying the sciences assiduously with the Jesuit experts, and fostering the spread of their knowledge in every possible way. Elsewhere in his book (p. 57), Verbiest was led to write, in a delightful passage which his fellow-Jesuit Bernard-Maître finds rather turgid: 'And now it was as if all the mathematical sciences, led by Astronomy in royal splendour, began together to enter the Imperial Palace—Geometry and Geodesy, Gnomonics and Perspective, Statics and Hydraulics, Music and the Mechanical Arts—all arrayed like the companions of an empress, each handmaiden more lovely than the next.'

[3] Thomas Pereira (A.D. 1645–1708) was also Portuguese, particularly well qualified in music. This chapel was the Nan-T'ang one.

[4] François Louis Stadlin (A.D. 1658–1740) was from Zug and a professional clock-maker. Chanting German canticles at his work he lived to the ripe age of 82; the Ch'ien-Lung emperor much deplored his loss. Part of the time he was assisted by a Frenchman, François Guetti, of the Missions Etrangères (1706).

[5] Jacques Brocard (A.D. 1661–1718) made scientific instruments as well as clocks.

[6] Pierre Jartoux (A.D. 1668–1720) was a cartographer as well as a mathematician, much concerned with the general map of China which the Jesuits made for the emperor, and for which he carried out numerous exhausting journeys. In horology he was assisted by the Czech Jesuit Charles Slaviczek (A.D. 1678–1735).

(Ch'ien-Lung) a fine clock which struck not only the equal day and night hours, but also the unequal night-watches. This was exactly the same problem which Su Sung had solved in his own simpler way 650 years earlier. Pelliot has reproduced an interesting letter written by Chalier at the time.

I have made [he says] a clock 4½ ft. high and 3 ft. wide which marks hours, minutes and seconds in Western style, and strikes quarters as a repeater, but besides this sounds the night-watches as they are sounded throughout China, marking on a special dial what watch it is, together with another giving the Chinese zodiacal sign according to which each night-watch has to be sounded or repeated as the seasons change. To understand all this, you must know that the night-watches begin regularly two hours after sunset and end two hours before dawn. This interval is divided into five equal watches, longer or shorter according to the length of the whole night.... The machine has been finished now four months, and the emperor is delighted with it. He keeps it in his own apartments. Nothing has been spared for the ornament of it and the case and dials are magnificent. While it was being made he often had me bring to him the new pieces, and seeing the multitude of arbours, cogs and pins he vowed that he would never have thought what he had asked me to do would have been so difficult. When it was finished he invited the princes and grandees to come and look at it, explaining its works as if it was something of his own invention. They complimented him highly, attributing to him all the execution of it, and the cream of the joke was that I had to speak in the same strain as all the rest of them.... As for clocks, the imperial palace is stuffed with them. Watches, carillons, repeaters, automatic organs, mechanised globes of every conceivable system—there must be more than 4000 from the best masters of Paris and London, very many of which I have had through my hands for repairs or cleaning.[1] I must know as much of the theory now as any clock-maker in Europe, for I am sure few have had as much experience. If only I had had more practice in my youth![2]

It is interesting to note that mechanised celestial globes, so important in the history of clockwork in China, were included in the imperial collection.

Later on the tradition continued with Gilles Thébault, a professional, one of the

[1] Many of the clocks and mechanical curiosities which were collected by the imperial court through the eighteenth and nineteenth centuries are still to be seen in the museums of the Imperial Palace in the centre of Peking, and the Summer Palace (I Ho Yuan [991]) some distance away towards the western hills. A catalogue of these was made by S. Harcourt-Smith and published at Peking in 1933; unfortunately it contains nothing of astronomical interest, though eighteen pieces by J. Cox are represented in it. Presumably all the scientific clocks were broken beyond repair or destroyed during the looting of the Yuan Ming Yuan [1004] in 1860 or that of the Forbidden City in 1900 if not later still. Nevertheless a number of the pieces still extant, ranging in date from 1760 onwards, embody moving figures similar to those in the great clock of the Yuan emperor Shun Ti (p. 140 above).

[2] Valentin Chalier (A.D. 1697–1747) was from Briançon. He suffered greatly from the adverse decision of the Vatican in the Rites Controversy, and his last years were saddened by failure to avert death sentences on Dominicans by the officials of Fukien province.

lay brothers,[1] and Jean Mathieu de Ventavon, one of the fathers,[2] who also made automata. At the end of the century, after the suppression of the Jesuit Mission, a Lazarist brother, Charles Paris,[3] took over, but the era of lay commerce was dawning, and the horological trade centred itself at Canton with English houses such as Cox & Beale, and Swiss horologists and merchants such as Petitpierre-Boy[4] and Charles de Constant de Rebecque.[5]

7 THE MERCHANTS OF THE 'SING-SONG' TRADE

Some reference to the astonishing history of the firm of Cox & Beale deserves insertion here.[6] Until as late as 1834, 'the sole exclusive right of trading, trafficking, and using the business of merchandise, into or from the dominions of the Emperor of China' was, so far as British subjects were concerned, legally monopolised by the Honourable East India Company. In 1781 the Company enforced its powers and expelled from China all British merchants not attached to its Factory. Except one—John Henry Cox—who was in China on a different basis from the others, as a dealer in English 'sing-songs'. This peculiar term was the coastal nomenclature for the trade in clocks, watches, and mechanical toys such as snuff-boxes containing jewelled birds which sang when the lid was opened (cf. Fig. 55). Such objects were one of the very few classes of goods which the West could offer, before the age of full industrial production, of interest to the East. Cox was finally forced to leave China, but came back before long as Cox & Beale, together with many other British merchants, all semi-disguised as Prussian, Austrian, Danish, and even Sardinian, consuls. We must not follow further here this entertaining story. The important points are that the trade in clocks was the first, and

[1] Gilles Thébault (A.D. 1703–66) was a Norman from St Malo. It was towards the end of his time (1764) that fountains representing the animals of the twelve double-hours, and playing in turn in accord with the time, were constructed by Michel Benoist (A.D. 1715–74) aided by Pierre Michel Cibot (A.D. 1727–80).

[2] Jean Mathieu de Ventavon (A.D. 1733–87) was helped by Hubert de Méricourt (A.D. 1729–74) who died only a year after reaching Peking.

[3] Charles (or Joseph) Paris (A.D. 1738–1804) reached Canton in 1784. For details on his work see Chapuis, Loup & de Saussure, p. 31.

[4] C. H. Petitpierre-Boy (A.D. 1769–*c.* 1810) was a lay Swiss clock-maker who accompanied the English embassy of Lord Macartney in 1793 and also the Dutch embassy which went to Peking two years later. The latter took two complicated clocks bought from Cox & Beale, but they arrived damaged. Petitpierre-Boy looked after the planetarium which the Macartney embassy took to present to the emperor, and set it up in Peking with the help of Dr Dinwiddie. Eventually he settled at Macao and then at Manila, married a Javanese lady, and was finally killed by pirates. For further details see Chapuis, Loup & de Saussure, pp. 45 ff.

[5] On Ch. de Constant (A.D. 1762–1835), see Chapuis, Loup & de Saussure, p. 51.

[6] See the interesting book of Greenberg (1). Cox & Beale were the lineal predecessors of the famous firm of Jardine, Matheson & Co.

Fig. 55. One of Cox's 'sing-songs' from the Imperial Palace collection; a mechanical bird-cage clock of 1783. When the clock chimes, the bird trills a little tune, moves its head, flaps its wings in a lively manner and hops from perch to perch; at the same time the spiral and star revolve, and various other motions take place (from Harcourt-Smith).

for a long time the only, provision of machinery from the Western European to the Eastern Asian area; and secondly that its volume was far too small to compensate in any degree for the great drain in silver bullion caused by the purchase of tea, silk and minor 'China goods' for the demands of the English home market. It was to counter this drain that the importation of Indian opium, though prohibited by the Chinese authorities, was undertaken. The Opium Wars (1839–42 and 1857–60) were the results. The 'sing-song' trade, it is interesting to note,[1] fell off very much after 1815, partly because of the growth of the Chinese clock-making industry, which produced time-pieces for half the cost.

Too little attention has been paid to the effects which this China trade had upon the designs of monumental clocks with astronomical adjuncts and various sorts of jack-work in seventeenth- and eighteenth-century Europe. There seem to have been two waves, in the latter parts of both these centuries, of European manufacture of astronomical clocks using not European principles only, but also certain features emanating from the ancient Chinese tradition of I-Hsing and Su Sung.[2] Trade with China thus acted back on European clock-making, stimulating a popularity of complex dials and jacks in an atmosphere of 'Chinoiserie'.

It will be remembered that the re-publication of Su Sung's book occurred in China, under the care of Ch'ien Tsêng [463] (cf. p. 12 above), probably about 1665. Within twenty years a German craftsman, Christopher Rad, following the designs of the famous Christopher Trechsler (or Treffler), made an astronomical clock called 'Automaton Sphaeridicum' (Fig. 56).[3] Vincenzo Coronelli saw it at Augsburg in 1683 and gave a description of it in his 'Epitome Cosmographica',[4] ten years later. The instrument was 7 ft. high, and carried a celestial globe provided with an automatic movement, together with dials and jacks which indicated the years, months, days, hours, and minutes, as well as eclipses for seventeen years

[1] Greenberg (1), p. 86.

[2] Cf. n. 1 on p. 23 above.

[3] See Stevenson (1), vol. 2, pp. 94 ff., 134. A careful study of this apparatus has been made by Bedini (1), pp. 488 ff., who believes that the real maker was Johann Philipp Treffler, and that it was fathered on the elder brother Christopher for reasons connected with the internal politics of Augsburg at the time. In any case the instrument was described in a pamphlet (now exceedingly rare) under the latter's name, published at Augsburg in 1679. From this we learn that the armillary sphere at the top rotated equatorially while the celestial globe within the open-work dome was pivoted upon the ecliptic poles. This was so constructed that it could be removed for manual demonstration and then replaced in such a way as to continue to receive the power-drive. Of the four clock-face dials below, one indicated hours and minutes in the usual way, another the time in various parts of the world, the third showed the days of the month and the signs of the zodiac, while the fourth gave the phases of the moon, etc. There were two clockwork movements, one for the sphere and the other for the globe and the dials. The whole structure was of silver, and there can be no doubt that it came duly to the court of the Emperor Leopold about 1685.

[4] Poletti, Venice, 1693, p. 333. On Coronelli and his book see Armão (1), p. 189.

in advance. Above the globe was mounted an armillary sphere. The whole design, with its crosswise framework base, and the relative positions of globe and sphere, is so reminiscent of Su Sung's clock tower as to invite the speculation that some knowledge of the latter had percolated to Europe, presumably through Jesuit channels, in time to inspire Trechsler's masterpiece of 1683.[1]

Certain descriptive catalogues and inventories of nearly a century later give us a glimpse of the kind of apparatus which was then being made in Europe for the Chinese market. James Cox was a clock-maker and jeweller who was deeply engaged in the trade, but in 1774 the 'great scarcity of money in the East Indies' obliged him to sell off his stock in a lottery. His catalogues[2] described a remarkable 'Chronoscope', the fellow piece of one which had been sent in the 'Triton' Indiaman to Canton in 1769, and which 'now adorns the palace of the Emperor of China'. The apparatus, 16 ft. in height, was of course made in gold studded with gems, but its interest lies in the extent to which it embodied features characteristic of the medieval clocks of China and Islam. Besides a mechanised armillary sphere and celestial globe, there were jack figures which struck bells at the hours and quarters, flying dragons which dropped (real) pearls into the mouths of other creatures waiting to receive them, and an elephant which paraded on a horizontal wheel waving its trunk and tail. Some of the clock casing was of 'crystal', permitting a view of the works, and one dial registered quarter-seconds.[3]

It is not generally known that a replica of the original 'orrery' was carried to China by the embassy of Lord Macartney in 1793. The first planetarium of this kind demonstrating the heliocentric system was made by George Graham[4] about

[1] At the same time the idea of having an armillary sphere or a celestial globe above geared to a mechanical clock below was not new in Europe. In A.D. 1556 Philipp Irmser had constructed a 'Planeten Prunkuhr' which had a four-faced clock surmounted by a rotating celestial globe with jack figures in between (Fig. 57). The figures made their circumperambulations like those of Su Sung. This clock was on display at the Ottheinrich Jubiläums Ausstellung, in the Castle at Heidelberg, September 1956. The panels of this beautiful apparatus show sixteenth-century astronomers using different kinds of instruments. Furthermore, in 1610 the masterpiece of Jobst Burgi, his rock-crystal clock, with its four-foliot cross-beat escapement and remontoire, incorporated a demonstrational armillary sphere (cf. Lloyd (3)).

[2] James Cox (1, 2).

[3] The English were not the only nation which presented works of horological virtuosity to East Asia. Takabayashi (1) figures (his fig. 27 and p. 50) a Japanese sketch of a clock sent from Russia by ship to Nagasaki as a present for the Imperial Court in 1804. Upon the base which contained the clock dial or dials there stood an elephant bearing on its back an armillary sphere, presumably mechanised, with a pavilion the spire of which bore—for some inscrutable reason—a wind-wheel. The elephant motif went back at least to al-Jazarī in A.D. 1200, for one of his designs shows an elephant bearing a mahout who periodically strikes a gong, and a rider who, pivoting, directs a pointer at the markings of a dial, the whole being surmounted by a pavilion of singing birds, etc. (Wiedemann & Hauser (1), p. 117).

[4] A good account of George Graham (A.D. 1673–1751) and his contributions to horology will be found in Lloyd (2).

PLATE XVb

Fig. 56. The 'Automaton Sphaeridicum' of Christopher (or John Philipp) Treffler, made at Augsburg in 1683 for the Emperor Leopold. With its celestial globe surmounted by an armillary sphere, and its dials below borne on a cross-shaped base and five columns, it is distinctly reminiscent of Su Sung's designs so long before (from Stevenson).

Fig. 57. The 'Planeten Prunkuhr' constructed in 1556 by Philipp Irmser, a clock which indicates day, zodiacal sign, and many other things besides hours and minutes; each quarter of an hour a new jack figure appears, as in Su Sung's clock. Height *c.* 4½ ft. The globe rotates automatically, and the panels show sixteenth-century astronomers using their instruments (photograph Ottheinrich (Kurfürst) Jubiläums-Ausstellung, Heidelberger Schloss, 1956; courtesy of Dr Arnold Koslow).

1706 for Prince Eugene of Austria.[1] It was immediately copied for the Earl of Orrery, whence the name of the instrument.[2] A magnificent replica was then ordered by the East India Company[3] in 1714, and described by John Harris a few years later.[4] Whether this was from the beginning destined for China we have not ascertained, but the instrument seems to have been the same as that taken there many years later. Staunton tells us[5] that in 1792 two Chinese students from the College at Naples came to England to accompany the mission as interpreters, and began by advising as to the presents which should be taken.

> Extraordinary pieces of ingenious and complicated mechanism, set in frames of precious metal, studded with jewels, and producing, by the means of internal springs and wheels, movements apparently spontaneous, had often borne (in China) excessive prices. They were indeed of no sort of use, but the imagination of the governing mandarines had been struck by them, and an intimation often followed to the native merchants to procure them, no matter at what price....

It was vain, he continued, to think of surpassing the 'sing-songs' which had already entered China, but it was decided to send whatever might serve to illustrate the sciences and arts.

> Astronomy being a science peculiarly esteemed in China, and deemed worthy of the attention and occupation of the Government, the latest and most improved instruments for assisting its operations, as well as the most perfect imitation that had yet been made of the celestial movements, could scarcely fail of being acceptable.

Accordingly, the list given to the officials who met the embassy at Tientsin included, besides the orrery, a reflecting telescope, a celestial and a terrestrial globe, a clock showing the lunar phases, an air-pump, and other instruments.[6] The high traditional value placed on astronomical science in China thus imposed itself on European diplomacy at the end of the eighteenth century, and Graham's masterpiece of astronomical clockwork found its way as a tribute, even if partly unconscious and unintentional, to the land of I-Hsing and Su Sung.[7]

[1] The circumstances of this were described by Desaguliers (1) not very long afterwards (1734); 1st ed., vol. 1, pp. 430 ff.; second ed., vol. 1, pp. 448 ff.; the planetarium itself is depicted in his pl. 31. Cf. Taylor & Wilson (1).
[2] Orrery (1), vol. 1, pp. xv ff.
[3] *Loc. cit.*
[4] In his 'Astronomical Dialogues' (1719).
[5] (1), vol. 1, p. 42.
[6] Staunton (1), vol. 1, p. 492. On the unpacking and installation of the planetarium at Peking, cf. vol. 2, pp. 165, 287.
[7] By the middle of the nineteenth century planetaria were being produced by East Asian clock-makers, e.g. Tanaka Hisashige in Japan (see Takabayashi (1), figs. 33, 34A, B). The first Japanese to construct a celestial globe rotated by clockwork was no doubt Nakane Genkei [681] (1661–1733), one of the founders of modern astronomy in that country (see Hayashi (1), p. 354).

We may end this little chapter by two agreeable pieces of information. In 1809 a Shanghai man, Hsü Ch'ao-Chün [647], who said in his preface that he came of a family which had been making 'European' clocks for five generations, issued an illustrated treatise on them entitled: *Tzu Ming Chung Piao T'u Fa* [800].[1] He may well have been a descendant of Ricci's friend Hsü Kuang-Ch'i [648], one of the finest scholars of his time (A.D. 1562-1634), a high official, mathematician and agriculturalist, also of Shanghai. It was he who prepared with Ricci the first Chinese translation of Euclid. And secondly, down to the end of the nineteenth century the clock-makers of Shanghai venerated Matteo Ricci as their tutelary deity or patron saint under the name of Li Ma-Tou P'u-Sa [992]—the Bodhisattva Li Ma-Tou. How Su Sung, and especially the Buddhist I-Hsing, would have laughed!

8 MING SAND CLOCKS

Turning now to the comments of Ricci and his companions on the methods of time-keeping which they found in late Ming China on their arrival, there is no doubt whatever that he regarded mechanical clocks as something absolutely new and unheard-of in the country. He says this several times in his memoirs. The clocks for the Cantonese officials in 1583, which struck all the hours automatically, were 'beautiful things never seen nor heard of before in China'.[2] The clock with three bells destined for the emperor, a piece 'which struck all the Chinese dumb with astonishment, was a work the like of which had never been seen, nor heard, nor even imagined, in Chinese history'.[3] The great clock of the Chao-ch'ing mission house 'with a hand visible from the street which indicated the hours, and a great bell which sounded them, was a thing previously unheard-of'.[4] Ricci's opinion is thus quite unmistakable,[5] and no other Jesuit thought differently, so far as we know.

[1] There was a parallel literature in Japan. The *Karakuri Zui* (Illustrated Schema of Horological (lit. Mechanical) Ingenuity), published by Hosokawa Hanzō Yorinao in A.D. 1796, has been reproduced in facsimile, with modernised transliteration of the text, in the book of Yamaguchi Ryuji (1).

[2] 'Un horiolo, che sonava per se stesso ad ogni hora, molto bello, cosa mai vista ne audita nella Cina.' D'Elia (1), vol. 1, p. 164.

[3] 'Horiuoli...che sonava le hore e li quarti con tre campanelle, che fece stupire a tutta la Cina, che ne aveva vista, ne udita, ne imaginato mai opra simile a questa.' D'Elia (1), vol. 1, p. 231.

[4] 'Un horiolo grande...con la mano che mostra le hore per fora nella strada, con una grande campana che sonava le hore, cosa mai avuta nella Cina.' D'Elia (1), vol. 1, p. 252.

[5] In other cases Ricci was probably quite right in his estimate of the novelty of some of his introductions. He said that the horizontal sundial which he made for the prince Chien An Wang [993] at Nan-ch'ang was a 'cosa mai vista nella Cina' (D'Elia (1), vol. 1, p. 366). This was because the characteristic form of the sundial was the equatorial form, with the plate in the plane of the equator and the gnomon pointing to the pole at right angles to it. Chinese geometry had never been adequate for the stereometric projections needed for pole-pointing gnomons on horizontal or vertical surfaces.

But as to the time-keeping methods in use in China beforehand, they spoke in rather mysterious terms. In the admirable description of China with which Ricci prefaced his memoirs, he wrote:

As for their clocks, there are some which use water, and others the fire of certain perfumed fibres made all of the same size;[1] besides this they make others with wheels which are moved by sand—but all of them are very imperfect. Of sundials they have only the equinoctial type, but do not know[2] well how to adjust it for the position (i.e. the latitude) in which it is placed.[3]

This strange, and for us very significant, statement about wheels was probably written after 1601 during Ricci's residence in Peking, and a dozen years later after his death it was considerably enlarged by Trigault. In the *De Christiana Expeditione* we find:[4]

They (the Chinese) have very few instruments for measuring time, and those which they do have measure it either by water or by fire. Those which use water are like large clepsydras, and those which use fire are made of a certain odoriferous ash very like that tinder which is used for torture among us.[5] A few other instruments are made with wheels rotated by sand as if by water, but all are mere shadows of our mechanisms and generally most faulty in time-keeping. As for sundials they know only that one which takes its name from the equator, but they are unable to set it up correctly according to the location (i.e. the latitude).[6]

[1] This reference, and the similar remark in the following passage, concern the practice of time-keeping by the burning of sticks or trails of incense. The method is old, and was used particularly in temples (see Needham (1), vol. 3, p. 330).

[2] This statement, repeated in the following passage, is quite erroneous. It could have applied only to the relatively ill-educated provincials with whom the Jesuits were first in contact.

[3] 'Gli loro horiuoli sino adesso furno di acqua e di fuoco con certe pipite odorifere, fatte tutte della stessa grandezza; fanno anco altri con ruote mosse di arena; cose tutte che di se tengono molta imperfettione. De' solari, solo hanno l'equinotiale, ma non lo sanno ben collocare conforme ai luoghi dove li pongono' (D'Elia (1), vol. 1, p. 33).

[4] Augsburg (1615), p. 22.

[5] This phrase puzzles us somewhat; we think Trigault may have been referring to medical cauterisation—but the practice of using tindery incense ash for this was characteristically Chinese, and came to Europe under the name of moxibustion. Moxa was not introduced into European medicine until 1674, by Buschoff (Garrison (1), p. 291).

[6] 'Horis metiendis vix habent instrumenta, quae habent, vel aqua vel igne mensurantur. Aquea sunt velut ingentes clepsydrae; Ignea ex odorifero cinere confecta, tormentorum nostrorum fomites imitantur. Pauca etiam alia conficiunt rotulis ab arena velut aqua circumactis, sed omnia ad nostratia artificia, sunt umbra, et fere multum peccant in ipsa metiendi temporis symmetria. E sciotericis solum norunt illud, quod ab aequatore nomen accepit, sed neque illud pro ratione locorum didicerant collocare.' Gallagher's translation of this seems to us particularly bad. He calls the clepsydras 'huge water-pots', and the tinder 'somewhat in imitation of our reversible grates through which ashes are filtered'. Regarding the wheels, he has 'operated by a kind of bucket wheel in which sand is employed instead of water', but though no doubt there were buckets or scoops they do not seem to be in the text. Regarding the sundials he has 'they (the Chinese) know that these take their name from the equator', evidently not appreciating the distinctions between the different types of sundial. Valuable though this book is as a draft translation useful in the rapid finding of a desired passage, careful checking is necessary.

We were thus obliged to believe that still at the time of Ricci and Trigault some remains were left of the old Chinese tradition of driving-wheel clocks equipped with escapements, but we ourselves could find no further information concerning them.[1] The use of sand seemed particularly curious, for the hour-glass is considered to have been an introduction from Europe.[2] We could only hope for enlightenment from further research.

Fortunately it was not long in coming, first from certain sources quoted in part by Yabuuchi Kiyoshi (1) in a paper on the history of horology in China, and then in admirable detail from Liu Hsien-Chou (3), who had the good fortune to find an elaborate description of the sand-driven wheel clocks of the Ming. First we must quote a few short passages from the opening astronomical chapter in the *Ming Shih* [829], the official history of the Ming dynasty, completed in A.D. 1739 by Chang T'ing-Yü [687] and others. We have already related the story of how the monumental striking water-wheel clock made by the last Yuan emperor perished in the wave of Confucian austerity which accompanied the rise to power of the nationalist dynasty of the Ming. The new emperor and his advisers evidently judged technological skill on a social and ethical basis. Gunnery was admissible as it aided the victory of the now popular political movement, but horological engineering (at any rate in some forms) was not, because it savoured of palatial luxury based upon undue oppression of the people. Hence we are interested, but not surprised, to read (ch. 25, p. 15*a*):

After T'ai Tsu [1002] had conquered the Yuan people, the Astronomer-Royal [935] presented him with a 'rock-crystal clock' [440], within which there were two wooden jacks automatically striking gongs and drums in accordance with the passing hours. But T'ai Tsu considered it to be a useless (extravagance) and had it broken up.

[1] One must have regard here to the psychology of the Jesuit missionaries. Not being interested in the history of science and technology, it would never have occurred to them to see in the more primitive types of escapement, even if they carefully examined any, an apparatus which had antedated, and perhaps even stimulated, the European clock-making with which they were familiar. Their chief concern was to point out its imperfections, in the interests of 'Western' clockwork, and hence by implication of the 'Western' religion which it was their primary aim to propagate.

[2] Wang Chen-To (1), who has gone into this question, concludes that the hour-glass, like the mariner's compass card (but not of course the compass itself) was transmitted to Chinese and Japanese use from Portuguese and Dutch trading ships towards the end of the sixteenth century. If then we find references to hour-glasses in translations of pre-Ming texts we can probably assume a technical inexactitude on the part of the sinologist. This is at any rate true in one case which we have checked—Sun & de Francis (1), p. 224, translating an account of Szechuanese farming by Kao Ssu-Tê [697] (fl. A.D. 1241). Here the 'hour-glasses' used to time the shifts at seasons of intensive work are only the usual portable farmers' clepsydras [401]. Cf. the picture and description in *Nung Shu* [827] (A.D. 1313), ch. 19, p. 20*a*.

Passing now from the first to the last years of the Ming dynasty, we find the Chinese astronomers who were the friends and collaborators of the Jesuits discussing the making of new equipment for the imperial observatories. In the course of this they reveal a whole tradition of sand-rotated wheel clocks starting from the end of the fourteenth century. In the same chapter (ch. 25, p. 18*b*) we find the following:

In the seventh year of the Ch'ung-Chên reign-period (A.D. 1634), Li T'ien-Ching [688] reported that Hsü Kuang-Ch'i [648] said that for determining time there had been anciently the clepsydra, but now there was the 'wheels-and-bells' (instrument). Both required human attention, and were therefore not so reliable as taking time from the movements of the heavens themselves. He therefore petitioned that three kinds of instruments, sundials, stardials, and telescopes, should be made. So the emperor authorised him to take charge of the matter.[1]

Then (p. 19*b*) the historian continues:

In the following year (1635) (Li) T'ien-Ching suggested that sand-clocks [441] should be made. At the beginning of the Ming (i.e. about 1360–80), Chan Hsi-Yuan [689], finding that in bitter winters the water froze and could not flow, replaced it by sand. But this ran through too fast to agree with the heavenly revolution, so to the (main driving-) wheel with scoops [442] he added four wheels each having thirty-six teeth [219]. Later on, Chou Shu-Hsüeh [690] criticised this design because the orifice was too small so that the sand-grains were liable to block it up, and therefore changed the system to one of six wheels (in all), the five wheels each having thirty teeth [219], at the same time slightly enlarging the orifice. Then the rotation of the machine really agreed with the movement of the heavens. What (Li) T'ien-Ching now petitioned for was surely this design deriving from (Chan and Chou).

Thus we gain a new name (hitherto unknown) of much importance for our history, Chan Hsi-Yuan, active about A.D. 1370. Chou Shu-Hsüeh, on the other hand, who flourished between 1530 and 1558, has long been quite well known as a mathematician, astronomer, cartographer and surveyor. Such is the general background of the sand-clock period. One might even be tempted to see the influence and encouragement of the first Ming emperor for a form of clock which could be easily constructed in all provinces and prefectures, rather than for those spectacular monumental types which had ornamented imperial palaces in former times.

The passage which throws most light on the Ming sand-clocks was found by Liu Hsien-Chou in a collection of the best essays of the dynasty. Its writer was

[1] The historian continues here with an account of the telescope, which he recommends as useful for artillerists as well as astronomers.

none other than Sung Lien [636], the famous scholar whom we have already met as the author or chief editor of the *Yuan Shih* (p. 134 above).

Ming Wên Ch'i Shang [830] (Collection of the Best Essays of the Ming Dynasty); the fourth part of *Ku Wên Ch'i Shang* [831] (Collection of the Best Essays of Former Times); machines section, ch. 2, pp. 53 ff. Ed. by CH'EN JEN-HSI [691] (A.D. 1581–1636)[1]

Wu Lun Sha Lou Ming Hsü [832] (Essay on an Inscription for a Five-Wheeled Sand-Clock), by SUNG LIEN [636] (A.D. 1310–81), *c.* 1375

The manner of the sand-clock is this; there is a container (*ch'ih* [42]) for very fine sand, from which it pours forth into the buckets (scoops, *tou* [442]). The movement consists of five wheels. The first axle is 2·3 ft. long, its circumference being 1·5 in. At the end of this axle, which is set horizontally [443], there is a wheel of circumference[2] 1·28 ft., having sixteen scoops on its rim, each of which is 0·8 in. broad and as much in depth.[3] The (other) end of the axle has (a small pinion wheel of) six teeth. Sand, pouring into the scoops, makes them fall one by one, whereupon these teeth turn also, and enmeshing with the second wheel, rotate it.

The axle of the second wheel is 1 ft. long, with a circumference the same as the former, and the wheel itself, vertically fitted [444], has a circumference of 1·5 ft. and thirty-six teeth. At the other end of this axle there is again (a small pinion wheel of) six teeth, which enmesh with the wheel, and rotate it.

The circumference of the third wheel and the length and circumference of its axle are identical with those of the second. Like the first axle, too, this bears at its end (a small pinion wheel of) six teeth, which enmesh with the fourth wheel, and rotate it.

The fourth is of the same dimensions as the third, but it is set up vertically in line with the second. It has again (on its axle) six (pinion) teeth at the (other) end, which enmesh with the 'middle wheel' and rotate it. (In dimensions) this is like the fourth, but while all the others are mounted vertically, this one alone turns horizontally. Its axle rises 1·6 ft. high and has no teeth at its end; instead of this it passes through the (centre of the) 'time-measuring dial' (*ts'ê ching p'an*, probably for *ts'ê ying p'an* [445]). On this dial are inscribed the twelve double-hours [143] and the 100 'quarters' [21]. The dial is carved round like the sun, and a pointer with cloud decorations is fixed to the axle. The five wheels enmesh with each other [446], transmitting the motion with progressive retardation [447]. In one day the 'middle wheel' makes one revolution around the dial, and we know that wherever the end of the cloud pointer stands, such and such an hour and quarter is indicated.

[1] Ch'en Jen-Hsi produced an encyclopaedia of some worth in 1632 and edited the *Comprehensive Mirror of History for Aid in Government* of Ssuma Kuang and Chu Hsi, with all its continuations and appendices. We are indebted to Dr Liu Hsien-Chou for a MS. transcript of this passage, for Ch'en's collection is not available in Cambridge.

[2] We are not sure that the writer did not mean 'diameter' here, and perhaps at some of the other points of his description.

[3] Perhaps 'length' was intended here.

All the other wheels are attached to the axles in pairs at measured distances apart (or, appropriately spaced [448]); only the 'middle wheel' is not. The wheels and the sand-tank are all concealed within the casing; only the dial and pointer are exposed. At the side (of the clock) there are two carved figures of boys dressed in yellow, one beating a drum and one sounding a gong. As for the movement of the sand, it ought to be checked each day.[1] Such is the general outline of the mechanism.

Formerly in Luan-yang [874][2] the water frequently froze, and even though often heated, it still would not drive water (-wheel) clocks. Thus it was that Chan Hsi-Yuan [689] of Hsin-an [875][3] brought his ingenuity to bear, and substituted sand for water. When his (prototype) clock had been completed, everyone said that such a thing had never before been heard of. Indeed it did not need to fear comparison with the Seven-Jewelled Illuminated Water-Clock[4] of Kuo Shou-Ching [635] (*c.* A.D. 1270), which automatically sounded the (double-) hours with bells and drums. There was also Chêng Chün-Yung [692] of P'u-yang [876],[5] who travelled to the capital with (Chan) Hsi-Yuan and therefore knew all the details of his design. After he returned he made (more of these clocks) and asked me for an inscription (for one of them). So I wrote the following words:

> 'The "Hoisters of the Water-Pots" were ancient men of State,
> And as of old the time they told by Water's constant rate,
> But wintry ice spoilt their device—the water turned to land—
> Till Chan by Earth did conquer Earth, and moved his wheels with sand;
> Which being neither firm nor flood, keeps faith with Heaven's round
> And makes the jacks of Master Chêng to beat their rhythmic sound.
> So now Stone flows where Water can't, ignoring fire and frost—
> Good people all, mark well the dial, and grudge each moment lost.'

This interesting description raises numerous questions. We also should mark well the dial. For it is something rather unexpected to find it appearing in fully modern form, that is to say, stationary with a circulating pointer, here in China in the latter part of the fourteenth century, just about the same time as it makes its appearance in Europe. Since clocks were still not very common in either civilisation it seems rather hard to believe that either transmitted the stationary dial-face to the other. Evidence has been given above (pp. 67 ff.) that the anaphoric clock with its rotating dial was not unknown in China in earlier medieval times, so

[1] There is something queer about this sentence, but we think that this is what it means.

[2] A Chin Tartar city in Hopei province north of Peking, perhaps a temporary capital at the beginning of the Ming dynasty.

[3] A city on the shores of the Pearl River estuary south of Canton. Perhaps significantly, it has long been famous for the manufacture of geomantic and navigational magnetic compasses.

[4] Undoubtedly the same as the Illuminated Water-Clock of the Ta Ming Hall [985].

[5] A town on the P'u-chiang river, a tributary of the Ch'ien-t'ang south of Hangchow, in Chekiang province. Southern Chekiang was always associated with the production and working of iron and steel in many small centres.

perhaps it is easier to suppose that the stationary dial-face was a parallel and independent development. Even more important is the more or less parallel divorce of the clockwork mechanism from the astronomical components, and its assumption of a purely time-keeping function. As we have seen (p. 135 above), the first Chinese mechanical clocks to abandon astronomical spheres and globes, confining themselves to the auditory and visual reporting of time, were those made for Khubilai Khan by Kuo Shou-Ching in 1262 and 1270. This was only a few decades before the first appearance of mechanical clocks 'as such' in Europe.[1]

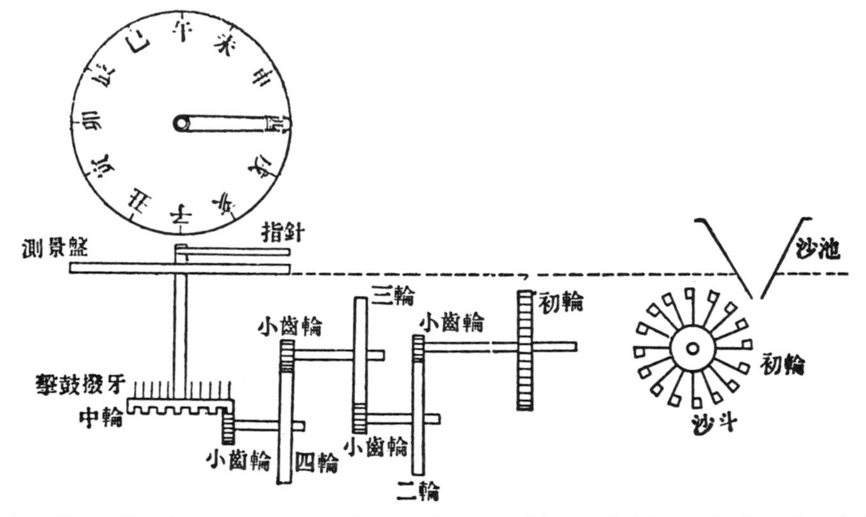

Fig. 58. Liu Hsien-Chou's reconstruction of the mechanism of the sand-driven wheel-clocks of Chan Hsi-Yuan (*c.* A.D. 1370). Its five wheels comprised the driving-wheel with scoops, three large gear-wheels, and one 'middle wheel' on the pointer shaft. The later modification by Chou Shu-Hsüeh (*c.* 1545) added a fourth large gear-wheel and changed the gear ratios. The jacks are not shown in the diagram, but they worked off the lugs on the 'middle wheel'.

Did the clocks of Chan Hsi-Yuan and Cheng Chün-Yung have an escapement? In his diagram of their mechanism, here reproduced as Fig. 58, which gives a very good idea of the arrangement, Liu Hsien-Chou (3) did not incorporate one, though he believed that one was present. The only words in the text by which he justified this view are in a phrase already noted [443], where the word *hêng* occurs just as it does in *t'ien hêng* [62], the upper balancing lever in Su Sung's clock. But

[1] It is true that the astronomical components had already been reduced to a very minor role in some of the Byzantine and Arabic striking water-clocks. In al-Jazarī's clock of 1206 (Fig. 34) the zodiacal signs were much overshadowed by the puppet orchestra, and in the Gaza clock of 510 Helios the sun-god alone, wending his way past the jack doors as the day waned, represented the astronomical interest.

the similarity is only very vague, and the passage seems to make much better sense if it is taken as we have rendered it, an 'axle set horizontally'. Grammatically this contrasts well with the parallel expression a little further on, a 'wheel set vertically'. We think, therefore, that the text gives no authority for assuming an escapement, and that perhaps what Chan Hsi-Yuan did was to invent something more original than Sung Lien realised, namely reduction-gearing clocks. In Chan's model each scoop (about thumb-size) must have filled in about 4·165 sec., and the scoop wheel turned the clock-face pointer by a gear train of 1296 to 1 using four sets of wheels at a 6 to 1 speed-ratio. In Chou's modification each scoop must have filled in about 1·75 sec., and the scoop wheel turned the pointer by a gear-train of 3125 to 1 using five sets of wheels at a 5 to 1 speed-ratio. Of course, with the level of technique available to mechanics in the sixteenth century the resulting accuracy may well have deserved the strictures of the Jesuits of 1610, and no doubt the introduction of weight- and spring-driven clocks was by then greatly to be desired, but for the fourteenth and fifteenth centuries the work of Chan Hsi-Yuan was most remarkable. Reduction gearing had been known in principle to the Alexandrians, and for time-keeping machinery it was of course a commonplace in fourteenth-century Europe—Leonardo, for example, made sketches of clocks which show it (*Codex Atlanticus*, fol. 348 *v*).[1] But the thought of relying completely upon it for slow motion relayed from a prime mover does not seem to have occurred to contemporary Europeans. It is to be hoped that the sand-clocks of the Ming may be illuminated much further by the discovery of texts or even perhaps of material evidence.[2]

9 A KOREAN ORRERY

In the meantime we are able to illustrate one exceedingly interesting piece of mechanism which, though of workmanship probably as late as the eighteenth century, still stands as the representative of 2000 years of the East Asian tradition of astronomical clockwork, embodying many of its characteristic features.[3] This

[1] See Ucelli di Nemi (1), p. 111, Entry 125.

[2] When one of us (J.N.) was in Peking in the summer of 1952, the Museum of the Imperial Palace had on show remarkably complete sets of costume jewellery belonging to Ming princesses and recovered from recent excavations of their tombs. Now (1958) the tomb of the Wan-Li emperor himself has been opened and studied. There would be nothing at all impossible in the finding of a sand-clock as archaeological work proceeds.

[3] Or at least still stood, twenty years ago, when Rufus was able to examine it. One can only express the sincerest hope that this wonderful piece has safely survived the dreadful destruction of the recent war in Korea.

is a demonstrational armillary sphere constructed in Korea,[1] the mechanical drive of which includes a time-telling and striking device to complete its clock function. It was described by Rufus (1), from whom we may quote the following passage in explanation of Fig. 59.

A *Syen-Kui Ok-Hyng* (i.e. *Hsüan-Chi Yü-Hêng* [162, 163], cf. p. 18), or astronomical clock, was ordered by king Hyo Chong (or Hyojong, i.e. Hsiao Tsung [994]) in A.D. 1657. Hong Chu-Yoon (Hung Ch'u-Yin [649]) attempted its construction, but the result was not a complete success. Afterwards a magistrate, Choi Yoo-Chi (Ts'ui Yu-Chih [650]) by name, was found, who had contrived an instrument of that kind, so he was ordered to make one for the king. According to the *Mun Hon Pi Go* it moved automatically by the motion of water, and indicated very accurately the degrees of movement of the sun and moon as well as the time of day. A somewhat similar clock which is however run by weights, and is probably of a latter date, is (1936) at the home of Mr Kim Sung-Soo (Chin Hsing-Shu [651]), who kindly permitted us to photograph it. The length is about 4 ft., the height of the main part 3 ft. 3 in., and the width 1 ft. 9 in. The astronomical sphere has a diameter of 16 in. and the terrestrial globe about three and a half. The time of day is given by discs carried round on a vertical wheel (i.e. a wheel on a vertical axis) and displayed through a window at the side. The position of the sun is given on the band representing the ecliptic, which marks the twenty-four solar divisions of the year. The motion of the moon is indicated on the ring representing its orbit (i.e. its apparent path), which is divided by pegs to mark the twenty-seven (i.e. twenty-eight) lunar constellations (i.e. mansions, *hsiu*). The mechanism is driven by two weights, one for the wheels and gears of the time-piece, which is regulated by a pendulum with a simple escapement; and the other for the striking device, which contains several iron balls and releases them to roll down a trough and trip the hammer that strikes the gong. Then they are lifted by paddles on a rotating wheel to repeat the process. The geography of the terrestrial globe includes some sixteenth-century voyages of discovery. The maps of the Old World and of South America are quite natural—but North America appears to have suffered from the effect of an explosion, as it consists of several islands with very irregular outline.

One stands amazed at the fidelity to ancient tradition manifested in this Korean work. Let us list the lineage of its main features. We see (*a*) that the middle nest of rings is made to rotate, just as in Su Sung's instrument of A.D. 1088 (Figs. 6, 24, 30 above), and (*b*) by means of a rotating polar axis, just as in all the mechanised armillary spheres before that time from Chang Hêng in the second century A.D. onwards (cf. Fig. 21 above).[2] (*c*) The sphere has a horizon circle, characteristic of all Chinese armillary spheres after Chang Hêng's time. Then there is (*d*) the earth

[1] Evidence will be presented elsewhere (Needham (1), vol. 3, p. 682) that of all the environing countries of the Chinese culture-area, the Koreans were for many centuries in the Middle Ages and later those who took greatest interest in all forms of science and technology. Medieval Korean embassies to China were always concerned with acquiring the latest books on medicine or the best astronomical equipment, and as late as the time of the Jesuits this tendency was still in evidence.

[2] Cf. p. 132, n. 5.

Fig. 59. An eighteenth-century Korean demonstrational armillary sphere, with clockwork. This instrument embodies numerous features characteristic of medieval Asian horological traditions (from Rufus).

model placed centrally, just as it was in the instruments of Ko Hêng and Liu Chih (third century A.D.; cf. Fig. 40 above), and now it is marked (*e*) with the chief continents like that terrestrial globe brought to Peking from Persia by Jamāl al-Dīn in A.D. 1267 for the consideration of Kuo Shou-Ching. We even find (*f*) a special ring for the path of the moon, like that which was incorporated in Li Shun-Fêng's design of A.D. 633, as well as in others afterwards. Then within the casing of the mechanism there is a periodical release of balls (*g*) reminiscent of those which regularly fell from the mouth of Wang Fu's 'candle dragon' of A.D. 1120, or earlier still from the seventh-century steelyard clepsydras of Antioch, described in the tenth-century *Chiu T'ang Shu* of Liu Hsü (cf. Fig. 34 above). Not only that, but there are the descendants of Su Sung's norias (*h*) waiting to raise the balls up again to their reservoir so as to repeat the process indefinitely (Figs. 19 and 20 above). Finally (*i*) there is a window at which discs (the descendants of Su Sung's jacks of 1088) present themselves to report the time (cf. Fig. 9 above). It would be an instructive thing to have a replica of this whole instrument, as well as that of Su Sung, together with suitable historical explanations, in every great museum of the history of science and technology in the world.

10 CHINESE OPINIONS OF JESUIT CLOCKWORK

Lastly we come to the converse Chinese opinions of the Jesuit importations. It is extremely interesting to see that there were a few scholars at the time who clearly recognised that something very similar to the new clocks had been discussed in Sung books.

If we open one of the great modern Chinese phrase dictionaries, the *Tz'u Yuan* [801], compiled by Lu Erh-K'uei [669] and many other scholars, and first published in 1915, we shall find, on p. 1444, an entry for the words *kun-t'an* [394]. It runs as follows:

> According to the *Shih Erh Yen Chai Sui Pi* [802] (Jottings from the Twelve Inkstones Studio, written by Wang Chün [652] about 1885) striking clocks and watches come from western foreign countries. But the *Hsiao Hsüeh Kan Chu* [727] encyclopaedia (of Wang Ying-Lin [501], written about A.D. 1270 but not printed until 1299) quotes Hsüeh Chi-Hsüan [653] (A.D. 1125 or 1134–73) as saying 'Nowadays time-keeping devices [390] are of four different kinds. There is the clepsydra (lit. the bronze vessels [391]), the (burning) incense stick [392], the sundial [393], and the revolving and snapping springs (*kun-t'an* [394]).' With these springs wheels were used [395]. When the hours arrived the springs were (automatically) released and thus sounded the time [396]. Was this not like a clock or a watch? Thus it was that Juan Wên-Ta [654] in his poem entitled *Hung Mao Shih Ch'en Piao* [803] (The Chiming Watches of the

Red-Haired People) wrote: 'Some say that these are the same sort of things, *kun-t'an*, which one can read about in Sung books.'

This information, so far as we have been able to verify it, is quite correct. The quotation from Hsüeh Chi-Hsüan is readily found in ch. 1, p. 42*b*, of Wang Ying-Lin's encyclopaedia, but unfortunately no further explanation is given. In the few books of Hsüeh which have been available to us we have not been successful in tracing the original passage, which perhaps would tell us more. The ancient meaning of *kun* was something like a well-polished axle bearing, or to rotate with little friction; *kun tzu* [397] later on meant a roller used on a threshing-floor.[1] The second word, if a noun, meaning a pellet- or bullet-bow, also ancient, is pronounced *tan*; if a verb, to snap or pluck a stringed instrument, or to shoot with a pellet-bow, is pronounced *t'an*. It is rather tantalising, however, here at the end of our investigations, to come across a new technical term which is neither among the copious engineering vocabulary used in Su Sung's book, nor to be found among the mass of other texts from which we have been able to construct the framework of a history of clockwork in China. The circumstances, however, make it overwhelmingly likely that in due course it will appear in some text or texts which will augment this history, and that the writer of the entry was quite justified in supposing that wheels were concerned. They would indeed have been Su Sung's jack-wheels, releasing the springs of the bells and drums operated by the time-announcing figures as they made their rounds. As for Juan Wên-Ta, who must have been a perspicacious reader of Sung books, perhaps Su Sung's among them, we have not been able to establish his date or identity either. In the light of all the data presented in the present monograph, his assumption was quite right, but it would be very interesting to know where his term *kun-t'an* came from.

The *Tz'u Yuan* was by no means the only dictionary which took account of Hsüeh Chi-Hsüan's important statement. It did not escape the eye of Wang Jen-Chün [655] who in 1895 was searching for evidences of old Chinese inventions to counter the all-inclusive and intimidating claims of occidental science at its most confident. His book, the *Ko Chih Ku Wei* [804] (Scientific Traces in Olden Times), which was printed in the following year, was far from being a scholarly production, yet it managed to quote (often correctly) many valuable passages. Talking of clepsydras (ch. 1, pp. 10*b* ff.) he gives the quotation from Hsüeh and then goes on:

[1] It is also to be noted that, according to the lexicographers, the word *kun* is interchangeable with *hun* [398, 399]. If this were to be adopted, the meaning would perhaps be 'the springs connected with the celestial globes and armillary spheres'.

The Western striking clock (*Hsi-Yang tzu ming chung* [400]) is derived from the clepsydra [401]. The *Ch'ou Jen Chuan* [770] says that the *kun-t'an* was in fact the same thing as the *tzu ming chung*, and that we had them already before the Sung.

Here again, in spite of considerable searching, we have not been able to find this passage, but there is no reason for doubting that Juan Yuan [656] (A.D. 1764–1849), the scholarly author and editor of the *Biographies of Mathematicians and Scientists*, did say this somewhere in his voluminous book or its appendices (1799). The remark that mechanical clocks went back not only to the Sung, but before the Sung, is particularly acute.

Wang Jen-Chün continues:

Fêng Shih-K'o [657][1] says in his *P'êng Ch'uang Hsü Lu* [805] (Continuation of the Records of the Weed-Overgrown Window)[2] that 'Li Ma-Tou [992] (Matteo Ricci) produced chiming clocks (*tzu ming chung* [214]) (or watches) which looked like small well-made gold boxes. They marked twelve hours in the day, and therefore struck twelve times. It was really something unique.' Then Wang Shih-Chên [658],[3] in his *Ch'ih Pei Ou T'an* [806] (Occasional Discussions North of Ch'ih-chow)[4] (1691), says that 'in Hsiang Shan Ao [873][5] there is a Standard Time Tower (*Ting Shih T'ai* [402]) which has a huge bell underneath it. On a platform with images of immortals at the corners (*Fei Hsien T'ai yü* [403]) there are jacks which strike the bell [404] and which are operated by revolving machinery [405]. ((It sounds according to the (double-) hours,[6] striking once at the end of *tzu* (11 p.m.–1 a.m.) and twelve at the

[1] Fêng Shih-K'o took his *chin-shih* degree in A.D. 1571 (*Ming Shih*, ch. 209, p. 7*b*; cf. d'Elia (1), vol. 2, p. 314).

[2] This book of his was probably a continuation of the *P'êng Ch'uang Lei Chi* [807], written by Huang Wei [659] in 1527, which was also concerned with strange things. Fêng's book must have been written within the first two decades of the seventeenth century, and he almost certainly knew Ricci personally as he speaks of the light folding fans, perhaps from Japan, which he used to present to his Chinese friends (d'Elia (1), vol. 1, p. 35; Forke (1), vol. 2, pp. 490 ff.).

[3] Wang Shih-Chên was a famous literary man, poet and official (A.D. 1634–1711). To read the biography of him by Fang Chao-Ying in Hummel (1), p. 831, is to savour all the atmosphere of the life of a charming scholar of the old style, rising to high rank and title, and then upon being deprived of all these honours by some administrative upheaval, quietly retiring to his home in Shantung to read and write during the last years. Nor was he without interest in techniques, as we see from the present passage, and from a description he gave of ship-mills which he saw working when he was on an official journey to Szechuan.

[4] The book contains miscellaneous notes, comments and criticisms, partly connected with things seen on the author's travels.

[5] This was a place just south of Ningpo on the coast. The context, not quoted by Wang Jen-Chün, shows that the tower clock with jacks was that of a Christian church, dedicated to the Blessed Virgin, where there was also a wind-organ and a telescope. Whether its existence and maintenance were due to the Jesuits we do not know, but this was a place where they were more than once in hot water (Pfister (1), pp. 369, 827).

[6] This phrase, apparently so unimportant, is nevertheless of much interest, for it shows how the occidental 24-hour system fitted perfectly the age-old Chinese system of 12 double-hours. Both were equal and uniform time-intervals, not depending on the season's varying lengths of day and night. Unequal 'canonical' hours had lasted till the fourteenth century in Europe, but apart from the night-watches, unequal hours had been abandoned far earlier in China.

beginning of *wu* (11 a.m.–1 p.m.), then going on from the end of *wu* when it strikes one, to the beginning of *tzu* when it strikes twelve. Day and night follow each other on the ring without the slightest error, a pointer indicating the twelve signs on a round dial. At every hour the bell sounds, the figures of the toads, etc. move, and the pointer indicates a particular position.'))[1]

Again, Wang Jen-Chün goes on:

Mr Yü Yüeh [660][2] in his *Ch'a Hsiang Shih San Ch'ao* [808][3] (Three Notebooks from the Tea-Perfume Arbour) says that 'if we understand rightly the (*P'êng Ch'uang*) *Hsü Lu* it seems that China had the beginnings of the striking (mechanical) clock (*tzu ming chung* [214]). But striking only twelve times in one day (and night) was different from what we have now. If we understand rightly the (*Ch'ih Pei*) *Ou T'an* it is clear that when the hours arrived (the machine) responded to them. When the end of *tzu* (11 p.m.–1 a.m.) is reached the clock strikes one, and when the middle[4] of the (midday double-hour) *wu* (11 a.m.–1 p.m.) is reached it strikes twelve. Then it begins (counting up) again from the end of *wu* till it comes to the middle[4] of *tzu*, the pointer having made (two) circuits of the twelve *ch'en* [151] of the dial [406]. At a given time the pointer arrives at a destined position. Such is the present striking clock. But it used to strike twelve at the beginning of *tzu* and the beginning of *wu*, which is a little different from the present system.'

These passages seem to throw some light on the adjustments which took place as the European 24-hour system mingled with the traditional Chinese double-hours. But there are many other things of interest in Wang Jen-Chün's book. Elsewhere, for example, he quotes (ch. 5, p. 28*a*) a scholar and official, Wang Chih-Ch'un [674], who about 1885 wrote an account of his embassy to Russia. Discussing the history of science and technology in East and West in his *Kuo Ch'ao Jou Yuan Chi* [818], ch. 19, he said: 'The automatically striking clock was invented by a (Chinese) monk, but the method was lost in China. Western people studied it and developed refined (time-keeping) machines. As for the steam-engine, it really originates from the (monk) I-Hsing of the T'ang (dynasty), who had a way of making bronze wheels turn automatically by the aid of rushing water—all that was added was the use of steam and the change of name....'[5] Wang Chih-Ch'un's perhaps somewhat exaggerated reaction was assuredly dismissed as chauvinistic by any westerners who ever came to hear of it, but we can now see what good justification he had for some of his ideas.

[1] Wang Shih-Chên's passage within double brackets is not quoted by Wang Jen-Chün.

[2] Yü Yüeh (A.D. 1821–1907) was a philological scholar who taught for more than thirty years at a college in Hangchow. See the biography of him by Tu Lien-Chê in Hummel (1), p. 944.

[3] This would have been written about 1890.

[4] Emending 'beginning' *ch'u* [407] to 'middle' *chung* [408] as the modern system is being described.

[5] Here Wang Chih-Ch'un undoubtedly had in mind primarily the paddle-wheel steam-boat, and confused the reciprocating engine with the steam turbine developed by de Laval and Parsons from 1882 onwards.

From all the foregoing quotations it can be clearly seen that during the first impact of the science and technology of Renaissance Europe brought to China by the Jesuits, there were some Chinese scholars who remembered enough of Sung literature to be sure that clockwork was not something fundamentally new. This is all to their credit. But the complicated situation of the time had its effects upon the thinking of Europeans in much later times, and in particular on that of European historians of science. The state of science and technology in Ming China was in fact decadent, and what was left of it the Jesuits were not at all disposed to emphasise. They wanted to be thought of as converting a highly cultured people, but the greater the superiority of the 'new, or experimental, science' of the West over the traditional empirical science of the Chinese, the greater, by implication, was the superiority of the Western religion. Not until the time of Antoine Gaubil,[1] and then (it must be admitted) only in a rather patronising way, did the Jesuits make a serious attempt to study the history of the sciences of the Chinese culture-area. Yet from the beginning, and in spite of all persecutions, they had been extremely welcome in their capacity as scientific men. In an interesting passage of Ricci's memoirs not included by Trigault in his book, he wrote:[2]

What with the great facility, commodity and freedom of printing, one can see wherever one goes in Chinese houses, how enthusiastically they collect books, much more than is the case with us. And by the same token they print far more books in any year than any other nation. And since they lack our sciences, they make many on other matters, useless and even harmful. But for them there was the greatest novelty in our tidings, telling so much of our own, and of all other nations, what with new laws (of religion), new science and new philosophy, so that much about us came to be printed in their books—partly concerning the arrival of the Fathers and the things we brought with us, pictures, clocks, books, descriptions and mechanical things, partly concerning the laws and sciences that we taught, partly concerning the printing of our books or quotations from them, partly concerning the many epigrams and poems which were composed in our honour. And stories true and false concerning us flew about to such an extent that there will be great memory of us in this kingdom for centuries to come—and what is better, good memory.

[1] Antoine Gaubil (A.D. 1689–1759) was a Jesuit missionary in China from 1723 onwards. He had had a considerable training in astronomy under Cassini and Maraldi at the Paris Observatory, and after his departure from France carried out what may truly be called titanic and indefatigable labours in acquiring an almost perfect knowledge of Chinese, collecting all possible texts bearing on astronomy, conferring with the few Chinese scholars of his time who were proficient in astronomy and mathematics, and himself making astronomical observations. So perfect was his knowledge of Chinese and other Asian languages that he was frequently called upon by the emperor to act as verbatim interpreter in State interviews. His chief works were the *Brief History of Chinese Astronomy* and the *Treatise on Chinese Astronomy*, both of 1729; a revision of the *History* finished in 1749, and a *Treatise on Chinese Chronology* also dating from this year but not published till 1814. Still today his works are valuable for sinological history of science.

[2] D'Elia (1), vol. 2, p. 314.

Thus novelty was the keynote of the Jesuit experience. Not unnaturally it became the keynote of the picture of the time formed by subsequent historians.[1] Towards the end of the note on clocks in his incomparably learned *Introduction to the History of Science*, George Sarton, one of the greatest among historians of science, crystallised the accepted view in the following words:

I may add that mechanical clocks were not discovered independently in the Far East. The Chinese (and Japanese) were of course familiar from early days with sundials, they had obtained some knowledge of clepsydras (from the Roman West?), and they had learned also to measure time with burning tapers or candles; but they never thought of mechanical clocks until some were shown to them by missionaries....The application of the clock to Far Eastern usage implied some difficulty, because the Chinese (and the Japanese) did not divide the day into twice twelve equal hours, but into two unequal periods (day and night) each divided into six equal divisions....[2] Hence the introduction of mechanical clocks necessarily created a deeper disruption of ancient usage in the Far East than in the Christian West.[3] The fact that the Chinese (and the Japanese)[4] did not invent them need not surprise us; it is more remarkable that they

[1] E.g. in 1956 Bedini (2) could write: 'A careful examination of the scant information that can be collected reveals, first of all, that the Chinese apparently did not invent or develop any special forms of time-pieces of their own at any time, and that they did not even produce good copies of the clocks of other countries. It is obvious that in spite of the many scientific achievements of this ancient civilisation, and its many contributions to the arts, the science and art of horology were not numbered among them and did not suit the Chinese temperament....The history of Chinese civilisation from its beginnings reflects a timelessness, indeed a lack of preoccupation with time in any form. Travellers in China in various ages reported that the natives felt no need to know the time except approximately, and this indifference continues in China to the present day. It is consequently not altogether surprising to learn that the Chinese never produced any competent clock-makers.' Apart from this devastating misjudgment, the article of Bedini includes some interesting first-hand information on modern Chinese clocks, though otherwise based mainly on McGowan (1).

[2] This is a complete misapprehension. The statement is true only of the Japanese. As d'Elia (1), vol 2, p. 128, has clearly and correctly stated, the Chinese from ancient times onward (at least from the middle of the first millennium B.C.) divided the whole of the day and night (from midnight to midnight) into twelve equal double-hours and one hundred 'quarters'. Unequal hours, varying in length with the seasons, had been in use before that time, in the Shang and early Chou periods, but did not persist. For further discussion of this, see pp. 202 ff. below. Enshoff (1) and Sarreira (1) have maintained that Ricci introduced mechanical clocks keeping unequal or canonical hours to China, but there is not the slightest foundation for this statement.

[3] Again this is an untenable proposition. The unequal hours were dominant only in Japan. And why should the introduction of equal hours into Japan have caused more disruption than the supersession of the canonical hours by equal hours in fourteenth-century Europe at the time of the first weight-driven mechanical clocks? The close association between these clocks and time-keeping by equal hours is well known; cf. Howgrave-Graham (1).

[4] On Japanese unequal-hour clocks see conveniently Ward (1), pp. 42 ff., from whom we reproduce two pillar clocks with adjustable scales (Fig. 60). Good Japanese sources include Takabayashi (1) and Yamaguchi (1), but from the considerable western-language literature we may also cite Rambaut (1) and Robertson (1). European clocks were taken to Japan by the Dutch at the beginning of the seventeenth century, and owing to the subsequent closure of the country they long retained, indeed until about 1870, the archaic verge-and-foliot escapement. Since at midsummer a day hour was about 2·33 times as long as a night hour, Japanese clocks were often equipped with two of these escapements, the foliots being so adjusted that one set beat much more rapidly during the night. The change-over was effected automatically by the striking mechanism, but the foliots had to be moved once a fortnight into new positions as the seasons changed. After the introduction of balance-wheels and springs, the variation had to be transferred to the method of indicating the time, as is seen in Fig. 60.

PLATE XVIII

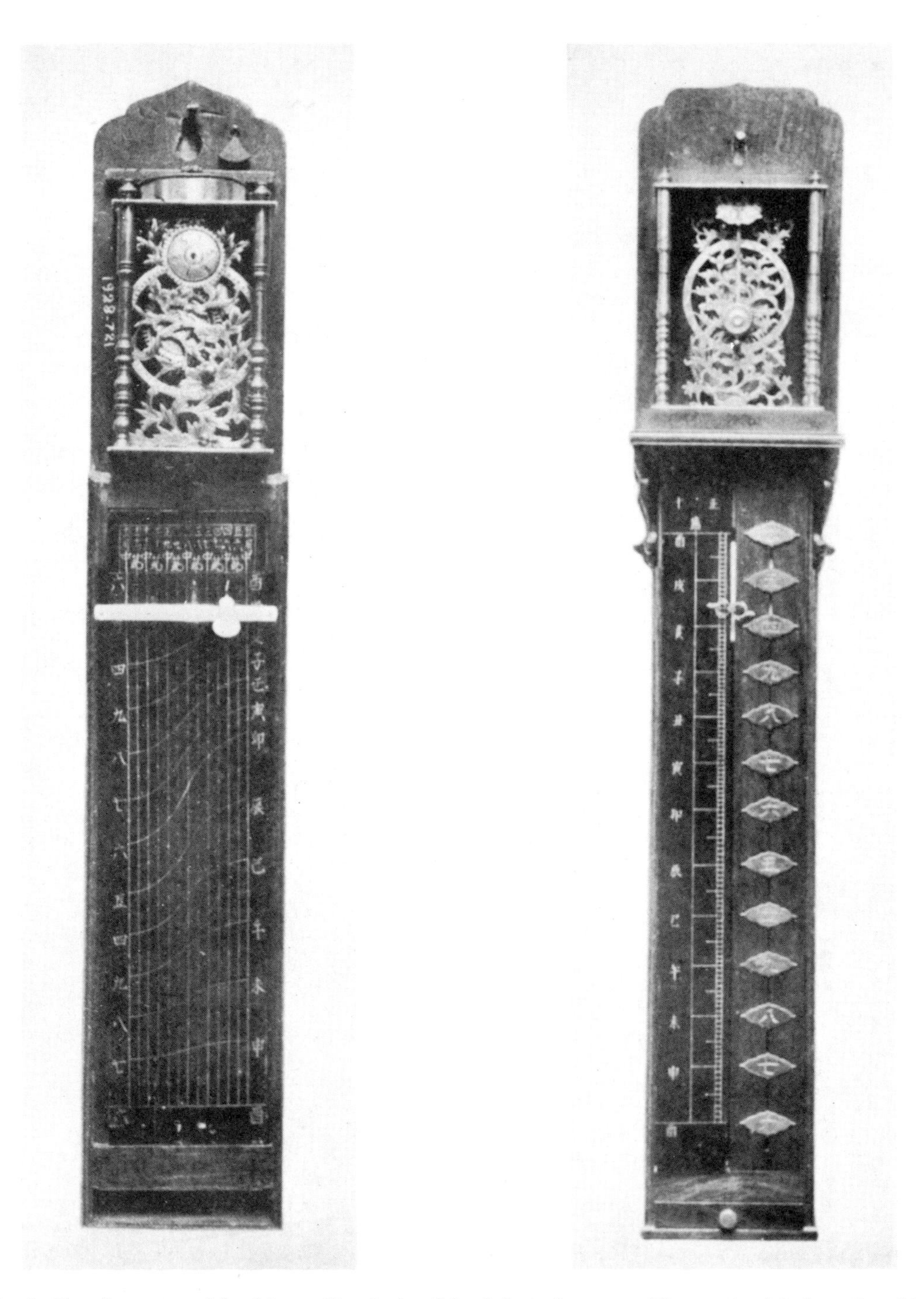

Fig. 60. Two Japanese weight-driven pillar-clocks of the eighteenth century. That on the right has adjustable 'hour'-marks, a separate graduated scale being provided for each pair of months. That on the left combines the time-scales for all the months on a single plate, the hour indicator being adjusted horizontally to bring it over the correct scale (from Ward). Only in Japan were unequal hours in use, and there they were retained until about 1870. In China equal hours had been universal since the middle of the first millennium B.C.

adopted them at all. This can be explained only by the necessity of adapting their ways to the Western ways forced upon them.[1]

Here out of eight statements or suggestions only three are right. This should be a warning to all of us of the provisional nature of all scholarship,[2] and of the danger of preconceived ideas about the comparative contributions of the cultures of East and West. Another scholar, equally learned, has written:[3]

> However far it may or may not be possible to trace back our Western mechanical trend towards the origins of our Western history, there is no doubt that a mechanical penchant is as characteristic of the Western civilisation as an aesthetic penchant was of the Hellenic, or a religious penchant was of the Indic and the Hindu.

In the light of the true history of clockwork this valuation seems most insecurely based.

And so we come back to Mr Fêng and Mr Yü talking about adjustments of clock time something analogous to our differences of an hour between summer and winter settings. They may seem rather an anticlimax after the bells and drums of twenty centuries, but there is nothing in it contrary to natural pattern. Before one could have 'summer time' it was necessary to have time itself. The prosaic necessarily follows upon the epic. The quiescence of the Yin follows upon the motion of the Yang—until the next cycle is unfolded by the timeless Tao of all things.

[1] Vol. 3, p. 1546. No Western ways were being forced upon the Chinese until two and a half centuries after Ricci's time. For the period of the Opium Wars and afterwards under treaty-port imperialism the phrase could well be used, but nobody forced the eighteenth-century emperors to accumulate thousands of clocks in their palaces, nor any of their subjects to use them. Disinterested curiosity reigned in good Father Ricci's golden days.

[2] One of us (J.N.) wrote in the first volume of *Science and Civilisation in China*, p. 243: 'The last important introduction (from the West to China) was clockwork, a distinctively European invention of the early fourteenth century.'

[3] Arnold Toynbee, in *A Study of History*, vol. 3, p. 386.

VIII

THE CONTEXT OF THE INVENTIONS

At the conclusion of this wide survey one question inevitably presents itself to the mind. Why did the invention of the first escapement, associated with water-wheel and linkwork, arise at the beginning of the eighth century and not at some other time? Why did the early part of the second see the attempts of Chang Hêng crowned with a preliminary success? Is it possible to point to any factors operating in their social environment at these times which would give some rationality to the achievements? For unless this can be done they remain pure freaks of chance, the actions of genius bombinating in a vacuum.

I THE VERMILION PENS OF THE LADIES-SECRETARIAL

It was, perhaps, the great medievalist Ducange who first proposed to explain the sudden interest in mechanical clocks at the beginning of the fourteenth century A.D. by the desire of the monks of the great abbeys to know more accurately the time of the night. So only could they adjust the hours of their lauds and matins. It is true that this explanation does not reveal why the desire was felt particularly at this time. But it was at any rate an attempt to set the invention of the weight-driven verge-and-foliot clocks in their social milieu, and find a reason for their rapid adoption. Exactly the same question can be asked about the invention of the water-wheel weigh-bridge escapement in T'ang China. What possible need could have been felt at such a period within the Chinese imperial palace—for it was always in close association with the emperor that these masterpieces of medieval engineering arose—for more accurate knowledge of the hours of the night, and for a means of following the march of the constellations even when foul weather rendered them invisible? It is true that time-keeping and stellar depiction were not the only functions of the water-wheel clocks; there was also the use of the mechanised globe as a calendrical computer. This was probably at least as important a determining factor as any other. We shall shortly speak of it, but first let us think only of the time-telling bells and gongs, and the slow (if rather jerky) rotation of the celestial globe with its map of the heavens.

It will be recalled that the Chinese emperor was a cosmic figure, the analogue here below of the pole-star on high. All hierarchies, all officialdom, all works and days, revolved around his solitary eminence.[1] It was therefore entirely natural that from time immemorial the large number of women attending upon him should have been regulated according to the principles of the numinous cosmism which pervaded Chinese court life. Ancient texts give us remarkable insight into the ranks of his consorts and concubines. Though their titles differed considerably during the two millennia which followed the first unification of the empire,[2] the general order comprised one empress [995], three consorts [996], nine spouses [997], twenty-seven concubines [998], and eighty-one assistant concubines [999]. The total adds up to 121, which (certainly by no coincidence) is one-third of 365 to the nearest round number. A classical passage in the *Chou Li* [764] (Record of the Rites of the Chou Dynasty) even gives us what might be called a duty rota. It runs (ch. 2, p. 20*b*):

> The lower-ranking (women) come first, the higher-ranking come last. The assistant concubines, eighty-one in number, share the imperial couch nine nights in groups of nine. The concubines, twenty-seven in number, are allotted three nights in groups of nine. The nine spouses and the three consorts are allotted 1 night to each group, and the empress also alone one night. On the fifteenth day of every month the sequence is complete, after which it repeats in the reverse order.

Thus it is clear that the women of highest rank approached the emperor at times nearest to the full moon, when the Yin influence would be at its height, and, matching the powerful Yang force of the Son of Heaven, would give the highest virtues to children so conceived. The primary purpose of the lower ranks of women was rather to feed the emperor's Yang with their Yin. In the ninth century A.D., Pai Hsing-Chien [670], the brother of the famous poet Pai Chü-I, complained that all these rules had fallen into disorder. In his 'Poetical Essay on the Supreme Joy' (*T'ien Ti Yin Yang Ta Lo Fu* [814]) he wrote:

> Nine ordinary companions every night, and the empress for two nights at the time of the full moon—that was the ancient rule, and the secretarial ladies [1000] kept a record of everything with their vermilion brushes....But alas, nowadays, all the three thousand (palace women) compete in confusion.[3]

[1] Cf. Creel (1); Granet (1); Soothill (1).

[2] Principal references: *Chou Li*, ch. 7 (tr. Biot (1), vol. 1, pp. 143, 154, 156); *Li Chi*, ch. 44, p. 42*a* (tr. Legge (4), vol. 2, p. 432); *Ch'ien Han Shu*, ch. 99 C, p. 23*b* (tr. Dubs (1), vol. 3, p. 438). These refer to the Han—for the T'ang see des Rotours (1), pp. 256 ff.

[3] Attention was first drawn to these texts, and many more translations made, by van Gulik (1), whose excellent monograph on the whole subject of sex physiology, custom and hygiene, in Chinese civilisation, should be consulted.

Now these secretaries were mentioned already in the *Chou Li* (ch. 2, see Biot (1), vol, 1, p. 158), i.e. in the second century B.C., and it is their activities which show us the relevance of these affairs to the history of clockwork.

What was at stake was the imperial succession. Chinese ruling houses did not always follow the primogeniture principle, and the eldest son of the empress was not necessarily the heir apparent.[1] Theoretically only her sons could be candidates, but this rule was constantly broken. Towards the end of a long reign, an emperor might expect to have quite a number of personable young princes from whom to choose his successor. In view of the importance of State astrology in China from the most ancient times, it may be taken as certain that one of the factors in this choice was the nature of the asterisms which had been culminating at the time of the candidate's conception.[2] Hence the importance of the records which had been kept by the ladies-secretarial, and the value of an instrument which not only told them the time, but from which one could read off the star positions at any desired moment.[3] Presages of stars, and of course also comets, novae, and other more unusual phenomena, would have been a factor of importance in the choice of the successor to the throne. To sum up the matter, careful records were a fundamental usage in the imperial family, and interest in recording machinery is therefore not at all surprising. But these facts, like the interest in horologia which the western monks doubtless had, fall short of giving a reason why the invention of the escapement, as the reply to a demand for more accurate mechanical time-keeping, should have come about just when it did, rather than at any other time.

[1] This has recently again been emphasised by the veteran German sinologist Ed. Erkes (1), commenting (p. 152) on the fact that the Yung-Chêng emperor (r. 1723–35) was a fourth son. It was never, he says, a foregone conclusion that the eldest son of the emperor should inherit; there had to be a choice when the time came.

[2] As is well known, ages of individual persons in China are still counted, not from birth, but from conception.

[3] The *Chin Shu* [775] (History of the Chin Dynasty), ch. 11, contains a remarkable corpus of Chinese State astrology. One star, for example, 'is the symbol of the commander-in-chief and the prime minister. It governs the readiness of the country to withstand attack, and the preparation of armaments.' Another 'denotes the officials in charge of ancestral worship, governing northern cities, temples, and all matters pertaining to ritual and prayer. It also governs death and lamentations.' One can readily see that in a dangerous time a prince whose conception had occurred under the aegis of the former asterism would have a good chance of nomination, other things being equal. A prince conceived under the latter star, however, would be somewhat handicapped at any time.

2 THE COMPUTING MACHINES OF THE CLERKS-CALENDRICAL

Let us see whether further light can be obtained from the function of the powered celestial globe as a calendrical computer. The nature of the procedure used has already been described in connection with the work of Chang Hêng. Comparison of the star transits in the heavens with those shown on the globe, the apparent positions of the sun, moon and planets being correctly shown on the model by means of appropriate movable studs or beads, afforded a means of detecting discrepancies between solar and stellar motions. Calendrical computations could thus be checked by a graphic method.

The promulgation of an official calendar was one of the most important acts of the Chinese emperor. It corresponded with the right of minting money in the West, and the acceptance of the official calendar implied recognition of Chinese imperial authority. About one hundred of these calendars were issued from the first unification of the empire in the third century B.C. until the end of the Ch'ing (Manchu) dynasty in the nineteenth century A.D. Each bore a specific name consisting of two or three characters, and there is a wealth of information about the dates of their introduction and the astronomers who compiled them. Unfortunately we lack any masterly survey of the whole in a western language[1] which would summarise the story, distinguishing between those calendars which simply involved new recensions of existing tables using new secular terms or radices in the calculations, and those which depended on new measurements so that new constants could be established and new tables made. However, the question may be asked whether there was any relation between the horological inventions and the frequency of introduction of new calendars. Were the inventions connected with what one might call periods of 'calendrical uneasiness'?

A preliminary answer is not at all difficult. There are several convenient lists of the Chinese calendars, notably one given by Chu Wên-Hsin (1). If we plot the number of new calendars introduced each century between 400 B.C. and A.D. 1900 we obtain the graph[2] shown in Fig. 61. The maximum can then be seen to have occurred in the sixth century A.D., when no less than fourteen new calendars were

[1] For references to the abundant literature, much of it in Chinese and Japanese, see Needham (1), vol. 3, p. 391.

[2] Cf. the analogous graph made by Price (3) for the intensity of production of astrolabes in Islam and Christendom between A.D. 900 and 1900; or the analysis by Zinner (1) of the categories of printed astronomical books between A.D. 1446 and 1630.

introduced.[1] But a correction is necessary. In order to approach our objective of true astronomical activity it is necessary first to exclude all new calendars which were merely changes of name, and secondly to exclude all those which were introduced to mark the inauguration of new dynasties. Since it was customary for each new ruling house to honour itself in this way, we cannot accept them as

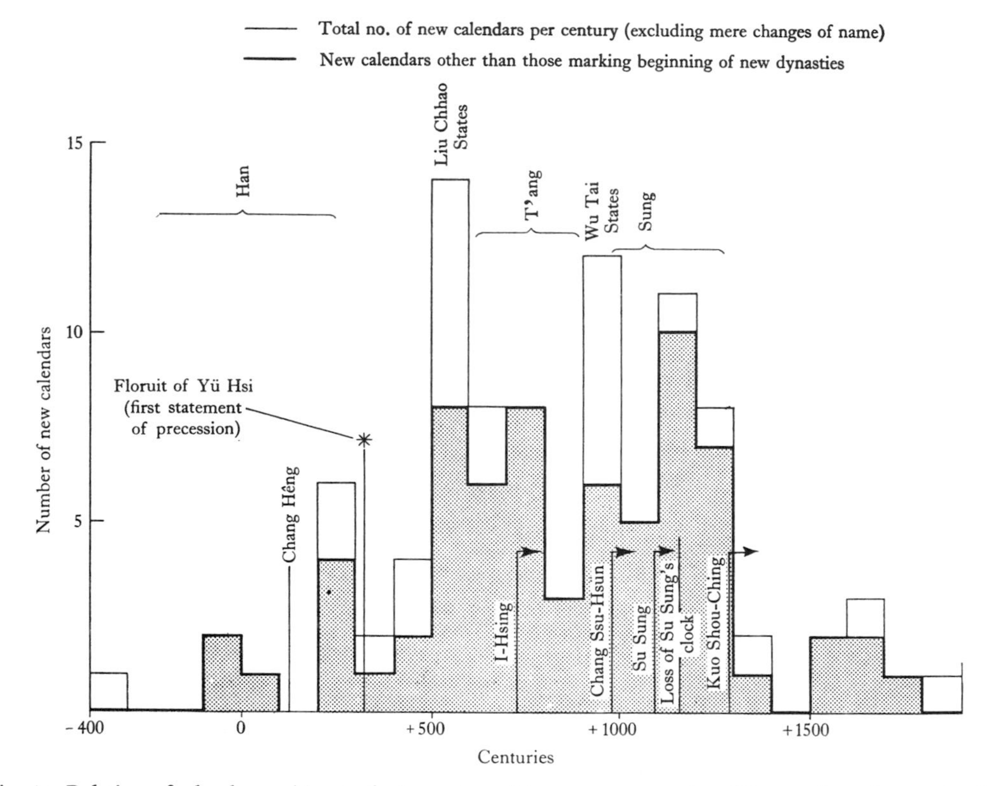

Fig. 61. Relation of calendar-making to the invention of the escapement; plot of the number of new calendars introduced in each century between 400 B.C. and A.D. 1900 (from Needham (1), vol. 4; data from Chu Wên-Hsin).

evidence of any unavoidable astronomical requirement. We thus have an array of smaller columns. Even these cannot be taken as the true index for which we are seeking, for new calendars were often introduced to signalise new individual accessions within the same dynasty or even changes of reign-period within the same reign. But the corrected columns may be regarded as a rough barometer of

[1] The calendrical discussions seem to have been most intense around the year A.D. 513. We have the names of some twenty calendar experts who held great debates at this time. Cf. *Ch'ou Jen Chuan* [770], pt. 4, ch. 3.

astronomical activity, for calendrical renewals at these shorter intervals by no means always occurred, and would indeed have been so inconvenient that one may assume that they often resulted from astronomical discussions and controversies. Frequently we know that this was the case—as in I-Hsing's own time.

The general picture seen in Fig. 61 is one of a rise to a double maximum (the sixth to the eighth centuries A.D. and the ninth to the thirteenth), followed by a fall. The obvious interpretation is that the problems were beginning to be posed in the Han, and that they had mostly been solved by the Ch'ing. Chang Hêng's invention occurs in the preparatory Han period, at a time of few calendars, but one must remember that he was setting in motion a technique which would have to be followed for many years to bring useful results. On the other hand, the invention of the escapement by I-Hsing comes just towards the latter part of the burst of 'calendrical uneasiness' and calendar-making activity of the Liu Ch'ao and T'ang periods. The great clocks of later times all come within the similar period of the Sung.

When the same material is plotted from A.D. 200 to A.D. 1300 in twenty-five-year periods (Fig. 62), the phases acquire further clarity. There is a relative quiescence to about A.D. 500, but from then onwards (apart of course from the spate of inaugurations) there is at least one calendar in each quarter-century, and once as many as four. The two or three preceding centuries had been a time of rapidly growing experimentation with astronomical instruments such as armillary spheres. Now also the foreign importations of Buddhism were beginning to take effect, and astronomical ideas were certainly among them. But the calendrical activity of this time belongs to a little-known period and deserves much further study. In any case, the uneasiness stimulated by foreign influences greatly increased in the T'ang, and at the beginning of the eighth century A.D. the merits of other calendars—Indian, Persian, Sogdian—were hotly debated at the capital. This was the very time at which I-Hsing and Liang Ling-Tsan, as if in answer to desperate demands for some more truly time-keeping machine, made their invention of the waterwheel link-work escapement.

Towards the end of the T'ang things were again quieter, but with the establishment of the Sung discrepancies must have become obvious once more, and we find the clock of Chang Ssu-Hsün (A.D. 979) coinciding with a quarter-century in which no less than three calendars were produced. Between 925 and 1225 no quarter-century passed without at least one new calendar, so the clock of Su Sung

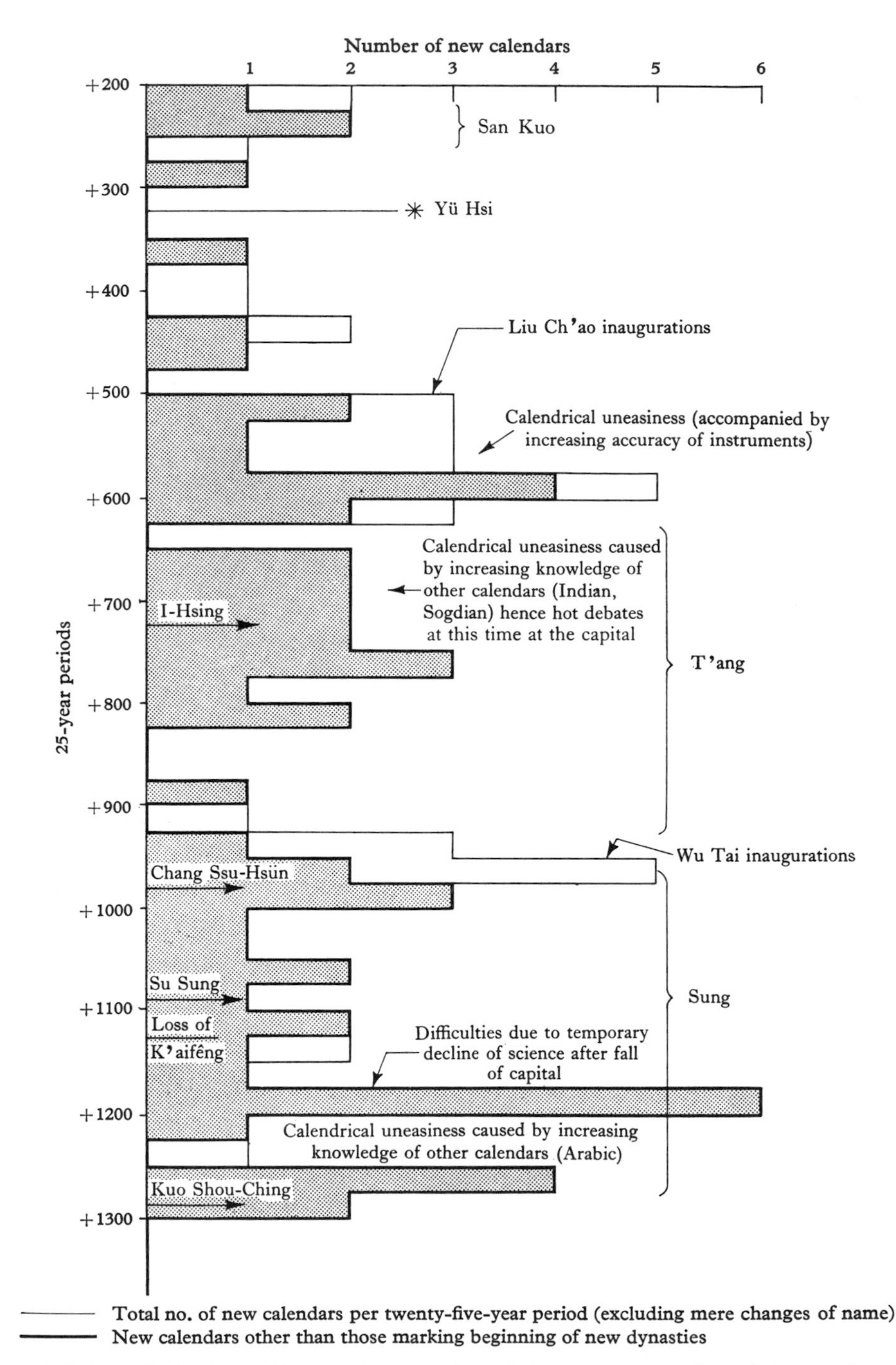

Fig. 62. Relation of calendar-making to the invention of the escapement; plot of the number of new calendars introduced in each twenty-five-year period between A.D. 200 and A.D. 1300 (from Needham (1), vol. 4; date from Chu Wên-Hsin).

(1088) finds its place naturally in this period. The all-time record of six calendars between 1175 and 1200 must almost certainly be a reflection of the harm done to astronomical science by the fall of the Sung capital to the Chin Tartars in 1126. Then in the last part of the thirteenth century A.D. there ensues a burst of activity which is readily explained by Arabic influence analogous to the earlier Indian and Sogdian. Eventually comes the lethargy of the Ming, and the beginnings of unified world astronomy with the Jesuits in the early Ch'ing.

Thus side by side with considerations arising from the private life of the imperial family we can place the needs of calendrical computation. It looks as if the invention of Chang Hêng about A.D. 130 derived from the growing doubts which had led Hipparchus to the discovery of the equinoctial precession in 134 B.C., and were to lead Yü Hsi [583] to state the same doctrine about A.D. 320. But by 700 the more accurate measurement of time had become a burning question, and the social need of a predominantly agrarian culture for an accurate calendar, as well as the intrinsic evolution of astronomical science itself, led to the answer found by I-Hsing. It is indeed a strange conclusion that an apparatus so deeply enrooted in western mechanical industrial civilisation as the clock should have originated partly in connection with the calendar required by an eastern agricultural people. But other perspectives no less remarkable must be mentioned. As we must often emphasise, Chinese astronomy was constructed on a polar and equatorial system, while Hellenistic astronomy was primarily ecliptic and planetary. Each had its peculiar advantages and its corresponding triumphs. If Hipparchus was able to state the fact of precession four and a half centuries before Yü Hsi, it was because he was measuring and comparing star positions on ecliptic co-ordinates, and therefore it became evident that their distances from the equinoctial points had changed. But if astronomical instruments were rotated mechanically by Chang Hêng fifteen centuries before the conception of the clock drive arose in Renaissance Europe,[1] and if in this I-Hsing with his first approximation to real mechanised time-keeping had a priority of nine or ten, it was because Chinese astronomers thought always in terms of equatorial co-ordinates and therefore of declination parallels. Along these lines indeed all stellar revolution proceeds, but following ecliptic latitudes no star ever moves. In China therefore it was an entirely natural thought to arrange the rotation of a celestial globe or a demonstrational armillary

[1] Cf. the interesting paper by Michel (2) on instruments ancestral to the planetarium. The earliest celestial globes rotated by clockwork date only in Europe from the sixteenth century A.D. Michel illustrates a fine one made by Christopher Schissler the younger about 1580; it shows automatically the motion of the sun on the ecliptic.

sphere if the plan promised to be useful.[1] What was not perhaps quite so easy was how to do it.

[1] How paradoxical it is, therefore, that traditionally the Greeks receive all the credit due to the first inventors of mechanically rotated astronomical instruments. 'It was necessary', writes Michel (2), discussing the limiting factors of this development, 'to await the geometrical genius of the Greeks, with their faculty of "seeing through space", in order to attain the resolution of the mechanics of the heavens as a logical combination of circular motions.' He is thinking, of course, of the Ptolemaic theory, with all its complexities, but in fact the Greek achievement in practice was confined to the rotation of flat discs by sinking floats, and it is unlikely that even the Antikythera planetarium or that of Archimedes (see p. 185 below) worked on any other principle. There is no doubt that the Greeks were the more successful theorists, but equally none that the Chinese, on the whole, were the better practical instrument-makers.

IX

GENERAL HISTORY AND TRANSMISSION OF ASTRONOMICAL CLOCKS

HAVING now traced the story of the great astronomical clocks in ancient and medieval China, we are in a position to review it against a larger perspective. It is of the greatest importance and interest to do this, for they belong to what was undoubtedly the earliest type of complicated machine, and probably also the most impressive of their times. Such devices of scientific technology have exercised not a little influence on the idea that the universe was a great mathematical machine whose workings could be comprehended by exact reasoning. Since astronomy and graphic representation are two of the most ancient of man's arts, it is no wonder that he should want to hold the cosmos in his hand by making a model of it—a painted ball, solid or made of hoops, which he might turn to illustrate the apparent rotation of the bowl of the sky.

Thus it is only to be expected that there is evidence in every ancient civilisation, almost from the beginning, of the making of astronomical models, and one may seek in vain for any which indicates the transmission of details or even of the idea itself. At the next level of complication two refinements appear; the simple ball becomes a more complex type of model designed to show more astronomical data, and some method is found to turn the model without crude handling by the operator. The first refinement comprises the whole history of armillary spheres, astrolabes and celestial globes; the second includes the development of mechanical time-keeping and its auxiliary arts of gear-work, jack-work and oscillating mechanism. Much has already been said throughout the historical section about the evolution of these separate parts of the problem. This evidence must now be augmented from sources outside China, and so drawn together as part of one single story which leads from the beginnings of astronomy to the most modern chronometer.

I THE EVOLUTION OF ASTRONOMICAL MODELS

There can be little doubt that the progress of astronomical model-making proceeded for the most part quite independently in China on the one side and the 'West' on the other. The 'West' must be understood here as comprising the successive high civilisations of the Hellenistic, Islamic and European regions. Somewhere between the two camps of astronomical horology one must place India—her position being complicated by the existence of an earlier tradition which, like the Hellenistic, probably derived in part from Babylonian culture, and a series of later influxes from Alexandria, Byzantium, Islam, and probably China as well.

In China, as we have seen, the armillary spheres seem to develop by easy stages from what appears to be the simplest possible type, having the minimum number of rings and of pieces of auxiliary apparatus.[1] This series of known armillary rings and later spheres stretches in unbroken lineage from the second century before our era unto the coming of the Jesuits, and in that lineage certain features consistently recur. The most striking characteristic is that almost all the spheres with axles and rotating parts (thus, all but the most primitive) set the axles along the polar axis. This is no accidental feature of mechanical design, but an inherent part of the format of Chinese astronomical theory. A second feature in the entire series is the use of a sighting-tube—a telescope without lenses—whenever an observing apparatus is required. In all other cultures a pair of pinnule plates was used instead, being either mounted on a straight bar so as to form an alidade (as in astrolabes), or mounted diametrally opposite each other (as in armillary instruments).

The difference in sighting techniques between the use of a tube and of a pair of holes seems unrelated to any exigencies of practice or theory, and the tube may well be associated with the ready availability of bamboo tubing in ancient Chinese civilisation. In any case, when one finds an ecliptic axis or pinnule sights in China there is good reason to presume that Western influence is at work; conversely, the sighting-tube found but seldom in the West may reasonably be thought to be imported from China. An example of the former possibility has already been discussed in the case of I-Hsing's armillary sphere of A.D. 725 (see p. 76), which had ecliptic pivots; the latter possibility is to be observed in the remarkable telescope-like sighting-tube used and described by Gerbert, *c.* A.D. 1000[2]—there is no

[1] Cf. Needham (1), vol. 3, pp. 342 ff.
[2] See Zinner (2), Eisler (1) and Michel (3).

material evidence to suggest that it derived from Chinese sources, but the device in this form is otherwise unknown in the West.[1]

Apart from these differences in points of detail, there is the historical consideration that the armillary sphere[2] was well developed and quite sophisticated in structure by the time of Ptolemy, *c.* A.D. 120, in Hellenistic Alexandria, at the very time when the evolution of the sphere in China was also advancing. One has the assertion of Ptolemy that he constructed the armillary *astrolabon*, but it may well be that Hipparchus (*c.* B.C. 150) knew some simpler form of the instrument and used it for his star observations, though positive evidence is lacking.[3] With this early start both in West and East it is rather unlikely that the Chinese series owes anything to transmission of ideas. If there had been transmission one would not find the gradual evolution from primitive beginnings in China.

The question is also linked closely with the different modes of astronomy in the Hellenistic and Chinese cultures. In the West, a series of happy accidents occurred soon after the arithmetically minded Babylonians had communicated their astronomy to the geometrically strongly-developed Hellenic scientists. These accidents of physical fact and mathematical structure had the effect of directing the best period of genius towards the mathematical analysis of planetary motions rather than to any other part of astronomy. From this there developed a subtle emphasis on motions in (and near) the plane of the ecliptic. This part of theory was more complicated, more beautiful and more impressive to them than any other part of the available corpus of astronomy, even perhaps more than any other part of their science. In China, on the other hand, the evolution was not disturbed by any comparable accident—they had no Babylonian race of calculators on their doorstep—and astronomy proceeded in what one must regard as a more 'normal' way, without overriding stress being placed on the importance of ecliptic measurements and ecliptic co-ordinates. In China the ecliptic was, of course, important as the course of the sun and the planets, but it was not a setting for the most important part of the theory.

Chinese armillary instruments necessarily show the ecliptic when it is required to indicate the paths of the sun and planets; their theory too takes it into account and develops to a good level the analysis of planetary periods and

[1] See Price (4).

[2] We are speaking here of the armillary sphere, not of the use of single armillary rings, the use of which must have long preceded the sphere in both civilisations (Greece and China). The ring must have been used at the beginning of the third century B.C. by Aristarchus and Timocharis, as in the fourth century by Shih Shen and Kan Tê [493, 494].

[3] Dicks (1) asserts it.

motions. It was, however, never central in their astronomical scheme as it was in the West. That is why the armillary sphere differs in detail and perhaps in rate of evolution in East and West. Even when astronomers like I-Hsing introduced ecliptic methods they could hardly become popular and generally accepted without a complete change in the framework and mathematical techniques of astronomical theory. This change did not occur in China until relatively modern times. That it occurred in the West at such an early date must be regarded as one of the most significant and crucial accidents of our whole civilisation; because of it Western science evolved in a matrix of mathematical theory favouring the physical sciences—without it Chinese science retained throughout a climate more like that of the biological and chemical sciences in the West.

What we have said about the armillary sphere applies equally to its observational and demonstrational forms, though in the West the purely demonstrational type may have been a quite late adaptation associated with the renaissance of astronomy during the second half of the fifteenth century. The Ptolemaic armillary spheres, intended principally for observation, were but little modified in Islam and they remained from first to last one of the most important of the larger instruments. In particular, descriptions of them were included in the corpus of auxiliary texts to the Toledan Tables of A.D. 1087 and the Alfonsine Tables of A.D. 1274. These became known in Europe by about 1200 and 1300 respectively, and the resulting double wave of transmissions brought almost all of what was known about astronomy before Renaissance learning made the original classical texts available. Not only did the armillary sphere return to Europe in this way, but the astrolabe and the globe too.[1] In addition, if the principle of the mechanical clock came to Europe from Islam it must have been transmitted in some such manner also.

From what little evidence we have it seems that the globe, like the armillary sphere, arose independently in China, or perhaps with only a minimal general stimulus transmission. The astrolabe, however, is peculiar to the West, and not a single mention, allusion or example occurs in medieval China; the reason is again the lack of any analytical techniques in their mathematical astronomy—the principle of stereographic projection used in the astrolabe constitutes quite a sophisticated example of the application of such techniques. This particular advantage seems to have misfired in the West, for although the astrolabe is so very

[1] It is curious that, although several medieval European astrolabes survive, there is not a single sphere or globe, celestial or terrestrial, older than the mid-fifteenth century.

ingenious, and perhaps the most useful calculating device known before the seventeenth century, it is no substitute for a globe. In the West it was always the wonderful astrolabe which was central in the great demonstrational clocks, and it was not until late in the Renaissance that anyone realised that the unsophisticated globe or solid model was more spectacular, more appealing to the lay mind, and just as useful mathematically as the flat map provided by the astrolabe or the anaphoric clock.

In India, the armillary sphere appears only in later texts of the twelfth century[1] and shows almost entirely the Islamic and therefore strictly Western features, pointing to an origin in common with the greater part of the rest of Indian astronomy of this period. It is, however, notable that the *Sūrya Siddhānta* describes the sphere as being sunk in a casing representing the horizon, thus following traditional Chinese practice in this respect. At such a late date, however, one must expect quite free mixing of Eastern and Western practice in India.

2 THE EVOLUTION OF POWER-DRIVEN ASTRONOMICAL MODELS

Now that we command so much acquaintance with the evolution of power-driven models in China, it is the Western history that must appear incomplete and confused. It is well known that the mechanical clock governed by an escapement was firmly established in Europe by the time of de Dondi's masterpiece of A.D. 1364, and that already even then it was found in association with highly complex astronomical dials. It is indeed strange that the earliest European clock of which we have a complete account should also be one of the most complicated pieces of mechanism ever recorded before quite recent times. Not until later, towards the end of the fourteenth century, can one find sure evidence of the crude and simple, purely time-keeping, clocks which, with all good reason, one would have expected to find at the head of the long historical procession.

Before the fourteenth century, it is agreed, there were all sorts of sundials and water-clocks—some of these latter having wheels and other working parts—but none of these things could be related, except by virtue of the purpose they served, to the ordinary 'weight and escapement' mechanical clock. Clepsydras and gear-wheels were known to the Greeks, but this has appeared to have little direct bearing on the course of horology in late medieval Europe. To historians of the subject, it is the invention of the escapement which has seemed to be not only fundamental but also curiously sudden and unexpected.

[1] In an insertion, perhaps dating from about A.D. 1100, in the *Sūrya Siddhānta* (orig. *c.* A.D. 500); and in the *Siddhānta Śiromaṇi* of 1150.

This situation was evidently most unsatisfactory, and it is clear that we must make every effort to bring the new Chinese evidence to bear on the recalcitrant problem. Even if one were satisfied that there had never been any sort of transmission in either direction between China and the West, the new material may afford a series of illuminating parallels where before there was only darkness. If transmission did occur, the former ignorance of the Chinese story might even explain why in spite of at least a century of intensive and scholarly research the mechanical clock still seemed to spring suddenly out of the ground in fourteenth-century Europe.

Upon examination, cross-cultural transmission and stimulus in the fields of astronomy and mechanical ingenuity may usefully be considered in two separate periods. It so happens that in the topics which most concern us, there was practically no progress in the West between the time of Ptolemy (second century A.D.) and that of the first great observatories and astronomers of Islam (*c.* 800 or even 900). In the interval between these dates, Hellenistic Alexandria declined and was overrun, while the barbarian invaders, slowly settling, assimilated its culture directly and also indirectly through Byzantium. Perhaps also, in this interval, some parts of that culture were handed on towards India. Because of this gap of some seven or eight centuries, one can say with confidence that instruments and techniques used before about A.D. 800 or 900 probably existed already in the Hellenistic period; one may equally well presume that things appearing after this date are probably new inventions, or transmissions from some culture other than the Hellenistic. It is therefore possible to consider on the one hand a Hellenistic corpus of knowledge and practice, and on the other hand, and quite separately, any changes made after a date in the ninth century.

3 THE HELLENISTIC BACKGROUND TO HOROLOGY

Summarising ground which we have already covered in detail, it is apparent from the works of Ptolemy, Philon, Heron, Vitruvius and similar writers[1] that many of the elements entering into the construction of later astronomical clocks were current and already in a high state of development in Hellenistic times. All the external trappings of the clocks were there—globes, spheres, 'astrolabes', jack-work and the love of impressive demonstrations of ingenuity by mechanics and natural philosophers. Moreover, the clepsydra had been brought to an

[1] See Drachmann (1).

equally high level of accuracy with devices to regulate the flow of water or change the scale so as to show the unequal hours of the day and night between the variable limits of sunrise and sunset.

But actual evidence of power-driven astronomical models is tantalisingly incomplete and unsubstantial. There are reports of Archimedes (*c.* 250 B.C.) having made a device which showed the planets and heavens rotating,[1] but although later Islamic texts describe 'Archimedean' clocks of more orthodox non-astronomical design[2] there are no details of the original device. One must also take into account the fantastically complex Antikythera machine with all its gearing, which seems very much like the mechanism of a planetarium; but it is so fragmentary after two millennia under the sea, that one can only conjecture about its nature.[3] There is also a curious passage in Heron's *Pneumatica*[4] where, in the midst of descriptions of various clepsydra-like performing vessels, there is inserted a chapter describing a model of the universe. It is merely a ball set in the space enclosed by a pair of larger transparent hemispheres and held there by being floated on water which half fills the enclosed space—it demonstrates the earth at the centre, the heavens surrounding, and the aether between. This is the only non-working model in the whole collection of Heron's works, and because its form is so garbled one cannot help thinking it may be but a fragment of what was once a description of a working model, rotated somehow by pneumatic or hydraulic force.

If it almost seems as if some mad dictator had contrived to expunge all details of Hellenistic astronomical models from the records, we can console ourselves with the knowledge that his influence did not extend to writings on the clepsydra. Fortunately there is a considerable corpus describing very sophisticated devices of this sort;[5] basically they use the flow of water to actuate mechanism in three different ways:

(*a*) By means of a float rising or sinking in the water, giving a linear motion to a cord or chain. This can be used directly or transformed into circular motion by winding the cord round a barrel axle or by using a rack and pinion. It will be recalled (pp. 64 ff., 80 ff.) that this expedient can only be employed when relatively little mechanical effort is required.

[1] See Ovid, *Fasti*, VI, 277, and Cicero, *De Re Publica*, I (i), xiv. There is a special study by Wiedemann & Hauser (2). This planetarium device must be distinguished from a striking 'alarm' clock with periodically dropping balls which has been variously ascribed to Archimedes, Plato, etc., on which see Drachmann (2), pp. 36 ff.; Wiedemann & Hauser (1), p. 25, fig. 4; and Kubitschek (1), p. 215.

[2] They seem mostly to stem from the Byzantine tradition which produced the famous clock at Gaza described about A.D. 510 by Procopius—see Diels (2).

[3] See the bibliography in Price (1).

[4] Ch. 46 (Woodcroft (1), p. 68).

[5] See M. C. P. Schmidt (1).

(*b*) By suddenly emptying a vessel filled by the clepsydra; this may be done either by using the movement of the centre of gravity of a bucket to tip it about its pivots, or by a siphon in which the water is allowed to reach the upper U-bend when the receiver is full, and so empty the whole accumulated supply of water into another vessel.

(*c*) By using a steady motion, obtained as in the first case, to work a trip-lever or some similar 'all-or-nothing' device; this is particularly frequently used in automata of the mechanical rather than the pneumatic variety.

It will be seen that the missing element is the combination of discontinuous action (tipping, tripping or siphoning) with a cumulative power-source which might be used to drive a clock—it is as if the Greeks had the pendulum (which of course they did not) but lacked the crown-wheel. The steady flow method could readily be employed to drive an astrolabic dial, and this important device, the anaphoric clepsydra, was probably the height of Western achievement in the matter of powered astronomical timekeepers.[1]

Because of a prevalent misconception it is perhaps necessary to emphasise that Hellenistic gearing did not extend to more than the simplest possible applications. Its greatest use was as a transmitter of power, the gear-wheels being employed in pairs. In the taximeter (or hodometer) the principle is that of a trip mechanism worked by a pinion-of-one on the carriage axle and on the subsequent shafts in series—this can hardly qualify as gearing.[2] Nowhere in the described automata does one see any ready use of gearing,[3] and nowhere at all, until later astronomical devices, was gearing used to provide differential rates of rotation.

[1] Fragments of plates belonging to two such anaphoric clocks have been excavated. They are both Roman, dating from about the second century A.D., and they were found in Salzburg and in the Vosges respectively. See Price (1), n. 14.

[2] As Liu Hsien-Chou (3), p. 5, has pointed out, the Chinese hodometer, which originated in the Han period if not before, is particularly relevant to the history of clockwork. For besides its method of construction, in which the gear-train driven by the road-wheel of the carriage included pinions of one or of three, its signalisation was auditory by means of jack figures beating a drum (cf. Fig. 43 above). For a full account of the history of the hodometer in China, see Needham (1), vol. 4, pt. 1.

[3] It is true that according to Vitruvius (IX, viii, 5 ff.) there were striking water-clocks in which an anaphoric drive was combined with gear-wheels, mainly used for jack-work. Ascribing these to Ctesibius of Alexandria (early second century B.C.), he says: 'First he made a hollow tube of gold, or pierced a gem; for these materials are neither worn by the passage of water nor so begrimed that they become clogged. The water flows smoothly through the passage, and raises an inverted bowl which craftsmen call the cork or drum (i.e. the float). The bowl is connected with a bar (i.e. a shaft) on which a drum (drum-wheel) revolves. The drums (i.e. gear-wheels) are wrought with equal teeth, and the teeth fitting into one another cause measured revolutions and movements. Further, other bars (shafts) and other drums (gear-wheels), toothed after the same fashion, and driven together in one motion, cause, as they revolve, various kinds of movement; therein

4 THE CHINESE BACKGROUND TO HOROLOGY (BEFORE *c.* A.D. 850)

It has already been noted that the armillary sphere appears to have evolved independently in China. These considerations apply almost equally well to the clepsydra, though here there may also have been transmission and stimulus at the very early, most crude, stages when it was little more than a simple leaking pot. The sloped sides of the cascaded pots in later Chinese illustrations (cf. Fig. 38) seem to indicate a derivation from early outflow clepsydras of the sort common in Egypt and probably Babylonia too during the second millennium B.C.[1]

Regarding the application of power to astronomical models, we know that by *c.* A.D. 850 there had arisen, not only the machine of Chang Hêng, but also that of I-Hsing. If the former might perhaps seem to have required nothing (except the design of the armillary sphere) that was not already present in Western, Hellenistic times, the latter is so far from that tradition that it must be regarded as an independent invention. New ideas had not come from Islam at this early date. If one grants this argument, the fairly steady evolution of Chinese clocks after I-Hsing leads one to think that in the main the development must have been quite independent until the beginning of the thirteenth century. It certainly indicates that the idea of using a water-clock with an escapement device originated in China.

5 ISLAM AND INDIA (BETWEEN *c.* A.D. 850 AND THE TRANSMISSION TO EUROPE)

After astronomical activity had got under way in Islam, clocks and similar devices seem to have attained a rapid popularity. At first there was merely a revival of the Hellenistic devices, later quite new and very important advances were made.[2] Probably the first notable contribution was that of using gear-wheels for

figures are moved, pillars are turned, stones or eggs let fall, trumpets sounded, and other side-shows (i.e. jack-work), activated.' The insertions bracketed in this passage are necessary because the technical terms of the translator in the Loeb Classics series differ from those which we adopt. Particularly noteworthy is the reference to falling balls, a form of chime so characteristic of the later Byzantine and Arabic striking water-clocks, and even mentioned in Chinese texts (cf. Fig. 34 and pp. 88, 97, 121 ff., 163). Noteworthy also is the absence of any reference to water-scoop driving-wheels even as part of the striking mechanism.

[1] See Borchardt (1). The flower-pot shape was a simple device to compensate somewhat for the loss of pressure-head as the water flowed out, and it was retained for two millennia in China long after the outflow clepsydra had been replaced by the inflow float clepsydra, and the pressure-head problem had been approximately solved by the insertion of a number of compensating tanks or an overflow device.

[2] For example, there was the striking water-clock presented by the King of Persia to Charlemagne in A.D. 807. This is described in Eginhard's *Annals* (in *Monumenta Germanicae Historiae*, ed. Pertz: *Scriptorum*, vol. 1, p. 194). Cf. Planchon (1), p. 21; Kubitschek (1), p. 217. The monarch was none other than Hārūn al-Rashīd (cf. Sarton (1), vol. 1, p. 527).

astronomical purposes, profiting by the differential rates of rotation they make possible. This was first done by al-Bīrūnī,[1] *c.* 1000, who used gears to provide calendar work and sun and moon indicators on the *dorsum* (back face) of an astrolabe—the special volvelle he described to show the phase of the moon is still commonly found in clocks. Actual specimens of astrolabes embodying devices similar to that described by al-Bīrūnī have fortunately been preserved—one made at Ispahan in A.D. 1221/2 (now at Oxford in the Museum of the History of Science), and a second made in France *c.* 1300 (now at the Science Museum, London).[2]

During the twelfth century there was much activity in designing still more complicated automata actuated by water from clepsydras.[3] Although these devices were far more complex than anything previously recorded in the West, they contained but few basic improvements on the Heronic designs. Noteworthy is the use of water-wheels[4] (Figs. 63, 64), but unlike the Chinese clocks the intermittent action is provided by a tipping bucket and the wheel drives jack-work only, without anything to give the cumulative effect that is essential for recording time. The time-measuring part of the apparatus still depends on the steady motion provided by an ordinary float clepsydra.

Great interest attaches to events in India during the twelfth century, and to the subsequent infiltration of the ideas current there into Islam and perhaps to Europe. In the *Siddhānta Śiromaṇi*, composed *c.* A.D. 1150, there is a detailed description of a Ptolemaic armillary sphere. In the same section occurs an account, quite intrusive in an astronomical text, which describes two perpetual motion wheels and a third (castigated by the author of the work) which helps its perpetuity by letting water flow from a reservoir and drop into pots around the circumference of the wheel.

Siddhānta Śiromaṇi, by BHASKĀRA (A.D. 1150), ed. L. Wilkinson; revised by Bapu Deva Sastri (Calcutta, 1861) [Bibliotheca Indica, n.s. nos. 1, 13, 28].

Ch. VI (pp. 151 ff.), 'Golabandha; on the Construction of an Armillary Sphere'.

1. Let the mathematician, who is as skilful in mechanics as in his knowledge of the sphere, construct an armillary sphere with circles made of polished pieces of straight bamboo; and marked with the number of degrees in the circle.

[1] See Wiedemann (1) and Price (1).
[2] See Gunther (1) and Price (1).
[3] On these the elaborate monograph of Wiedemann & Hauser (1) should be consulted.
[4] Wiedemann & Hauser (1), p. 143.

The text continues, describing a three-shell, demonstrational or calculating armillary; later this instrument is also used for observation (p. 210), being turned to such a position that the disc marking the place of the sun throws a shadow on the central globe representing the earth. The middle shell of the armillary contains the three mutually perpendicular circles of meridian, prime vertical and horizon, the two great circles bisecting the angle between meridian and prime vertical (i.e. N.E. to S.W., N.W. to S.E.), an equator circle and a novel addition, similar to the equator but set at an angle to the horizon equal to the latitude instead of the co-latitude of the place of observation. The outer (?) shell, mounted on a polar axis, has the usual three mutually perpendicular circles, a set of three small circles of equal declination above the equator and another three below, the ecliptic and a circle which can be slightly tilted from the ecliptic to correspond to the plane of planetary orbits. The inner devices comprise a simple circle mounted on a vertical axis—probably for marking the places of planets and perhaps also for sighting altitudes, and a set of rings carrying representations of the seven planets and the sphere of fixed stars. At the centre is a globe representing the earth. There seem to be no ecliptically-mounted rotating circles, and no special sighting device.

Ch. XI (pp. 227 ff.), 'On the Use of Astronomical Instruments'

50. Make a wheel of light wood and in its circumference put hollow spokes all having bores of the same diameter, and let them be placed at equal distances from each other; and let them also be placed at an angle somewhat verging from the perpendicular (i.e. as in the noria): then fill these hollow spokes with mercury: the wheel thus filled will, when placed on an axis supported by two posts, revolve of itself (cf. Fig. 65).

Or scoop out a canal in the tire of the wheel and then, plastering leaves of the *tala* tree over this canal with wax, fill one half of the canal with water and the other half with mercury, till the water begins to come out, and then cork up the orifice left open for filling the wheel. The wheel will then revolve of itself, drawn round by the water (cf. Fig. 66).

Make up a tube of copper or other metal, and bend it into the form of an elephant hook, fill it up with water and stop up both ends. And then putting one end into a reservoir of water, let the other end remain suspended outside. Now uncork both ends. The water of the reservoir will be wholly sucked up and fall outside (i.e. a simple siphon).

Now attach to the rim of the before described self-revolving wheel a number of water-pots, and place the wheel and these pots like the water-wheel so that the water from the lower end of the tube flowing into them on one side shall set the wheel in motion, impelled by the additional weight of the pots thus filled. The water discharged from the pots as they reach the bottom of the revolving wheel, should be drawn off into the reservoir before alluded to by means of a water-course or pipe. (Presumably one would also need a pump or water-raising device).

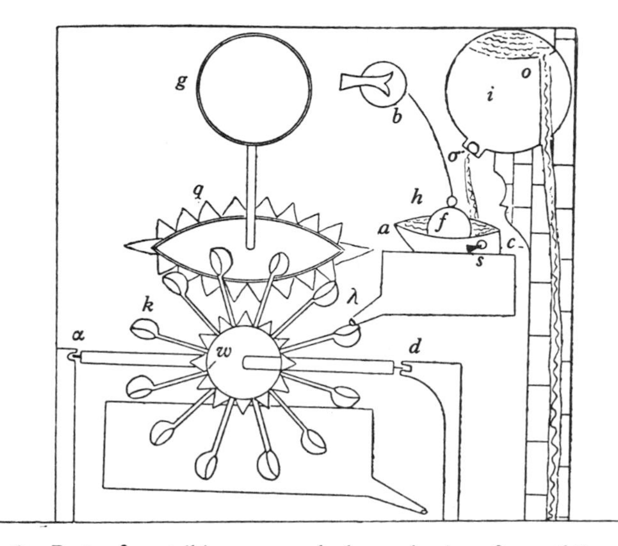

Fig. 63. Part of a striking water-clock mechanism from al-Jazarī's book on mechanical contrivances (A.D. 1206). The periodically tipping bucket activates a water-wheel geared to a disc or globe (from Wiedemann & Hauser).

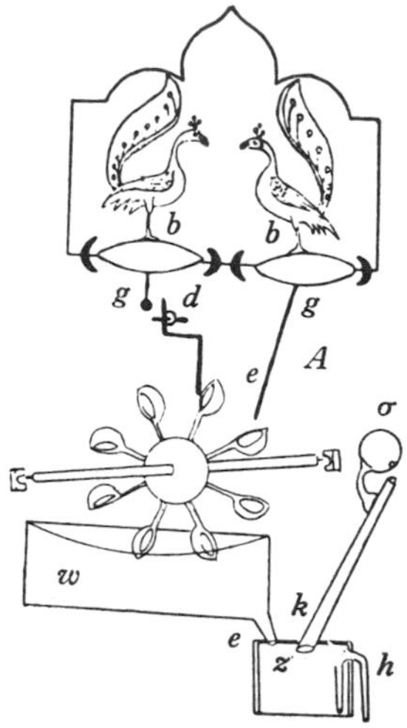

Fig. 64. Another of al-Jazarī's mechanisms. The water-wheel, periodically turned, trips the lugs *d*, *e*, thus moving the spindles *g*, *g*, and the peacock jacks *b*, *b*, while the water escaping through the sump *w* into the tank *z*, causes the pipe *k*, *σ*, to sound, and eventually siphons off through *h*. It is noteworthy that in both these devices the water-wheel is part of the striking, not the going mechanism.

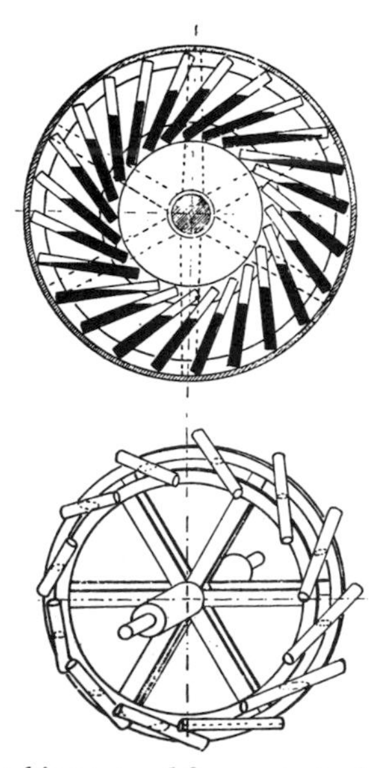

Fig. 65. An Arabic proposal for a perpetual motion wheel of noria type with the buckets filled with mercury, similar to that described in Indian texts (from Schmeller).

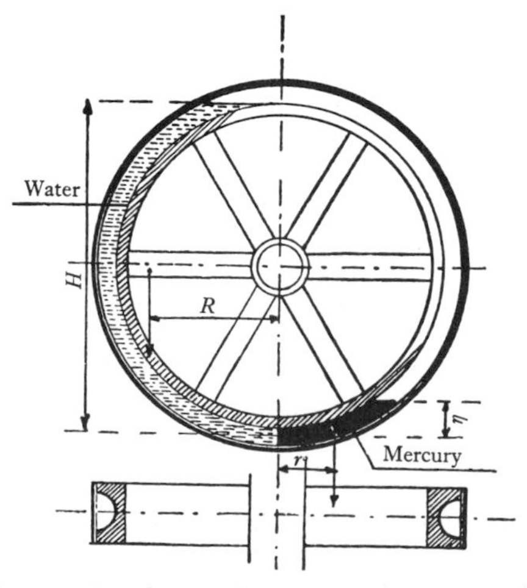

Fig. 66. Another Arabic and Indian proposal for a perpetual motion wheel (from Schmeller).

The self-revolving machine (mentioned by Lalla, etc.) which has a tube with its lower end open is a vulgar machine on account of its being dependent, because that which manifests an ingenious and not a rustic contrivance is said to be a machine (i.e. it is made like a perpetual motion machine but needs to be driven around by applied power).

And moreover many self-revolving machines are to be met with, but their motion is procured by a trick. They are not connected with the subject under discussion. I have been induced to mention the construction of these, merely because they have been mentioned by former astronomers.

Other instruments include the equinoctial armillary, the staff, the gnomon, the *ghati* clepsydra 'of copper like the lower half of a water-pot, which should have a large hole bored in its bottom...', circle, semi-circle, quadrant, etc.

Curiously enough, almost identical explanations and descriptions are given in Islamic texts which occur in association with the treatises on automata by Ridwān and by al-Jazarī (*c.* 1200), with accounts of the 'clock of Archimedes', and in a complex corpus of the Heron–Philon–Archimedes type.[1] The manuscripts found in this interesting set of associations clearly refer to exactly the same forms of perpetual motion wheels as in the *Siddhānta Śiromaṇi* and to other types as well. Some of the details of these wheels have a distinctly Chinese flavour,[2] while some of the forms not included in the Indian sources are found again later in Europe. Thus there is a wheel with partitioned chambers of mercury arranged in just the same 'escapement' fashion as occurs in the Alfonsine astrolabe clock (Fig. 67),[3] and there is a bucket-chain water-raiser in which the power is provided by a weight drive exactly like that of the mechanical clock and in which gearing is used to transmit power.

Yet again, in the *Sūrya Siddhānta* (*c.* A.D. 500) there is a comparable, though rather garbled, account of armillary spheres and perpetual motion;[4] but almost certainly the entire section is an interpolation dating from the eleventh or twelfth century, and hence probably of about the same period as the *Siddhānta Śiromaṇi*. The armillary sphere described in the *Sūrya Siddhānta* embodies the typically Chinese feature of a casing which represents the horizon, and it also includes a model of the earth, placed at the centre of the sphere. In addition the text mentions the use of sand and of 'quicksilver-holes' for time-keeping, and refers cryptically to another

[1] Some of these texts have been edited and translated in an interesting monograph by Schmeller (1).

[2] For example, some of the Arabic perpetual motion wheels are described as being fifteen spans (about 400 cm.) in diameter, made of teakwood, with twenty-four or forty-eight iron or copper scoops round the rim, each one five spans by four fingers (125 × 16 cm.).

[3] See the edition of Rico y Sinobas (1), and the note of Feldhaus (4).

[4] See the edition of Burgess (1), pp. 305 ff. in ch. 13.

horary device with the 'man, monkey and peacock'—it is strikingly significant that these animals are amongst the favourite devices for jacks in the clocks of al-Jazarī.[1]

Thus in these Indian texts, as well as in those from Islam which seem so closely connected with them, one finds the curious running together of ideas about the armillary sphere and about perpetual motion wheels. One finds also several other associated features which are most significantly connected with the history of the mechanical clock—there is the Alfonsine mercury drum, the weight-drive and other things from the West, and there are also features which look more specifically Chinese.

At this date then, India shows a development closely allied to that taking place in China and in Islam. Here the curious association between perpetual motion and armillary spheres may possibly give an important clue. Perhaps some observer, seeing the Chinese clocks imperfectly from the outside and marvelling at them, innocently invented that myth of perpetual motion which was to dog the science of mechanics for so long. An intervention of rationality produced the compromise of 'perpetual motion' wheels powered by scoops fitted to the rim and receiving a stream of water to help things along. Perhaps the perpetual motion fable first originated from an unprecise account of a Chinese clock like Su Sung's. At all events, from the middle of the twelfth century onwards, there is considerable likelihood of the transmission of such Chinese ideas to India and Islam and thence (after 1200 and 1300) to Europe.

6 EUROPEAN CLOCKWORK AFTER THE TRANSMISSION FROM ISLAM

Having reached this point, we are now in a position to consider the developments that took place in Europe when it received the transmissions from Islam. The transmission took place, as has been said, in two waves which may be approximately identified with the receipt of the Toledan Tables (*c.* A.D. 1087) and the Alfonsine Tables (*c.* 1274) with their associated 'canons' explaining astronomical theories, techniques and instruments. Not only did Hellenistic learning reach Europe in this way, but at the same time there came all the subsequent improvements which had been made in Islam and elsewhere—the use of gearing, the more complicated jack-work, perhaps also the Chinese ideas on clocks which may have reached India and Islam in some more or less distorted form.

[1] Wiedemann & Hauser (1), p. 145.

PLATE XIX

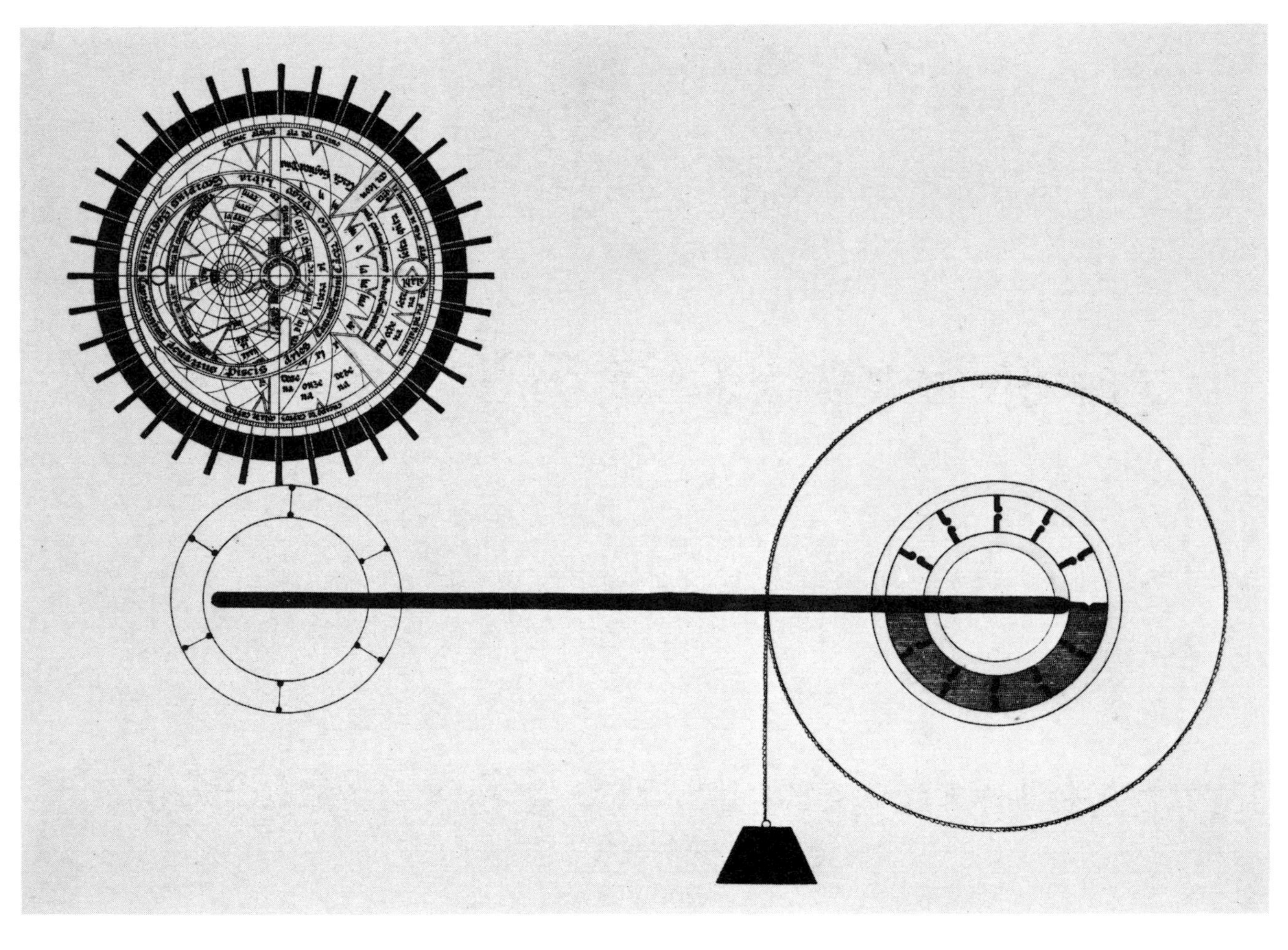

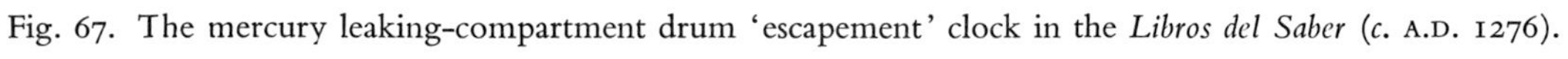

Fig. 67. The mercury leaking-compartment drum 'escapement' clock in the *Libros del Saber* (*c.* A.D. 1276).

It must certainly be more than coincidence that after the Toledan transmission one begins to find in Europe the same curious juxtaposition of the armillary sphere and the perpetual motion machine. Many have noticed that before the twelfth or thirteenth century it is hard to detect in the West any trace of the idea of perpetual motion as a practical possibility.[1] Speaking of the time around 1250 Sarton says[2] that one of the new approaches to mechanical problems was that of engineers trying to improve their machines. 'The dream of perpetual motion was beginning to engross their imagination.' And indeed the notebook of Villard de Honnecourt, written just about this date (1257), contains a famous design for a wheel with a series of mallets fixed round its rim which would, if once properly made, continue to rotate indefinitely.[3] 'Many a day', wrote the craftsman in his album, 'have skilful masters debated the construction of a wheel that shall turn of itself; here is a way to make such a one, by means of an uneven number of mallets, or by quicksilver.' And Leonardo da Vinci himself, 250 years later, was not above elaborating on the same design with sketches and notes.[4]

Even more telling is the fact that these two things—the armillary sphere and the perpetual motion mechanism—soon became connected with a third, the attraction of magnets and lodestones. In part this association could be explained by supposing that ideas of perpetual motion were a natural outcome of the observed eternal and unceasing motion of the apparent diurnal rotation of the heavens. The magnet with its directivity was the source of a mystical orientation which might very naturally be connected with such celestial motion. But it seems on the whole more likely that the first association derived from an incomplete knowledge of Chinese mechanical clocks, and the second from a transmission through the same Islamic channels—perhaps indeed at the same time—of accounts of the magnet and compass as the Chinese had developed it through previous centuries.[5] In other words, what travelled was a complex of true facts and misconceptions.

[1] On this subject the reader may be referred to the interesting compilation of Dircks (1).

[2] Sarton (1), vol. 2, p. 764.

[3] See the discussion in Dircks (1), vol. 2, pp. 1 ff.

[4] See the discussion in Dircks (1), vol. 2, pp. 4 ff. The life of Villard's mallets, which he may have derived from certain wheels used in medieval churches as a substitute for the luxury of bell-ringing on Good Friday, proved extraordinary. They appeared in various modifications down to the end of the nineteenth century (e.g. Jeremiah Mitz in 1658, Alessandro Capra in 1678, C. E. Neumann in 1766, George Linton in 1821, Charles Colwell in 1867 and many others). John Wilkins exploded them in his *Mathematical Magic* (1648) but it is remarkable to find how many worthy scientific men of the seventeenth century believed in the possibility. For details see Dircks (1).

[5] See the account of the history of the magnetic compass in China and its transmission to the West, in Needham (1), vol. 4, pt. 1.

As is well known, Petrus Peregrinus, who was working between A.D. 1245 and 1275, and to whom we owe the first and greatest western medieval work upon the magnet,[1] believed that by means of magnets it would be possible to construct a *perpetuum mobile*. Speaking of the polar properties of the magnetic compass, Taylor (1) says that Peregrinus

conceived of it as presenting a model of the celestial sphere, part corresponding to part, and believed that if correctly poised it would rotate in sympathy with the rotating firmament. He proposed therefore to make an armillary sphere with a freely turning axis which held a perfectly shaped and balanced lodestone. Once the axis was elevated so as to correspond exactly with the celestial axis, and the apparatus patiently set in motion by hand until it took up the motion of the sky, the astronomer would possess an instrument which would serve as a perpetual time-piece and give indeed all the celestial data he required.

Then in Petrus Peregrinus himself (*Epistola...de Magnete*, A.D. 1269) we find:

Part II, ch. 3. Of the construction of a certain wheel which will move continually and perpetually.

Part I, ch. 5. Of the inquiry whence the magnet receives the virtue which it has....That being done, arrange the stone on the meridian circle on its pivots fixed lightly in the poles of the stone that it may move in the manner of armillaries, in such wise that the elevation or depression of its poles may be in accordance with the elevation and depression of the poles of the heavens in the region in which you are. Now if the stone then move according to the motion of the heavens you have arrived at a secret marvel. But if not let it be ascribed rather to your own want of skill than to a defect of Nature. For in this position or mode of placing I deem the virtues of this stone to be properly conserved, and I believe that in other positions or parts of the sky its virtue is dulled rather than preserved. By means of this instrument, at all events, you will be relieved from every kind of clock (dial?) for by it you will be able to know the ascendant at whatever hour you will, and all the other dispositions of the heavens which astrologers seek after.[2]

Roger Bacon refers to the work of Peregrinus (which he had himself personally seen) in the *De Secretis*, *Opus Majus* and *Opus Minus* (1267)—the first mention must be some time before 1250. In the *Opus Minus* he finally says that the wonderful diurnal motion of the armillary would be attained by the use of a magnet. He describes the Ptolemaic armillary sphere, following the *Almagest*, and says that it cannot be made to move naturally by any mathematical device, but adds: 'A faithful and magnificent experimenter is now straining to make one out of such material and by such a device that it will revolve naturally with the diurnal

[1] Cf. Sarton (1), vol. 2, pp. 1030 ff.

[2] The final sentence here is reminiscent of the preoccupations of the astronomers of the Imperial Palace at the Chinese capital (cf. p. 172 above), and would not have been strange to Petrus Peregrinus' contemporary, Kuo Shou-Ching.

heavenly motion.' The idea lived on,[1] but within half a century or little more, horological engineering of a purely mechanical kind had been so successful as to consign the dream of Peregrinus to a lengthy oblivion. Only in our own time, and obliquely, has the application of electric power to these devices, and the use of radioactivity clocks, brought a justification far beyond the imagination of the thirteenth century.

During the latter half of the thirteenth century there is clearly evident in Europe a preoccupation with the ideas of water-clocks, simulated diurnal rotation, and the basic elements of clockwork. The notebooks of Villard de Honnecourt, *c.* 1257, provide an illustration of a clock housing,[2] though no clue is given as to the mechanism of the clock that was to go inside.[3] A fine illumination in a manuscript of about 1285 in the Bodleian Library (see Fig. 68), shows a water-clock with a wheel seemingly divided into compartments, but again the details of the machinery cannot be made out. Its discoverer, Drover (1), discusses the device, and brings forward a certain amount of textual evidence showing that the measurement of time in twelfth- and thirteenth-century European monasteries was effected by water-clocks of some kind or other. What the Bodleian MS. shows is that at least one of them involved a water-wheel. Even after the invention and spread of mechanical clockwork in the fourteenth century, the use of water-power for clocks or astronomical models continued in mind—indeed Leonardo himself wrote of this. In relation to drawings in the *Codex Atlanticus*, fol. 401 *r*, we read[4] that he

> studied various arrangements for the automatic filling and emptying of vessels of particular capacities. Apart from functioning as meters for water supplies, such combinations of tanks and valves could serve for the perfecting of hydraulic clocks, and especially for making a kind of alarm-clock 'for those who are thrifty of their time'. Then with improved types of wheels with blades or scoops which he designed, Leonardo expected to be able to utilise water-power

[1] Indeed with surprising tenacity. Some time before A.D. 1579 Johannes Taisnierus wrote on 'continuall motion by the stone Magnes'; his proposal will be found in Dircks (1), vol. 1, pp. 18 ff., though exactly what he intended is not very clear. In 1612 Thomas Tymme, an English divine, wrote a 'philosophicall dialogue' in which he described a perpetual motion clock presented by the well-known Dutch technician Cornelius Drebbels to King James. The description in Dircks (1), vol. 2, pp. 10 ff., is worth reading, for the apparatus took the form of a globe of brass and glass, with a smaller globe superimposed on it. Besides the time, astronomical information of various kinds could be read off from dials. But the motive power was a certain imprisoned 'fierie spirit', not a magnet.

[2] See the facsimile edition of Hahnloser (1).

[3] The opinion formerly entertained that the notebooks of Villard de Honnecourt contain a drawing of an early form of an escapement is now no longer acceptable.

[4] In the catalogue of the Milan Museum of Science and Technology (edited by G. Ucelli di Nemi), p. 90, Entry 99.

to set in motion systems of gearing, however complicated, and to attain, for instance, in the last little wheel of such a system 'so great a velocity that every other speed, whether of the sun and the other planets (in their daily motion round the earth), would appear in comparison very sluggish'.[1]

Meanwhile, in his commentary on the *Sphere* of Sacrobosco, Robertus Anglicus in 1271 mentioned incidentally that horological

> artificers are trying to make a wheel which will pass through one complete revolution for every one of the equinoctial circle, but they cannot quite perfect their work. If they could, it would be a really accurate clock, and worth more than any astrolabe or other astronomical instrument for reckoning the hours....

He goes on to say that it would be necessary to drive such a wheel by means of a falling weight, but though by this time such a drive seems to be becoming fairly familiar, Robert makes no mention of any sort of escapement.[2] Very shortly afterwards comes the astrolabe mercury compartment-wheel clock described and illustrated in the Alfonsine corpus.[3] This has a weight-drive, and for the first time in Europe there is certain mention of a regulating device—a drum wheel of the 'perpetual motion' type with mercury leaking through its chambers.

By about A.D. 1300, when the flood of new astronomical texts arrived in Europe, all the elements of the mechanical clock were known—excepting only the principle of the escapement. The crucial period is, of course, the transition from this stage to that of Richard of Wallingford's reputed clock (*c.* 1335)[4] and the astronomical masterpiece described in such superb detail by Giovanni de Dondi (1364),[5] who however disappoints us by merely mentioning and drawing the escapement, saying that it is 'common' and therefore not necessary to describe in detail. Thus there is a gap of little more than half a century between no knowledge of the escapement at all, and its dismissal as commonplace. Without the discovery of the ever-elusive European texts which might fill this gap one can only guess at what happened.

The whole situation could be clarified by the single assumption that some incomplete and garbled idea of the Chinese water-wheel escapement was transmitted to the West together with an equally muddled notion connecting the

[1] This is a curious remark. One would expect Leonardo to be pondering over the problem of inducing wheels to rotate extremely slowly. But grammatically the words must mean the rotation of wheels faster than the diurnal headlong rush of the sun and planets round the earth.

[2] See the note of Thorndike (1).

[3] See Rico y Sinobas (1) and Feldhaus (4). Fig. 67.

[4] See Price (5); Price (6), pp. 127 ff.; Gunther (2), vol. 2, pp. 48 ff.

[5] See Lloyd (1).

Fig. 68. A monastic water-driven wheel-clock of thirteenth-century Europe; the illumination in the biblical MS. (Bodleian Library 270*b*, fol. 183*v*). The incident depicted is the moving back of the sun 10 degrees by the clock for King Hezekiah during his illness by the prophet Isaiah (2 Kings xx. 5–11 and Isaiah xxxviii. 8). The mechanism of the clock is hard to make out but see pp. 195 and 215 (from Drover (1)).

Fig. 69. The simplest type of early fourteenth-century mechanical clock with weight-drive, crown-wheel, pallets, and verge-and-foliot escapement; an example, probably of later date, from the tower of St Sebald's Church at Nürnberg (from Zinner).

armillary sphere with the (apparently) perpetual motion wheels of the same Chinese clocks. Perhaps such ideas travelled in caravan company, as it were, with similar stories about the magnetic compass and the Cardan suspension.[1] India and doubtless Islam were its way-stations. Now it is altogether reasonable to think that in such a case the complicated lever system of the escapement would be most likely to suffer delay and distortion in transmission. By the time it arrived in clear form, the weight-drive would have replaced the water-wheel, and perhaps the Alfonsine type of governor had been evolved and rescued from the tangle of notions about *perpetuum mobile* wheels. Or alternatively it may have become known to some of the monastic horologers early in the thirteenth century, so that the Bodleian MS. clock might have had a Chinese escapement. One cannot tell. But it is interesting to set this possible transmission side by side with others which are more or less certain. If we may choose the date 1310 as a focal one for the first mechanical clocks in Europe, we may remember that 1325 is equally focal for gunpowder, the original home of which had been tenth-century China. Towards 1380 we find the first blast-furnaces producing cast iron in Europe—in China this technique had been developed a century or so before the beginning of our era. Towards 1375, and also in the Rhineland, comes the first European block-printing, an art which had been current in China since the ninth century. Still closer to clockwork in time are the great segmental arch bridges of Europe, the first being about 1340, though in China structures of this kind had first appeared in the seventh.[2] Such is the context of the clockwork transmission—if one there was.

Perhaps we can imagine some early fourteenth-century scholar-craftsman puzzling over the tale that in the East they used a set of oscillating levers, tripping and holding back a wheel so to regulate its turning that it would agree with the heavens like the two halves of a tally, never deviating by a hair's-breadth from that motion. So he made his own oscillating device—taking it, in all probability, from the weights at the end of the arms of those screw-presses so characteristic of Hellenistic and European technology (Fig. 69). Then he applied it, in alternating opposition, to a wheel driven by the fall of a weight, and found that it worked,

[1] On this see Needham (1), vol. 4, pt. 1.

[2] These thoughts crowded to the mind one evening in Florence in the summer of 1956 when gazing down at the arches of the Ponte Vecchio from the Piazzale Michelangelo. For the detailed evidence concerning the transmissions see Needham (1)—for gunpowder, vol. 4, pt. 3, for iron and steel technology, vols. 4, pt. 3 and 5, for printing, vol. 4, pt. 3, and for bridge-building, vol. 4, pt. 2. Besides the remarkable series of European acceptances during the fourteenth century, one should remember the coming of the magnetic compass about 1190 and that of the stern-post rudder about 1200.

perfecting thus his machine which should show forth the glory of God in the heavens. Alas that we do not know his name. And he in his turn assuredly never knew the names of those who long before, away at the farthest end of the known world, had built successful machines which faithfully served the same purpose.

APPENDIX

CHINESE HORARY SYSTEMS

THREE systems for subdividing the day into smaller units have co-existed in China since ancient times:

(1) A full day (midnight to midnight) divided into 12 'double-hours' (*shih* [143]), sometimes with each double-hour divided into two halves so as to give twenty-four divisions, as in the modern practice.

(2) A full day (midnight to midnight) divided into 100 'quarters' (*k'o* [21]), each of them being equivalent to 14 min. 24 sec. of modern reckoning. From time to time efforts were made to modify this practice so as to reduce the number of *k'o* to ninety-six or to increase them to 120, so as to make the system commensurate with the duodecimal hours. But these reforms were never long successful.[1]

(3) A night (sunset to sunrise, or some similar division variable throughout the year and dependent on the latitude) divided into five 'night-watches' (*kêng* [360]), these being further subdivided each into five *ch'ou* [139].

The first two methods are especially interesting by virtue of their long co-existence and by the obvious difficulties in reconciling them. The duodecimal system, from which modern practice is derived, goes back at least as far as the Old Babylonian rule which divided the day into twelve *kas.bu* (double-hours)[2] and may well have been transmitted to China from this source at an early date.[3]

[1] Maspero (1), pp. 208 ff., made a beginning of writing the history of this subject, but much more could be said. For example, there was Hsia Ho-Liang [682], whose story is told in the *Ch'ien Han Shu*, ch. 11, pp. 5*b* ff., and ch. 26, p. 34*b*. He was a candidate for official rank about 15 B.C., and a disciple of the astrologer Kan Chung-K'o [683]. Kan claimed that the mandate of the Han dynasty had expired and that he had revelations which could ensure its re-creation or continuance. The division of the day and night into 120 *k'o* was one of the things which it would be advisable to do. Impressed by Hsia Ho-Liang's advocacy of Kan's ideas, the emperor Ai in 5 B.C. changed the name of the reign-period and ordered that the horary reform should be adopted. But later, his illness not improving, the policy was reversed, and Hsia and his friends were executed. Cf. Dubs (1), vol 3, p. 93. A later adjustment occurred under the Liang dynasty, when in A.D. 507 the emperor Wu reduced the number of *k'o* in the day and night to ninety-six. But this improvement, which made them exactly equivalent to modern quarters, lasted only a very short time, since no justification for the number could be found in the classics, and in 544 the number 108 was substituted. Later still, the emperor Wên of the Ch'en dynasty, between 560 and 565, returned to the old figure of 100. A convenient collection of excerpts giving an introduction to the story of these vicissitudes may be found in *T'u Shu Chi Ch'êng* (*Li fa tien*), ch. 98.

[2] See Ginzel (1), vol. 1, p. 122.

[3] As was urged by Bilfinger (1) long ago.

Perhaps by this association or possibly independently, the Chinese double-hours are linked in one-to-one agreement with the astronomical succession of signs of the zodiac—starting with the first double-hour (i.e. 11 p.m. to 1 a.m.) designated by the Rat (=Aries).[1]

TABLE

DAY AND NIGHT DOUBLE-HOURS

(from *Hsiao Hsüeh Kan Chu*, ch. 1, pp. 23*a*, *b*; A.D. 1299)

1	Tzu	子	11 p.m.–1 a.m.	*yeh pang*	夜半	rat
2	Ch'ou	丑	1 a.m.–3 a.m.	*chi ming*	雞鳴	ox
3	Yin	寅	3 a.m.–5 a.m.	*p'ing tan*	平旦	tiger
4	Mao	卯	5 a.m.–7 a.m.	*jih ch'u*	日出	hare
5	Ch'en	辰	7 a.m.–9 a.m.	*shih shih*	食時	dragon
6	Ssu	巳	9 a.m.–11 a.m.	*yü chung*	隅中	snake
7	Wu	午	11 a.m.–1 p.m.	*jih chung*	日中	horse
8	Wei	未	1 p.m.–3 p.m.	*jih tieh*	日昳	sheep
9	Shen	申	3 p.m.–5 p.m.	*pu shih*	晡時	monkey
10	Yu	酉	5 p.m.–7 p.m.	*jih ju*	日入	cock
11	Hsü	戌	7 p.m.–9 p.m.	*huang hun*	黃昏	dog
12	Hai	亥	9 p.m.–11 p.m.	*jen ting*	人定	boar

The third, fifth, eighth, ninth and tenth double-hours are also listed in *Ch'ien Han Shu*, ch. 26, p. 22*a*, with the variant *tieh* 跌 for the eighth. The ninth, 'dinner-time', can also be written *pu* 餔 as in *Ch'ien Han Shu*, ch. 26, p. 33*b*.

A further period, a late afternoon 'post-prandium', *hsia pu* 下晡 or 下餔, is listed in ch. 26, p. 22*a*, as coming between the ninth and tenth double-hours. It may therefore be thought of as approximately the period 4 p.m.–6 p.m. This expression occurs in *Ch'ien Han Shu*, ch. 99, p. 32*a*, to designate the time of day at which the soldiers of the Han army in A.D. 23 finally fought their way up the tower where Wang Mang [684] had taken refuge, and killed him. See Dubs (1), vol. 3, p. 465.

In medicine, the hours of the day were analogised with the seasons of the year, and divided into the following four periods:

chao 朝 3 a.m.–11 a.m. (4 double-hours) corresponding to spring
chou 晝 11 a.m.–5 p.m. (3 double-hours) corresponding to summer
hsi 夕 5 p.m.–11 p.m. (3 double-hours) corresponding to autumn
yeh 夜 11 p.m. –3 a.m. (2 double-hours) corresponding to winter.

[1] See Chavannes (1). In the Han period the personified spirits of the 12 double-hours were often represented by pottery or terra-cotta figurines having the heads of animals and dressed in dignified robes. One such set, taken from a tomb on the outskirts of Sian in Shensi province, is depicted in the journal *Wên Wu Ts'an K'ao Tzu Liao*, 1954 (no. 10), pl. 80.

An illness was likely to follow a diurnal cycle. In the first of these periods the patient's condition would probably be good, in the second quiet, in the third worsening, and in the fourth bad. The *locus classicus* for this information is the *Huang Ti Nei Ching, Ling Shu* [833], ch. 44, p. 7*a*, essentially a work of the Han dynasty. For the reference we are indebted to Dr Lu Gwei-Djen.

In the early Shang period, before about 1270 B.C., there were no equal subdivisions of the day and night, and a set of some half-dozen terms sufficed for dawn, midday, dusk, etc.[1] These were *mei* [412] just before dawn, *hsi* [413] first lightening, *chao* [414] morning, *jih* [415] midday, *mu* [416] just before dusk, and *hun* [417] just after dusk. But from that time onward the day was divided into six unequal (i.e. varying) double-hours by the use of seven special terms, and there was one also for midnight.[2] These were *ming* [418] dawn, *ta ts'ai* [419] (greater assembly) at the end of the first double-hour, *ta shih* [420] (greater meal) at the end of the second, *chung jih* [421] midday, *chê* [422], a character derived from a pictogram which showed the declining sun, to indicate the end of the fourth double-hour, *hsiao shih* [423] (lesser meal) at the end of the fifth, and *hsiao ts'ai* [424] (lesser assembly) at dusk, the end of the sixth. Midnight was termed *hsi* [425]. These usages have been established by the researches of Tung Tso-Pin, Kuo Mo-Jo and others on oracle-bone and bronze inscriptions.

The full cycle of 12 equal double-hours is found stabilised in the Han period (from the beginning of the second century B.C. onwards), but it probably goes well back into the Chou. A passage in the *Tso Chuan* under date 534 B.C.[3] was taken by Wang Ying-Lin [501] at the end of the thirteenth century A.D. to mean[4] that at that time only 10 double-hour names were used, two of the periods lasting 4 hours each.[5] A passage in the *Kuo Yü* [823],[6] giving what purports to be a discussion of 519 B.C., indicates that by the third or fourth century B.C., at least, the twelve cyclical characters, *chih* [230], had become attached to the twelve equal double-hours. The time of adoption of the Babylonian equal duodecimal system

[1] *Yin Li P'u*, pt. 1, ch. 1, pp. 4*b* ff.

[2] *Yin Li P'u*, *loc. cit.*

[3] Duke Chao, seventh year; see Couvreur (1), vol. 3, p. 129.

[4] *Hsiao Hsüeh Kan Chu*, ch. 1, p. 23*b* [727].

[5] It would seem that the two *chih* [230] omitted were not the midnight ones to allow for night-watches, as might have been expected, but *mao*, 5–7 a.m., and *ssu*, 9–11 a.m. So apparently *yin* went on from 3 to 7 a.m. and *ch'en* from 7 to 11 a.m. This may be just why Wang Ying-Lin ranked them in importance immediately after the midday double-hour.

[6] *Chou Yü*, ch. 3, pp. 36*a*, *b*. The 12 equal double-hours are also mentioned in the *Chou Li* [764], ch. 6, p. 44*b* (Biot (1), vol. 2, p. 112); which points to the same time.

must therefore be placed not much later than the middle of the first millennium B.C., and may have been nearer its beginning than its end.[1]

From an early time it was customary to bisect each double-hour into separate halves, equivalent to our hours, the first being known as the *ch'u* portion [407] and the second as the *chêng* portion [411]. It should be noted that the Chinese double-hour system was so arranged that the first period straddled midnight (running from an hour before midnight to an hour after), and correspondingly the seventh straddled noon.[2]

If one concedes that the institution of the equal double-hours was imported into China from Babylonian or some other predominantly sexagesimal culture, it seems reasonable to suppose that the centesimal division into *k'o* ('quarters') represents an indigenous Chinese system. We have already noted the perennial malaise about the reconciliation of the two, and several times in the history of the astronomical clocks it seems apparent that part of the motive of their construction is the demonstration and marking of both types of period (cf. Fig. 70). The clear predilection of Chinese computers and mathematicians for decimal systems from Shang times onwards[3] speaks in favour of an autochthonous character of the 100 quarters. And it is interesting to find that during the centuries some of the Chinese calendars divided the *k'o* into 100 *fên* [364] while others divided them into sixty.[4] These were thus even less like our minutes than the *k'o* were like our quarters.

[1] Allusion has already been made (p. 168 above) to the mistaken idea of Enshoff (1, 2) that the unequal hours persisted through historical times in China. A recent coincidence has thrown light on the origin of this misconception. The Deutsches Museum at München exhibits in its horological section, as Chinese, a clock of 'Gothic' type crowned with a large bell. This clock has the adjustable double verge-and-foliot system for unequal hours (cf. p. 168, n. 4 above), and is unmistakably Japanese. But the greater part of Enshoff's argument was based upon this very clock, of which he gives a photograph. P. D. Enshoff was a worthy Benedictine missionary of advanced years who lacked sinological knowledge and gathered his data on Ch[illegible]a from very secondary sources; it was unfortunate that he devoted so much time to tilting at this particular windmill. One of us (J.N.) noted the München clock on a visit in the summer of 1957, before studying in the autumn the Enshoff papers in the Jäger Collection.

[2] We have been unable to verify a suggestion which Ginzel (1), vol. 1, p. 464, attributes to Gaubil (see p. 167 above), that in very ancient times the first double-hour commenced with midday instead of midnight, as was always the case in historical times. Tung Tso-Pin seems to think that the Shang day probably began at sunrise. Against this Dubs (2) argues strongly in favour of extrapolating Han practice backwards and assuming that the Shang people also began their day at midnight. Only in this way, he says, can certain eclipse records on the oracle-bones be satisfactorily interpreted. But it would take us too far to consider the controversy between Tung and Dubs on the dating of the Shang eclipses. In any case we can be sure that the Babylonian and Graeco-Jewish system of beginning the day at sunset was never used in China. It is curious that the Roman and the Han empires, alike in so many other respects, agreed in taking midnight as the bound between days.

[3] On this, see the evidence in Needham (1), vol. 3, pp. 82ff.

[4] Cf. Maspero (1), p. 211.

A system of night-watches is also to be found in Old Babylonian literature,[1] where mention is made of the *bararītu* (rising of stars), *kablītu* (midnight), and *namarītu* (dawn or twilight) watches. Such divisions, being of length variable with the changing duration of darkness throughout the year, correspond to the 'unequal hours' of ancient and medieval European practice. As we have seen, this system died out in China very early, but for the watches of the night it persisted throughout the ages. We are free to assume that it was also inspired by Babylonian precedent.

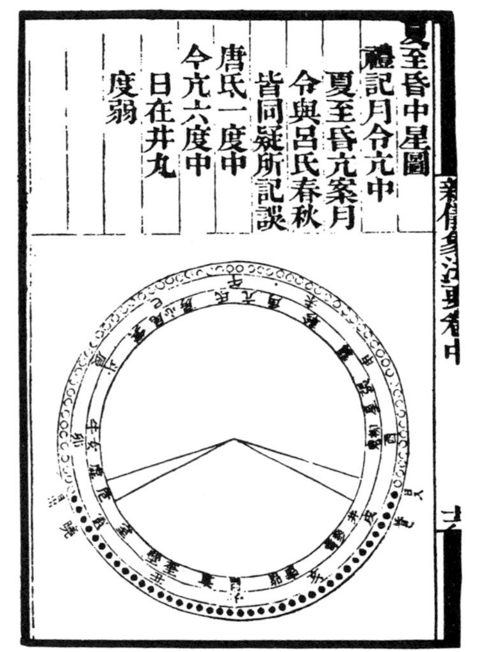

Fig. 70. A diagram from Su Sung's *Hsin I Hsiang Fa Yao* (ch. 2, p. 16*b*) showing the number of quarters occupied by light and darkness respectively at the summer solstice at K'ai-fêng in A.D. 1088; the former are indicated by white discs, the latter by black ones. The thin wedges demonstrate the twilight periods. Time runs clockwise round the circle, the inner ring of which marks the equatorial constellations (lunar mansions), and the middle ring of which carries the cyclical signs for the twelve double-hours of the day and night. The text above gives the equatorial constellations culminating at dusk at this date for several periods: at the time of (*a*) the *Yüeh Ling* in the *Li Chi* (*c.* fifth century B.C.); (*b*) the *Lü Shih Ch'un Ch'iu* (239 B.C.); (*c*) during the T'ang period (*c.* eighth century A.D.); and (*d*) in his own time. Su Sung doubts some of the older statements, but in his age the precession of the equinoxes had long been well known. Cf. p. 27 above.

Here is Wang Ying-Lin's statement on the night-watches about A.D. 1299:[2]

The night has five watches. In the *Chou Li* [764] (Record of the Rites of the Chou Dynasty; an early Han compilation) the night-watchmen (*ssu wu shih* [1005]) have charge of the times of

[1] Ginzel (1), vol. 1, p. 122.

[2] *Hsiao Hsüeh Kan Chu*, ch. 1, p. 24*a*.

> the night.[1] One of the commentators says that these went from *chia* to *wu* (i.e. that the first five of the ten cyclical characters *kan* [229] were used to denote them). The chapter of the *Ch'ien Han Shu* [824] on the Western Regions[2] says that the sentries (*hou chih shih* [1006]) beat the time with wooden spoons. The commentary says that the night is divided into five watches, and that according to the old Han system, the palace officials (*huang mên* [1007]) kept count of them. The *Ssuma Fa* [825] (Master Ssuma's Art (of War); a military treatise of the third or fourth century B.C.) says that the *hun ku* [426] or 'darkness drum' marks the fourth period which is therefore called *ta ts'ao* [427], 'Big Drum beaten in Camp at Night'. The *yeh pang* [428], midnight, is the third period, which is called also *ch'en chieh* [429], 'Warning of Dawn'. The *tan ming* [430], or 'first light', is the fifth period, which is therefore also called *fa hsü* [431], 'First touch of the Sun's Warmth'.

Thus from this encyclopaedist we learn the principle and a number of interesting archaic military terms, but not much more.

The general character of the Chinese method of dividing the period of night into five watches is clear enough, but there is considerable difficulty in obtaining detailed definitions of these watches in the various schemes which appear to have been current from time to time. Above all it is seldom clear whether the name of an interval refers to the period of the whole watch or only to the point which marks its beginning or its end. Furthermore, the terminal points involved in the scheme not only include sunset (*jih ju* [133]) and sunrise (*jih ch'u* [137]) but also dusk (*hun* [134], $2\frac{1}{2}$ *k'o* after sunset), a point 10 *k'o* after dusk (*ch'u kêng* [138]), another—'waiting for dawn'—(*tai tan* [135], 10 *k'o* before dawn), and a fourth, dawn (*hsiao* [136], $2\frac{1}{2}$ *k'o* before sunrise). It is, at times, difficult to tell whether these intervals are included in the night-watches or whether the watches are five equal divisions of some interval between a pair of these fundamental points.

The original system, which remained in use in country districts, was to let the night-watches run from dusk to dawn. In the latitude of K'ai-fêng (34° N.) the variation in length of the night is such as to give a period from sunset to sunrise of about $(50-10)$ *k'o* at midsummer to $(50+10)$ *k'o* at midwinter; with the convention of a crepuscular interval of $2\frac{1}{2}$ *k'o* (cf. the medieval European convention of 10°, i.e. $2\frac{7}{9}$ *k'o*) this would result in the total length of night-watches varying from 35 to 55 *k'o*. Dividing this into five equal parts, one would have watches varying from 7 to 11 *k'o* each in summer and winter respectively, in comparison with the standard double-hour and its equivalent of $8\frac{1}{3}$ *k'o*. One should note the ambiguity that the middle (i.e. the third) night-watch is referred to as 'The Night-Watch' in some texts.

[1] See Biot (1), vol. 2, p. 380.

[2] Chs. 96 A and B.

A very interesting variant of the night-watch system seems to have arisen as a result of living in an urban capital with miscellaneous pleasures and duties to be attended to by lamplight. Lu Yu [673] (+1125 to +1210) in his *Lao Hsüeh An Pi Chi* [817], ch. 7, p. 3*a*, comments as follows:

In former dynasties the five night-watches ceased at dawn, and all the provinces still use this system, but in the palace the night-watches stop 10 *k'o* before dawn; this is called 'waiting for dawn'. This is because after the night-watches are ended, the emperor's ablutions must be accomplished so that at dawn he may preside over his court. How diligent our forbears were in instituting such a custom.

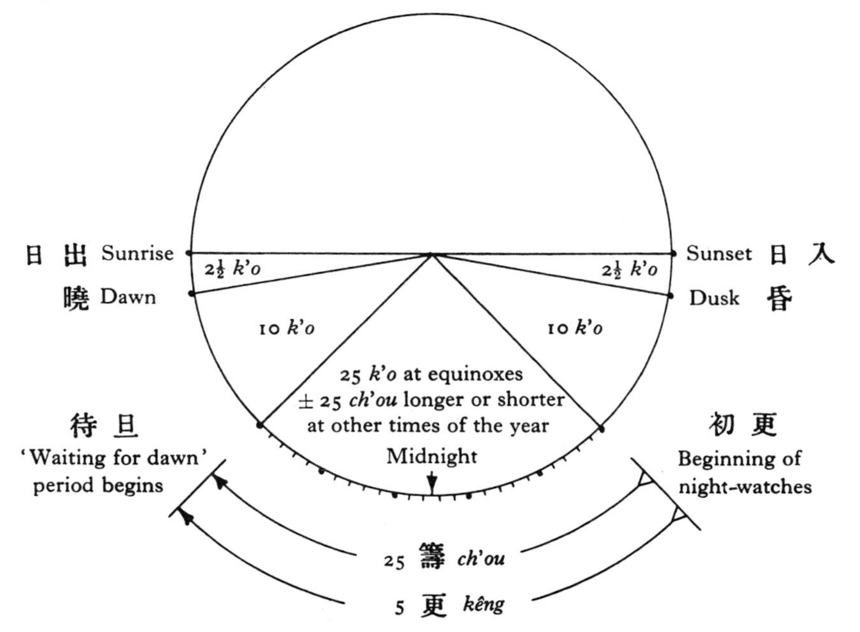

Fig. 71. Scheme of the time-divisions of the night in ancient and medieval China—equal quarters (*k'o*), and unequal night-watches (*kêng*) each divided into fifths (*ch'ou*).

It is, however, not stated whether the new system also modified the beginning of the night-watches so that they started a similar 10 *k'o* after dusk; if this were not done, the resulting unsymmetrical arrangement could not set midnight at the middle of the night-watches, and such a scheme would therefore be inconvenient. One is led therefore to suppose that the modified night-watches would run from 10 *k'o* after dusk until 10 *k'o* before dawn, a period varying in length from 15 to 35 *k'o* in latitude 34° in summer and winter respectively. Each night-watch would therefore vary from 3 to 7 *k'o*, being just 5 *k'o* long at the equinoxes, as is shown in Fig. 71.

SUPPLEMENT

I CHINESE ASTRONOMICAL CLOCK-TOWERS

THE 1960 edition of *Heavenly Clockwork* included the pregnant observation {p. 58 note 2} that 'No doubt the more subtle aspects of Su Sung's mechanism will not reveal themselves until a working model has actually been constructed.'[1] This proved to be indeed true.

Model construction began in 1961[2] with sand-driven experimental versions of the time-keeping mechanism.[3] It continued in 1962 with an improved sand-driven version,[4] together with a $\frac{1}{48}$-scale skeleton model of the astronomical clock-tower of A.D. 1086–9;[5] and in 1963 with a water-driven $\frac{1}{6}$-scale model of the time-keeping mechanism.[6] In 1967 engineering designs were prepared for a half-scale working-model reconstruction of the mechanism (Fig. 72) for the Merseyside County Museums.[7] Additional half-scale working-model reconstructions were built to these designs in 1969[8] and 1984.[9]

The key to the satisfactory operation of the time-keeping water-wheel linkwork mechanism described and incompletely illustrated in the book *Hsin I Hsiang Fa Yao* [702], published in A.D. 1172, was seen in 1961 to lie in the fact that a water-wheel

[1] In this Supplement, braces { } enclose references to pages or figures in the original text. Square brackets enclose references to the Tables of Chinese Characters {reprinted on pp. 229–42}.

[2] Anon (1); Anon. (2); Needham (4), p. 459, note a.

[3] Combridge (1), Fig. 2; Needham (4), Fig. 659; Needham (5), Fig. 5.

[4] Combridge (1), Fig. 3. (The name 'celestial balance', ascribed {on p. 55} to the time-keeping linkwork mechanism as a whole, was re-used in the title and figure-captions of Combridge (1). We now read the characters *t'ien hêng* [62], from which it was derived, as a simple technical term 'upper balance-lever' {cf. the explanation on p. xii, note 5} for a particular component of the mechanism (Fig. 73, item 11). We think that, having been inserted as an annotation against this component in an early version of the illustration {Fig. 18}, they may have been miscopied as its main caption during a medieval redaction.)

[5] For a fine museum-made derivative of the home-made prototype, see Needham (5), Fig. 64; Needham (6), Fig. 18. In the light of the studies described below, the skeleton-model reconstructions of the clock-tower incorporate changes, in proportions and constructional details, from the 1956 pictorial reconstruction {Fig. 1}. They also include internal stairs for access to the upper floor; we take this opportunity to withdraw the statement {p. 28 note 2} that a staircase is not mentioned in the text, and to correct the translation of the term *hu t'i* [2] {p. 229}, which should be read as 'staircase with balustrade'.

[6] Combridge (3), Fig. 2. (For a museum-made working model to the same scale see Ward (3), Fig. (2); Needham (6), Fig. 20.)

[7] Robinson (2); Combridge (4), Fig. 19; Combridge (5), Fig. 226/1.

[8] Made by Barnett and Walls Ltd., London E12, for The Time Museum, Rockford, Illinois, U.S.A. See Turner (1), pp. v–vi, 59–65, Figs. 32–4 (also Fig. 30 for a pictorial clock-tower reconstruction).

[9] Made for The International Exposition, Tsukuba, Japan, 1985, by Kyoritz Model Engineering Co. Ltd., represented by Masaru Nagae, Tokyo, in consultation with John H. Combridge, London.

PLATE XXII

Fig. 72. Time-keeping water-wheel: half-scale working-model reconstruction. The outer framework of the model is 9 ft high. An electrical contact on the model operates the modern wall-dial. (Photo.: Merseyside County Museums)

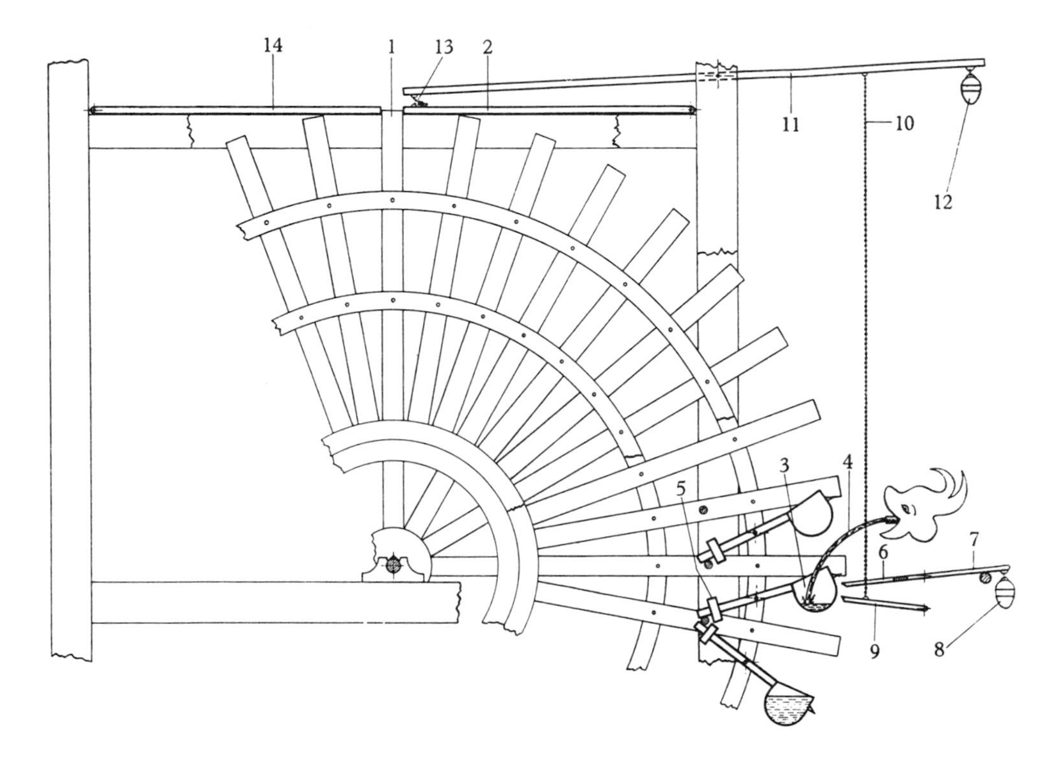

Fig. 73. Time-keeping water-wheel: explanatory drawing, roughly to scale. At the beginning of each 24-second time-unit interval the water-wheel is stationary, being held so by the engagement of its top spoke 1 with the right-hand upper lock 2. As water 4 from the constant-level tank enters the bucket 3, the counter-weight 5 is soon overcome, and the excess weight of water than causes the lug at the side of the bucket to engage with the checking fork 6 at the end of the lower balance-lever 7. When there is sufficient water to overcome the lower balance-weight 8 the lever 7 is suddenly tripped. It releases the lug onto the trip-lever 9, which is connected by the long chain 10 to the upper balance-lever 11. The upper balance-weight 12 was by itself insufficient to raise the left-hand end of this lever, but the excess weight of the water in the filled bucket now sets both levers in motion. Momentum is gathered during the swing of the bucket and levers, near the end of which the short chain of the upper link 13 suddenly becomes tight, so that the right-hand upper lock 2 is jerked out of the way of the spoke 1. The wheel now makes a quick clockwise step under the gravitational driving force of the filled buckets in its lower right-hand quadrant, while the left-hand end of the upper balance-lever 11 and the right-hand upper lock 2 fall under their own weight to arrest the following spoke. The left-hand upper lock 14 is lifted by the passage of this spoke, and falls again to prevent any recoil as the wheel is stopped. Meanwhile levers 7 and 9 regain their normal places ready for the next cycle of operations.

In the above account, the terms 'bucket', 'trip-lever', and 'upper link' replace the terms 'scoop', 'stopping-tongue', and 'upper stop', used in 1960. We would now also refer to the 'pivots' or 'hinges' of the levers, rather than to 'shutting axles' {as on pp. 41ff}.

The explanatory drawing, which is based on Combridge (1), Fig. 4, shows the buckets as having rounded bases. During the later construction of water-driven working models it was concluded that good evidence for this shape is lacking. Buckets of rectangular cross-section, of the proportions stated in the description of the astronomical clock-tower of A.D. 1086–9, were therefore used, as shown in the photograph of the Merseyside County Museums' half-scale working-model reconstruction, Fig. 72.

with fixed buckets {Figs. 1, 23, 25, 26} would (if indeed it could in real life have been made to drive the mechanism of a clock-tower[1]) certainly not have permitted weighing of the water in the buckets with the precision needed for accurate time-keeping.

The 'scoop-holders' *hung* [30][2], postulated in 1960 {p. 31 note 1} as simple shelves, were accordingly interpreted as pivoted and counterbalanced levers,[3] each of which supported one of the thirty-six 1-gallon buckets upon the wheel in a manner which allowed its water content to be weighed accurately against the balance-weight carried by the central balance-lever *shu hêng* [69] (Fig. 73, no. 7).[4]

Experimental modification of a 1961 sand-driven version also confirmed the essential function of the upper balance-lever *t'ien hêng* [62] (Fig. 73, no. 11) in accumulating enough kinetic energy, during the downward swing of a single filled bucket, to give the right-hand upper lock *t'ien so* [72] (Fig. 73, no. 2) a sudden jerk sufficiently powerful to overcome static friction between its extremity and the front face of the top spoke of the heavily loaded water-wheel.[5] The elegance of this device may be seen as an item of evidence in favour of our opinion that the time-keeping water-wheel may well have been one of the inventions of Chang Hêng [480], whose expertise in the use of inertial and kinetic-energy effects is also apparent in his famous second-century earthquake-annunciator.[6]

The 'star-shaped gadget' {p. 58 note 2} at the right-hand centre of the (probably tenth-century, if not earlier) schematic drawing {Fig. 18} was in 1961 seen to be not part of the mechanism but a representation of a traditional Chinese dragon's-head gargoyle, from the mouth of which the outlet-pipe from the constant-level tank projected towards the buckets of the water-wheel, as shown in Fig. 73.[7] The presence in Cambridge University Library[8] of an 1844 quarto edition of *Hsin I Hsiang Fa Yao* allowed the later preparation of a photographic

[1] Burstall (1), Fig. 69, shows 'essential features of a working model' with fixed triangular-section buckets. See also note 1, on p. 208.

[2] The etymology of the technical term *hung* [30] is obscure. In present Chinese usage the character signifies 'in flood', etc. We suspect that in the Sung Dynasty it had to do with some device associated with fish-weirs.

[3] This interpretation may have been prompted unconsciously by the reference {p. 57 note 2} to 'a series of balancing clepsydras mounted on the periphery of a wheel'.

[4] Combridge (1), p. 85 and Fig. 4.

[5] *Ibid.* pp. 84–5.

[6] Needham (3), pp. 626–32; Needham (4), p. 484. (Lorch (1), pp. 291–3, discusses the alternative possibilities that Chang Hêng's mechanisms may have been powered by descent of a clepsydra float {cf. Fig. 33 (*a*)} or of a weight resting on granular material in a leaking container (cf. Needham (5), Fig. 61). We remark that the second of these appears to be excluded by the evidence {p. 108} for the use of water.)

[7] Combridge (1), Fig. 4.

[8] Cambridge University Library: Class-mark FC.55·84/19 (Wade B 1258).

enlargement revealing the outlet-pipe to have been drawn on an alignment differing from that of the balance-lever, which can plainly be seen {cf. Fig. 18} as terminating in front of the dragon's-head rather than crossing behind it.[1]

The need for the clock-tower's external time-annunciator to indicate visibly and audibly both the twenty-four half-double-hours and the Hundred Intervals *k'o* [21] of the day-and-night, and other considerations, led to the conclusion that the water-wheel's time-unit interval was 24 seconds, that is $\frac{1}{3600}$ and not $\frac{1}{100}$ {p. 3, p. 39 note 1, p. 49, and p. 56 note 1} of a day-and-night.[2] Thus the wheel revolved precisely a hundred times in a day-and-night, and consumed in each double-hour a quantity of water needing about five man-minutes of work at the norias for its recirculation to the storage tank.

More recent study has suggested that the Hundred Intervals were not in fact reckoned and announced consecutively but in twelve groups of eight, each group being reckoned and numbered anew from the mid-point of a double-hour, and completed by a 'remainder fraction' of $\frac{1}{3}$ of an Interval before the mid-point of the next double-hour.[3] This is now seen as the true explanation of the fact that there were in all only 96 Interval jacks {p. 37 note 2} instead of a full 100.[4] The method of reckoning and numbering sequences of $8\frac{1}{3}$ Intervals from and to the mid-points (rather than the beginnings) of double-hours is also now believed to account for a medieval remark {p. 93 note 5} that the double-hours 'overlapped with each other almost by half'.[5]

Earlier, a study of the text and illustrations of *Hsin I Hsiang Fa Yao* had shown the 1172 edition by Shih Yuan-Chih [464] to have been an editorial conflation of an official report of 1092 by Su Sung [450] with texts embodying technical descriptions relating to three astronomical clock-towers, of the periods 976–9, 1078–85, and 1086–9, with illustrations the majority of which can be ascribed to the first of these periods.[6] Analysis allowed the identification of the principal

[1] Burstall *et al.* (1) described an experimental model in which the 'star-shaped gadget' was interpreted as an 'interrupter cam' for reversing the rotation of a wheel fitted with shallow trays from each of which the water was quickly spilled.

[2] Combridge (1), p. 83.

[3] Needham *et al.* (1), Chapter 2 note 47.

[4] *Ibid.* (This finding supersedes the suggestion in Combridge (1), p. 83.)

[5] *Ibid.*

[6] Combridge (4), pp. 291–5. (Combridge (4), Figs. 1 and 2, reproduces, from the 1969 octavo reprint of *Hsin I Hsiang Fa Yao* (see Shih Yuan-Chih (ed.) (1) in Supplementary Bibliography, p. 218) two illustrations which in 1960 were reproduced {Figs. 5 and 8} from the 1922 octavo reprint. Combridge (4), Figs. 3–15 and 17, reproduces, from the 1844 quarto edition (see note 8 on p. 207), six illustrations of which five {Figs. 7, 10, 21, 22, 28} were reproduced in 1960 from the 1922 octavo reprint, and eight illustrations (Combridge (4), Figs. 4, 7, 9, 10, 11, 12, 14, 17) which were not reproduced in 1960. Combridge (4) Fig. 16 was reproduced from the 1922 octavo reprint {cf. Fig. 16}.)

features of each clock-tower, and revealed the principal objective of successive designers to have been the improvement of calendrical studies by the provision of a clock-drive to the sidereal-reference component *san ch'en i* [4] {Fig. 24, no. 4} of an observational armillary sphere, rather than to its sighting-tube *wang t'ung* [132], {p. 24 note 3}.[1]

The results of the earlier practical work and textual studies were published in a series of articles,[2] and the later conclusions are noted in a currently published joint work.[3] We reproduce, as Fig. 73 of the present edition and with a rewritten caption, an explanatory drawing first published in 1961, and reprinted with a descriptive caption in 1965.[4]

2. TIME-KEEPING IN EAST ASIA BEFORE AND AFTER SU SUNG

2.1 *Steelyard clepsydras*

In their survey of Chinese clepsydra technique {pp. 85–94} the authors of *Heavenly Clockwork* distinguished in 1960 between a small steelyard clepsydra *hsing lou* [279] (literally 'Travel Clepsydra'[5]) {Fig. 38, Type C} for timing short intervals by weighing the liquid (water or mercury) contents of an inflow-vessel suspended from a balance-beam having a counterweight shifted manually according to the time elapsed; and a large steelyard clepsydra *ch'êng lou* [204, 287] or *ch'üan-hêng k'o-lou* [292, 401] {Fig. 38, Type D} in which the balance-beam was then thought to have been used in monitoring the rate of water inflow to the stationary receiver of an inflow-float clepsydra by means of an intermediate 'compensating tank' suspended from the balance-beam and counterbalanced by a weight positioned according to the flow-rate desired. In supposing that the flow must have been adjusted to run fast during the short summer nights and winter days[6] they temporarily overlooked the fact that in all Chinese clepsydras an unvarying flow was needed for accurate timing of the seasonally invariable twelfths, twenty-fourths, and hundredths of the day-and-night.[7]

[1] The illustration reproduced in 1960 as Fig. 6 (from Maspero (1) {see p. 224}, Fig. 8, cf. Needham (3), Fig. 159, Needham (6), Fig. 10) is from a drawing, of a manually adjusted armillary sphere of 976–9, into which a representation of a clock-drive gear-ring was inserted later; see Combridge (4), pp. 290–2, with Figs. 3 and 7. Combridge 4, Fig. 18, reproduces a postage-stamp drawing of an armillary sphere (cf. Needham (3), Figs. 156, 163, and Needham (6), Fig. 11) believed to resemble the improved clock-driven version of A.D. 1086–9 by Han Kung-Lien [457].

[2] Combridge (1), (2), (3), (4), (5), (6).

[3] Needham *et al.* (1), Chapter 2, notes 29 and 47.

[4] Combridge (1), Fig. 4, reprinted in Needham (4), Fig. 658, and Needham (6), Fig. 21.

[5] Needham *et al.* (1), Chapter 2, pp. 84–6.

[6] Needham (3), p. 327 note j.

[7] Combridge (8), p. 530.

A review of Chinese clepsydra literature led to the publication in 1981 of an alternative interpretation[1] in which the distinction between small and large steelyard clepsydras was abandoned, and both were seen as straightforward adaptations of Graeco–Arabic time-keeping devices to Chinese horary requirements and traditional techniques.[2]

Revision of the translation of the key Chinese text-description of a steelyard clepsydra {pp. 92–3} led to the suggestion that the characters *t'ien ho* [295], previously read as a name, 'Celestial River', for the whole system {p. 92 note 1} should be read as a name, 'Sky River', given to a chase cut into the upper surface of the wooden balance-beam to accommodate interchangeable time-scale rods {cf. the reference, p. 92 note 1, to graduated scales inscribed on the beam}. We now read the characters *t'ien ho* [295] as a simple technical term, 'upper channel', for the chase; cf. the technical terms 'upper flume' *t'ien ho* [51] {for which see pp. 31 and 42} and 'upper balance-lever' *ti'en hêng* [62] (on which see note 4 on p. 206). The sixteen faces of the four time-scale rods provided would have been more than sufficient to accommodate thirteen different seasonal time-scales graduated not only for the equal and invariable double-hours, half-double-hours, and hundredths of a day-and-night, but also for the equal though seasonally variable night-watches *kêng* [360] and subdivisions *chou* [139] or fifths-of-a-night-watch {Fig. 71}.[3]

Translation of an extract from the *Kuan-Shu K'o-Lou T'u* [757], written by Wang P'u [547] about 1135, which accompanies the illustration of clepsydras reproduced in Fig. 39,[4] confirmed that, in some Chinese steelyard clepsydras, auxiliary movable weights were used for measuring small fractions of the whole running period, as in Islamic steelyard clepsydras.[5]

The 'iron arch' *t'ieh hu mên* [310] (literally 'iron foreign doorway') {p. 92} was perceived to be a traditional Islamic balance-suspension, comprising a metal 'yoke' to which the cross-piece of the balance-beam was linked by rows of silk threads.[6]

[1] *Ibid.* pp. 530–2. Cf. the similar interpretation of Lorch (2), published independently at about the same time. Sleeswyk (1) gave a markedly different interpretation of the steelyard clepsydra. A critique by John S. Major of the differing approaches is in preparation; see Needham *et al.* (1), Chapter 2, note 89.

[2] Graeco–Arabic steelyard clepsydras are described and illustrated by al-Khāzinī. The translation, in part, by Khanikov (1) {see p. 88 note 7, and p. 223} is now superseded by that, in full, of Hill (1), pp. 47–68.

[3] For the designations of individual night-watches and fifths see Needham *et al.* (1), Chapter 2, p. 39, and note 29.

[4] From *Liu Ching T'u* [758] by Yang Chia, a rare *c.* 1155 edition of which survives in Taiwan; see Combridge (8), p. 535 note 39.

[5] Combridge (8), p. 535 note 39. (Cf. Lorch (2), Fig. 1 A.)

[6] *Ibid.* note 42; Hill (1), Fig. 23.

We believe there is now adequate evidence that Chinese steelyard clepsydras resembled Islamic steelyard clepsydras in most respects, apart from the incorporation in the former of seasonally interchangeable time-scale rods.[1]

2.2 *Striking Clepsydras*

The term 'Striking Clepsydra' is a shortened translation of the Chinese–Korean name *tzu-chi lou* (literally 'self-striking clepsydra'; cf. the name *tzu-ming chung* [214], literally 'self-sounding bell', for a striking clock {pp. 142–3}) which was given to a monumental time-keeping installation built in Seoul on the orders of King Sejong in A.D. 1432–4.[2] We extend it here to embrace the whole class of time-keepers automatically giving audible and/or visible time-announcements, and in some cases also rotating celestial globes, by means of mechanisms which we now think were probably similar to those of the A.D. 1432–4 installation.

The time-keeping elements of King Sejong's Striking Clepsydra comprised a pair of inflow-float clepsydras, the float-rods of which released, from seasonally interchangeable vertical racks, small metal balls which in turn released larger metal balls to work audible and visible effects, including a horizontally rotating jack-wheel.[3] It is now thought that similar methods were used in Kuo Shou-Ching's 'Precious Mountain Clepsydra' and 'Lantern Clepsydra' of A.D. 1262–70 {pp. 134–6}, Shun Ti's 'Palace Clepsydra' of A.D. 1354 {p. 140}, and other East Asian Striking Clepsydras built after the eclipse of time-keeping water-wheel technology following the fall of K'ai-fêng in 1126 {pp. 126–31}.[4] Their origin is ascribed to the adaptation of Graeco-Arabic ball-release techniques to East Asian horary requirements.[5]

Special considerations arise from T'ang Dynasty reports of a monumental 'steelyard' clepsydra in a Western city {p. 88 note 7}. Our present view is that this may have been a Striking Clepsydra in which metal balls, released by a mechanism probably of Graeco-Arabic anaphoric type,[6] operated external jack-work arranged to represent the horizontal balance-beam and other features of a manually attended steelyard clepsydra.[7] We think the traditional Graeco-Arabic

[1] Interchangeable time-scale rods continued in use in Japanese weight-driven pillar-clocks {Fig. 60, right-hand} until about 1870.

[2] Needham *et al.* (1), Chapter 2, pp. 23, 26–44.

[3] *Ibid.* Chapter 2, pp. 29–30, 38–9.

[4] *Ibid.* Chapter 1, note 26; Chapter 2, note 30.

[5] *Ibid.* Chapter 2, pp. 41–4.

[6] Hill (1), Fig. 3.

[7] Combridge (8); Hill (1), pp. 47–68.

form of visible time-annunciation, by the successive opening of twelve doors arranged in a horizontal row,[1] may have originated from jackwork thus designed.[2]

Some Korean Striking Clepsydras included 'advisory inclining vessels' *wo ch'i* [275] {pp. 84, 89, 94, 103}. These vessels were, however, filled by the non-uniform overflows of water from the clepsydras' constant-level tanks. Thus they cannot have served any useful time-keeping function, and presumably were there only to impress onlookers.[3]

2.3 *Ming sand-clocks and water-clocks*

Further studies of the Ming sand-clocks described by Sung Lien [636], and of several Ming water-clocks described by Chou Shu-Hsüeh [690], {pp. 157–9} have been published by Li Ti and Pai Shang-Shu.[4] They seem to confirm that the sand-clocks had no escapements, but relied solely upon gear-train friction. We can only suppose that the inescapable equivalence between the rate of energy-extraction by the bucket-wheel {Fig. 58} from a fairly constant stream of falling sand, and the rate of energy-consumption by friction in the gear-train, was found to yield a sufficiently constant going rate.[5] At least one of the water-clocks described by Chou Shu-Hsüeh is shown to have included a clock-driven celestial globe.[6]

2.4 *Wang Chêng's verge-and-foliot clock*

Reconsideration of the information available on the verge-and-foliot clock {pp. 146–7 and Fig. 53} of Wang Chêng [696], in the light of the knowledge gained from the study of Song Iyŏng's armillary clock (Section 2.5 below), has led us to believe that the function of its lead 'cross-bow bullets' was only to sound the drum and bell, either by direct percussion or by the operation of simple hammer-levers, and that after coming to rest in the drawers at the base of the clock case they had to be reloaded at the top by hand.[7]

[1] Hill (1), Plate 1 (in colour) {see also Fig. 34, opposite p. 68}.

[2] The attribution of the reported 'steelyard' clepsydra to Antioch {p. 88 note 7} was that of Hirth, whose attention was focussed on the Roman Orient. When read in the light of later publications by Enoki (1), Hirth and Rockwell (1), and Shiratori (1), the *T'ang Shu* texts appear consistent with the probability, in our view, that this clepsydra was one in the triple-walled original Round City of Baghdad, founded in A.D. 762 as described by le Strange (1). It was from a Caliph of Baghdad, Harūn ar-Rashīd, that Charlemagne received in A.D. 807 a twelve-door Striking Clepsydra; see Needham (4), p. 499 note d, and Section 3 below.

[3] Needham *et al.* (1), Chapter 2, pp. 77–9.

[4] Li Ti & Pai Shang-Shu (1).

[5] This opinion supersedes that expressed in Needham (4), p. 512 note a.

[6] Pai Shang-Shu & Li Ti (1).

[7] Needham *et al.* (1), Chapter 4, p. 124.

2.5 *Song Iyŏng's armillary clock*

A detailed study of the Korean armillary clock {illustrated in Fig. 59, and briefly described on p. 162} was undertaken in 1964,[1] and is currently published with a full set of photographs and explanatory drawings.[2] The clock includes an armillary sphere designed to model the instrument depicted in a Sung Dynasty illustration[3] which was also the source for the manuscript drawing reproduced in Fig. 30. The twenty-seven pegs on the moon-path ring of the armillary sphere {p. 162} are now believed not to mark the twenty-eight lunar mansions (or lodges), but to have operated a phase-of-moon mechanism (reconstructed schematically in the current publication[4]) which resembled in function that of which a prototype was described by Wang Fu [599] as having been made in A.D. 1124 {pp. 119–20}.

2.6 *Hsü Ch'ao-Chün's encyclopedia of astronomy and horology*

Eight pages of illustrations from the horological section, published in 1809 under the title *Tzu Ming Chung Piao T'u Fa* [800], of the encyclopedia *Kao Hou Mêng Ch'iu* of Hsü Ch'ao-Chün [647] {p. 154} have been published by Bedini, after Bonnant.[5]

Our own studies of Hsü's encyclopedia led us to the fact, recorded by Wylie,[6] that the publication in 1807 of its astronomical section was accompanied by that of a large and handsome twin-planispheric star-map on ecliptic co-ordinates, updated from those which in the first half of the seventeenth century had been published by Hsü's famous namesake (and likely ancestor) Hsü Kuang-Ch'i [648] in collaboration with the Jesuit astronomers at the emperor's court in Peking.[7]

In another section of the encyclopedia, published about 1815, Hsü Ch'ao-Chün described the construction of celestial and terrestrial globes and included a full set of equatorial half-gores printed with reversed star-maps based on the ecliptic planispheres of 1807. We believe the publication of Hsü's encyclopedia provided the inspiration for the production, under the supervision of Ch'i Mei-Lu,[8] and by other makers around the years 1828–30, of a series of clock-driven celestial globes engraved with reversed star-maps according to these gores and planispheres.[9]

[1] (Robinson) (1).
[2] Needham *et al.* (1), Chapter 4.
[3] *Ibid.* Fig. 4.10.
[4] *Ibid.* Chapter 4, note 37 and Fig. 4.19.
[5] Bedini (4), after Bonnant (1).
[6] Wylie (3), rev. ed., p. 124. (The Chinese identification-numbers of the stars in Wylie (2) follow those of Hsü's planispheres, as copied by Ch'ien Wei-Yüeh in 1839; see Wylie (2), p. 110.)
[7] Pelliot (1) {see p. 224}, p. 67 note 2; Needham (3), p. 447; Needham (4), p. 532 note a.
[8] Courtesy-name of Ch'i Yen-Huai, for whom see Needham (4), pp. 527–8 and Fig. 670.
[9] Combridge (9) (in preparation: meanwhile see Needham *et al.* (1), Chapter 4, note 14).

In his encyclopedia, Hsü also described the construction of sun-, moon-, and star-dials, planispheric astrolabes, and other instruments, examples of which may be expected to come to light.

3. TIME-KEEPING IN MEDIEVAL EUROPE

The contemporary description of Charlemagne's striking water-clock {p. 187 note 2} appears to be that of an eye-witness, and it tallies closely with that of a Striking Clepsydra of the kind {cf. Fig. 34} having twelve time-announcing doors arranged in a horizontal row.[1]

By the middle of the fourteenth century, weight-driven clocks with mechanical escapements were well known. Efforts have been made to determine a precise date for the first wholly mechanical clock, but we believe there has been unnecessary academic timidity in admitting the probability that such clocks existed at dates well before those for which firm documentary proof is available.

The 1960 authors' perception {p. 58} of an analogy between the time-keeping water-wheel linkwork mechanism and the seventeenth-century anchor escapement may have been, with hindsight, a little rash, but we still think the suggestion {pp. 196–8} that there was a stimulus to Western inventors by rumours from East Asia should not be altogether dismissed.[2]

Robertus Anglicus reported in A.D. 1271 {p. 196} that 'horological artificers' were then trying to make a wheel keep sidereal time accurately, but could not 'quite perfect their work'. Lynn Thorndike thought in 1941 that this implied there were no mechanical clocks in 1271, but soon would be.[3] We, however, tend to see it as suggestive of the existence of an established clockmaking trade, the workers in which were already seeking the time-keeping accuracy desired for astrological purposes. We think there is no justification for dismissing their products as necessarily 'water-clocks'. We think Robertus may merely have wished to suggest that a well-made compartmented cylindrical clepsydra might regulate a weight-driven mechanism more accurately than available types of mechanical escapement.[4]

A study of compartmented cylindrical clepsydras has been published byBedini,[5]

[1] Hill (1), p. 14. (The Latin text of the description is quoted by Kurz (1), p. 7 note 2. For an English version, of unstated provenance, see Goaman (1), p. 22, but read 'clepsydram' in lieu of 'elepfydram' and 'sort of bells' in lieu of 'fort of bells'.)

[2] See, e.g., Cipolla (1), p. 40; Landes (1), p. 23.

[3] Thorndike (1) {see p. 227}, p. 242.

[4] Combridge (6), p. 607.

[5] Bedini (3).

whose examples include the one illustrated in Fig. 67 from the *Libros del Saber* of *c.* 1276.[1] Following Lynn White, we now believe that the 'King Hezekiah' clock, illustrated in Fig. 68 from a Paris *Bible Moralisée* of *c.* 1250,[2] was also a compartmented cylindrical clepsydra, which was weight-driven by the heavy rope-suspended water-container visible in the illustration.[3]

The clock-housing seen before 1235 by Villard de Honnecourt {p. 195} has been persuasively interpreted by Hahnloser[4] and Simoni[5] as designed to accommodate the mechanism of a weight-driven carillon clock, similar to the one of which the mechanism was well depicted in an anonymous (and now lost) drawing copied later (in three separate portions, for artistic reasons) by the artist of the so-called 'Brussels Miniature' of *c.* 1450.[6]

[1] Bedini (3), Fig. 1, reproduces another example of the same drawing.

[2] White (1), p. 120 and Fig. 10; Drover (1) and (4).

[3] Combridge (6), p. 607. (While that article was in the press, a different interpretation of the 'King Hezekiah' water-clock was published by Sleeswyk (2), who was unaware of the evidence for the presence of water in one of the compartments of the wheel; see Combridge (6), note 35 (misprinted as note '34'), and Combridge (7).)

[4] Hahnloser (2), pp. 29–32, 349.

[5] Simoni (1) and (2).

[6] Michel (5), Fig. 1; Michel (6), p. 289; Michel (7), pp. 124–5; Drover (2) and (3); Combridge (6), p. 606–7, Figs. 2 and 3.

BIBLIOGRAPHY TO SUPPLEMENT

ANON. (1). 'Model of Su Sung's Escapement'. *Horological Journal*, August 1961, **103**, 481.

ANON. (2). '"Heavenly Clockwork" – a Sequel' (by B.L.H.). *Antiquarian Horology*, March 1962, **3** (10), 297.

BEDINI, S. A. (3). 'The Compartmented Cylindrical Clepsydra.' *Technology and Culture*, Spring 1962, 115–41.

BEDINI, S. A. (4). 'Oriental Concepts of the Measure of Time', in Fraser and Lawrence (eds.) (1), *q.v.*, pp. 451–84.

BONNANT, G. (1). 'L'Introduction de l'horlogerie occidentale en Chine.' *La Suisse Horlogère*, 27 August 1959, 767–78; English tr. 'The Introduction of Western Horology in China', *La Suisse Horlogère*, international ed., April 1960, **75** (1), 28–38.

BURSTALL, A. F. (1). *A History of Mechanical Engineering*. Faber and Faber, London, 1963, repr. (paperback) 1965.

BURSTALL, A. F., LANSDALE, W. E. & ELLIOTT, P. (1). 'A Working Model of the Mechanical Escapement in Su Sung's Astronomical Clock Tower.' *Nature*, 1963, **199**, 1242–4.

CIPOLLA, C. M. (1). *Clocks and Culture 1300–1700*. Collins, London, 1967.

COMBRIDGE, J. H. (1). 'The Celestial Balance: a Practical Reconstruction.' *Horological Journal*, February 1962, **104** (2), 82–6; extracts in *Bulletin of the National Association of Watch and Clock Collectors, Inc.*, August 1975, **17** (4), 338–41.

COMBRIDGE, J. H. (2). 'Chinese Water Clock.' *Horological Journal*, November 1963, **105**, 347.

COMBRIDGE, J. H. (3). 'The Chinese Water-Balance Escapement.' *Nature*, December 1964, **204**, 1175–8.

COMBRIDGE, J. H. (4). 'The Astronomical Clocktowers of Chang Ssu-Hsun and his Successors.' *Antiquarian Horology*, June 1975, **9** (3), 288–301.

COMBRIDGE, J. H. (5). 'Clockmaking in China: Early History', in Smith, A. (ed,) (1), *q.v.*, pp. 225–6.

COMBRIDGE, J. H. (6). 'Clocktower Millenary Reflections.' *Antiquarian Horology*, Winter 1979, **11** (6), 604–8.

COMBRIDGE, J. H. (7). 'The 13th Century "King Hezekiah" Water Clock: Addenda.' *Antiquarian Horology*, Autumn 1980, **12** (3), 300.

COMBRIDGE, J. H. (8). 'Chinese Steelyard Clepsydras.' *Antiquarian Horology*, Spring 1981, **12** (5), 530–5.

COMBRIDGE, J. H. (9). 'The clock-driven celestial globes of Qi Mei-lu and others' (in preparation).

DROVER, C. B. (1). 'A Medieval Monastic Water-Clock.' *Antiquarian Horology*, December 1954, **1** (5), 54–63; partly repr. *ibid.*, Summer 1980, **12** (2), 165–9.

DROVER, C. B. (2). 'The Brussels Miniature.' *Antiquarian Horology*, September 1962, **3** (12), 357–61.

DROVER, C. B. (3). 'The Brussels Miniature.' *Antiquarian Horology*, September 1965, **12** (4), 381–2.

DROVER, C. B. (4). 'The 13th century "King Hezekiah" Water Clock.' *Antiquarian Horology*, Summer 1980, **12** (2), 160–4.

ENOKI, K. 'Some remarks on the country of Ta-ch'in as known to the Chinese under the Sung.' *Asia Major*, 1954, new series, **4** (1), 1–19.

FRASER, J. T. & LAWRENCE, N. (eds.) (1). *The Study of Time*, vol. II. Springer-Verlag, New York, Heidelberg, and Berlin, 1975.

GOAMAN, M. (1). *English Clocks*. The Connoisseur, and Michael Joseph, London, 1967.

HAHNLOSER, H. R. (2). *Villard de Honnecourt*. 2nd rev. and enlarged ed. Schroll, Vienna, 1972.

HILL, D. R. (1). *Arabic Water-Clocks*. University of Aleppo Institute for the History of Arabic Science, Aleppo, Syria, 1981.

HIRTH, F. & ROCKHILL, W. W. *Chau Ju-Kua: his work on the Chinese and Arab trade in the twelfth and thirteenth centuries, entitled Chu-fan-chï*. St Petersburg, 1911: repr. Taipei, 1970.

KURZ, O. (1). *European Clocks and Watches in the Near East*. Warburg Institute, London, and E. J. Brill, Leiden, 1975.

LANDES, D. S. (1). *Revolution in Time: Clocks and the Making of the Modern World*. Belknap Press, Cambridge, Mass., and London, England, 1983.

LE STRANGE, G. *Baghdad under the Abbasid Caliphate*. University Press, Oxford, and Humphrey Milford, London, 1900; repr. 1924.

LI TI & PAI SHANG-SHU. (1). 'Ch'ao-Pen "Shen-Tao Ta-Pien Li-Tsung Tung-I".' *Nei Mêng-ku shih-fan hsüeh-yüan hsüeh-pao (Inner Mongolia Teacher-training College Journal)*, 1981 (2), 85–90.

LORCH, RICHARD P. (1). 'Al-Khāzinī's "Sphere That Rotates by Itself".' *Journal for the History of Arabic Science*, 1980, **4**, 287–329.

LORCH, RICHARD P. (2). 'Al-Khāzinī's Balance-Clock and the Chinese Steelyard Clepsydra.' *Archives Internationales d'Histoire des Sciences*, June 1981, **31**, 183–9.

MICHEL, H. (5) 'L'Horloge de Sapience et l'histoire de l'horlogerie.' *Physis*, 1960, **2** (4), 291–8.

MICHEL, H. (6). 'Some New Documents in the History of Horology.' *Antiquarian Horology*, March 1962, **3** (10), 288–91.

MICHEL, H. (7). *Images des Sciences*. Visscher, Rhode-St-Genèse, Belgium, 1977.

NEEDHAM, J. (3). *Science and Civilisation in China*, vol. III. University Press, Cambridge, 1959.

NEEDHAM, J. (4) *Science and Civilisation in China*, vol. IV, part 2. University Press, Cambridge, 1965.

NEEDHAM, J. (5). *Clerks and Craftsmen in China and the West*. University Press, Cambridge, 1970.

NEEDHAM, J. (6). 'Astronomy in Ancient and Medieval China.' *Phil. Trans. Roy. Soc. London*, 1974, **A. 276**, 67–82.

NEEDHAM, J., LU GWEI-DJEN, COMBRIDGE, J. H. & MAJOR, J. S. (1). *The Hall of Heavenly Records: Korean Astronomical Instruments and Clocks, 1380–1780*. University Press, Cambridge, 1986.

PAI SHANG-SHU & LI TI (1). 'Chou Shu-Hsüeh tsai chi-shih-ch'i fang-mien-ti kung-hsien.' *Tsü-jan k'o-hsüeh-shih yen-chiu (Studies in the History of Natural Sciences)*, 1984, **3** (2), 138–44.

(Robinson, T. O.) (1). 'A Korean 17th Century Armillary Clock.' (Notes on a lecture given by J. H. Combridge on 27 November 1964.) *Antiquarian Horology*, March 1965, **4** (10), 300–1.

R(obinson), T. O. (2). 'The Astronomical Clock of Su Sung.' *Antiquarian Horology*, June 1968, **5** (11), 414–15.

Shih Yuan-Chih (ed.) (1). *Hsin I Hsiang Fa Yao*, 1172. Repr. in *Ts'ung-shu Chi-Ch'eng* (no. 1302), Shanghai, 1935–7, and in *Jen-jen Wen-ku* (no. 1248), Taipei, 1969.

Shiratori, K. 'A New Attempt at the Solution of the Fu-lin Problem.' *Memoirs of the Research Department of the Toyo Bunko (The Oriental Library)*, 1956, no. 15, 165–329.

Simoni, A. (1). 'Un orologio a cembalo in una miniatura quattrocentesca.' *La Clessidra*, November 1965, 40–2.

Simoni, A. (2). 'Un nuovo documento per la storia dell' orologeria.' *La Clessidra*, April 1968, 18–21.

Sleeswyk, A. W. (1). 'The Celestial River: a Reconstruction.' *Technology and Culture*, July 1978, **19** (3), 423–49.

Sleeswyk, A. W. (2). 'The 13th Century "King Hezekiah" Water-Clock.' *Antiquarian Horology*, Autumn 1979, **11** (5), 488–94.

Smith, A. (1) (ed.). *The Country Life International Dictionary of Clocks*. Country Life Books, c/o Hamlyn Publishing Group, Feltham, Middlesex, 1979.

Turner, Anthony J. (1). *The Time Museum Catalogue, vol.* 1, *part* 3: *Water-clocks, Sand-glasses, Fire-clocks*. The Time Museum, Rockford, Illinois, 1984.

Ward, F. A. B. (3). *Science Museum Descriptive Catalogue of the Collection Illustrating Time Measurement*. H.M.S.O., London, 1966.

White, L., Jnr (1). *Medieval Technology and Social Change*. University Press, Oxford, 1962; paperback 1964.

White, L., Jnr (2). *Medieval Religion and Technology*. University of California Press, Berkeley, Los Angeles, and London, 1978.

Wylie, A. (2). 'List of Fixed Stars' (*c.* 1850). Repr. in his *Chinese Researches*, Shanghai, 1897, repr. Taipei, 1966; part 3, 'Scientific', pp. 110–39.

Wylie, A. (3). *Notes on Chinese Literature*. Shanghae [*sic*] and London, 1867; rev. ed. Shanghai 1901, reissued 1902.

GENERAL BIBLIOGRAPHY

Armão, E. (1). *Vincenzo Coronelli, Cenni sull'Uomo e la sua Vita, Catalogo Ragionato delle sue Opere, Lettere....* Bibliopolis, Florence, 1944 (Biblioteca di Bibliografia Italiana, no. 17).

Baillie, G. H. (1). *Clocks and Watches; an historical Bibliography.* Nag Press, London, 1951.

Baillie, G. H. (2). *Watches.* Methuen, London, 1929.

Bazin, M. (1). 'Recherches sur l'Histoire, l'Organisation et les Travaux de l'Académie Impériale de Pékin.' *Journ. Asiat.* 1858.

Beck, T. (1). 'Der altgriechische u. altrömische Geschützbau nach Heron dem älteren, Philon, Vitruv und Ammianus Marcellinus.' *Beitr. z. Gesch. d. Technik u. Industrie,* 1911, **3**, 163.

Beckmann, J. (1). *A History of Inventions, Discoveries, and Origins.* (1st ed. 1797); 4th ed. 2 vols., Bohn, London, 1846; enlarged ed. 2 vols., Bell & Daldy, London, 1872.

Bedini, S. (1). 'Johann Philipp Treffler, Clockmaker of Augsburg.' *Bull. Nat. Assoc. Watch and Clock Collectors,* 1956, **7**, 361, 415, 481 *et seq.* Sep. pr. pub. 1957.

Bedini, S. (2). 'Chinese Mechanical Clocks.' *Bull. Nat. Assoc. Watch and Clock Collectors,* 1956, **7** (no. 4), 211.

Benndorf, O., Weiss, E. & Rehm, A. (1). 'Zur Salzburger Bronzescheibe mit Sternbildern.' *Jahreshefte d. österr. Archäol. Institut* (Vienna), 1903, **6**, 32.

Bernard-Maître, H. (1). 'Ferdinand Verbiest, Continuateur de l'Œuvre scientifique d'Adam Schall.' *Monumenta Sinica,* 1940, **5**, 103.

von Bertele, H. (1). 'Precision Time-Keeping in the pre-Huygens Era.' *Horol. Journ.* 1953, **95**, 794.

Bilfinger, G. (1). *Die Babylonische Doppelstunde; eine chronologische Untersuchung.* Wildt, Stuttgart, 1888.

Biot, E. (1). *Le 'Tcheou-Li' ou 'Rites des Tcheou'.* 3 vols. Imp. Nat. Paris, 1851; photographically reproduced, Wêntienko, Peiping, 1930.

Biot, E. (2). 'Traduction et Examen d'un ancien Ouvrage intitulé *Tcheou-Pei....*' *Journ. Asiat.* 1841 (3e sér.), **11**, 593; 1842 (3e sér.), **13**, 198.

Borchardt, L. (1). *Die altägyptische Zeitmessung.* Pt. 1 of *Die Geschichte d. Zeitmessung u. d. Uhren,* ed. E. von Basserman-Jordan; de Gruyter, Berlin and Leipzig, 1920.

Brunet, P. & Mieli, A. (1). *L'Histoire des Sciences (Antiquité).* Payot, Paris, 1935.

Burgess, E. (1). *Translation of a Textbook of Hindu Astronomy; the 'Sūrya Siddhānta',* ed. P. Gangooly. Univ. Press, Calcutta, 1860, repr. 1935.

Carter, T. F. (1). *The Invention of Printing in China and its Spread Westward.* Columbia Univ. Press, New York, 1925, revised ed. 1931; 2nd edition, revised by L. Carrington Goodrich, Ronald, New York, 1955.

de Caus, I. (1). *Nouvelle Invention de lever l'Eau plus Hault que sa Source, avec quelques Machines movantes par le moyen de l'Eau, et un Discours de la Conduite d'ycelle.* London, 1644. Eng. tr. by J. Leak; Moxon, London, 1659.

CHANG YÜ-CHÊ (1). 'Chang Hêng; a Chinese Contemporary of Ptolemy.' *Popular Astron.* 1945, **53**, 1.

CHANG YÜ-CHÊ (2). 'Chang Hêng; Astronomer.' *People's China*, 1956 (no. 1), 31.

CH'ANG WÊN-CHAI (1). 'Shansi Yung-chi Hsien Hsüeh-Chia-yai Fa-hsien-ti i p'i T'ung Ch'i' ('On a Group of Bronze Objects (of the Warring States Period) discovered at Hsüeh-Chia-yai near Yung-chi city in Shansi province'), in Chinese. *Wên Wu Ts'an K'ao Tzu Liao* (Reference Materials for History and Archaeology), 1955 (no. 8), 40.

CHAPUIS, A., LOUP, G. & DE SAUSSURE, L. (1). *La Montre 'Chinoise'*. Attinger, Neuchâtel, n.d. (1919).

CHAVANNES, E. (1). 'Le Cycle Turc des Douze Animaux.' *T'oung Pao*, 1906, **7**, 51.

CHU WÊN-HSIN (1). *Li Fa T'ung Chih* (History of the Chinese Calendars), in Chinese. Com. Press, Shanghai, 1934.

CORONELLI, V. (1). *Epitome Cosmographica*. Poletti, Venice, 1693.

COUVREUR, F. S. (1). *'Tch'ouen Ts'iou' et 'Tso Tchouan'; Texte Chinois avec Traduction Française.* 3 vols. Mission Press, Hochienfu, 1914; repr. Belles Lettres, Paris, 1951.

COX, JAMES (1). *A Descriptive Catalogue of the Several Superb and Magnificent Pieces of Mechanism and Jewellery exhibited in Mr Cox's Museum at Spring Gardens, Charing Cross.* Cox, London, 1772.

COX, JAMES (2). *A Descriptive Inventory of the Several Exquisite and Magnificent Pieces of Mechanism and Jewellery comprised in the Schedule annexed to an Act of Parliament made in the thirteenth year of H.M. King George III, for enabling Mr James Cox of the City of London, Jeweller, to dispose of his Museum by way of Lottery.* Cox, London, 1774.

CREEL, H. G. (1). *Sinism; a Study of the Evolution of the Chinese World-View.* Open Court, Chicago, 1929.

CRONIN, V. (1). *The Wise Man from the West* (biography of Matteo Ricci). Hart-Davies, London, 1955.

DESAGULIERS, J. T. (1). *A Course of Experimental Philosophy.* 2 vols. Innys, Longman, Shewell & Hitch, London, 1734; 2nd ed. 1745.

DICKS, D. R. (1). 'Ancient Astronomical Instruments.' Inaug. Diss. London, 1953; abridged in *Journ. Brit. Astron. Assoc.* 1953, **64**, 75.

DIELS, H. (1). *Antike Technik.* Teubner, Leipzig and Berlin, 1914; enlarged ed. 1920.

DIELS, H. (2). 'Über die von Prokop beschriebene Kunstuhr von Gaza, mit einem Anhang enthaltenden Text und Übersetzung d. *Ekphrasis Horologiou* des Prokopios von Gaza.' *Abhdl. d. preuss. Akad. Wiss. Berlin (Phil.-Hist. Kl.)*, 1917, no. 7.

DIRCKS, H. (1). *Perpetuum Mobile; or, a History of the Search for Self-motive Power from the thirteenth to the nineteenth Century....* Vol. 1, Spon, London, 1861; vol. 2, Spon, London, 1870.

DRACHMANN, A. G. (1). 'The Plane Astrolabe and the Anaphoric Clock.' *Centaurus*, 1954, **3**, 183.

DRACHMANN, A. G. (2). 'Ktesibios, Philon and Heron; a Study in Ancient Pneumatics.' *Acta Histor. Scientiarum Naturalium et Medicinalium* (Copenhagen), 1948, **4**.

DROVER, C. B. (1). 'A Mediaeval Monastic Water-Clock.' *Antiq. Horol.* 1954, **1**, 54.

DUBS, H. H. (with the assistance of P'an Lo-Chi & Jen T'ai) (1). *History of the Former Han Dynasty, by Pan Ku, a critical Translation with Annotations*, 3 vols. Waverly, Baltimore, 1938–.

DUBS, H. H. (2). 'The Date of the Shang Period.' *T'oung Pao*, 1951, **40**, 322; 1953, **42**, 101.

EBERHARD, W. & MÜLLER, R. (1). 'Contributions to the Astronomy of the Han Period, III. Astronomy of the Later Han.' *Harvard Journ. Asiatic Studies*, 1936, **1**, 194.

EBERHARD, W. & MÜLLER, R. (2). 'Contributions to the Astronomy of the San Kuo Period.' Includes a translation of Wang Fan's *Hun T'ien Hsiang Shuo. Monumenta Serica*, 1936, **2**, 149.

EISLER, R. (1). 'The Polar Sighting-Tube.' *Arch. Internat. d'Hist. des Sciences*, 1949, **2**, 312.

D'ELIA, P. (1). *Fonti Ricciane*. 3 vols. Libreria dello Stato, Rome, 1942, 1949.

ENSHOFF, P. D. (1). 'Pater Ricci's Uhren.' *Die katholische Missionen* (Bonn), 1937, **95**, 190.

ENSHOFF, P. D. (2). 'Hatte China die ungleichen Stunden?' Typescript paper in the Jäger Collection.

ERKES, Ed. (1). *Geschichte Chinas, von den Anfängen bis zum Eindringen des ausländischen Kapitals.* Akademie Verlag, Berlin, 1956.

FAVIER, A. (1). *Pékin, Histoire et Description.* Desclee & de Brouwer, Lille, 1900.

FELDHAUS, F. M. (1). *Die Technik d. Antike u. d. Mittelalters.* Athenaion, Potsdam, 1931.

FELDHAUS, F. M. (2). *Technik d. Vorzeit....* Engelmann, Leipzig & Berlin, 1914.

FELDHAUS, F. M. (3). *Die geschichtliche Entwicklung d. Zahnräder.* Stolzenberg, Berlin–Rickendorf, 1911.

FELDHAUS, F. M. (4). 'Die Uhren des Königs Alfons X von Spanien.' *Deutsche Uhrmacher Zeitung*, 1930, **54**, 608.

FERGUSON, J. C. (1). 'The Southern Migration of the Sung Dynasty.' *Journ. Roy. Asiat. Soc. (North China Branch)*, 1924, **55**, 14.

FERGUSON, J. C. (2). 'Political Parties of the Northern Sung Dynasty.' *Journ. Roy. Asiat. Soc. (North China Branch)*, 1927, **58**, 36.

FISCHER, J. (1). 'Fan Chung-Yen (989–1052); das Lebensbild eines chinesischen Staatsmannes.' *Oriens Extremus*, 1955, **2**, 39, 142.

FITZGERALD, C. P. (1). *The Empress Wu.* Cresset Press, London, 1956.

FORKE, A. (1). '*Lun Hêng*; Philosophical Essays of Wang Ch'ung.' *Mitt. Seminar f. Orient. Sprachen*, Beibände **10** and **14**; sep. pub. 1907 and 1911, Harrassowitz, Leipzig; Kelly & Walsh, Shanghai; and Luzac, London.

FRÉMONT, C. (1). *Études Expérimentales de Technologie Industrielle*, No. 47; *Origine de l'Horloge à Poids.* Paris, 1915.

GALLAGHER, L. J. (1). *China in the Sixteenth Century; the Journals of Matteo Ricci, 1583–1610.* Random, New York, 1953. A complete translation, preceded by inadequate bibliographical details, of Nicholas Trigault's *De Christiana Expeditione apud Sinas* (1615) which was an expanded version of Ricci's diary and notes. Based on an earlier publication: *The China that was; China as discovered by the Jesuits at the close of the Sixteenth Century—from the Latin of Nicholas Trigault.* Milwaukee, 1942.

GARRISON, F. H. (1). *An Introduction to the History of Medicine.* Saunders, Philadelphia, 1913; 4th ed. 1929.

GAUBIL, A. (1). Contributions to Vol. 1 of *Observations Mathématiques, etc.*, ed. E. Souciet. Rollin, Paris, 1729.

GAUBIL, A. (2). *Histoire Abrégée de l'Astronomie Chinoise*, forming vol. 2 of the former. Rollin, Paris, 1732.

GAUBIL, A. (3). *Traité de l'Astronomie Chinoise*, forming vol. 3 of the former. Rollin, Paris, 1732.

GINZEL, F. K. (1). *Handbuch d. mathematischen u. technischen Chronologie.* 3 vols. Hinrichs, Leipzig, 1906–14. Vol. 1, *Zeitrechnung d. Bab., Ägypt., Moh., Perser, Inder, Südostasiaten, Chinesen, Japaner, und Zentralamerikaner.* Vol. 2, ...*Juden, Naturvölker, Römer u. Griechen.* Vol. 3, ...*Makedonier, Kleinasier u. Syrier, de. Germanen u. Kelten, des Mittelalters, der Byzantiner (und Russen), Armenier, Kopten u. Abessinier.... Zeitrechnung d. neueren Zeit....*

GRANET, M. (1). *La Pensée Chinoise.* Albin Michel, Paris, 1934.

GREENBERG, M. (1). *British Trade and the Opening of China, 1800–1842.* Univ. Press, Cambridge, 1951.

VAN GULIK, R. (1). *Erotic Colour Prints of the Ming Period, with an Essay on Chinese Sex Life from the Han to the Ch'ing Dynasty.* 3 vols. in case. Priv. pr., Tokyo, 1951.

GUNTHER, R. H. (1). *The Astrolabes of the World.* 2 vols. Univ. Press, Oxford, 1932.

GUNTHER, R. H. (2). *Early Science in Oxford.* 14 vols. Priv. pr., Oxford, 1923–45.

HAHNLOSER, H. R. (1). *The 'Album' of Villard de Honnecourt.* Schroll, Vienna, 1935.

HARCOURT-SMITH, S. (1). *A Catalogue of various Clocks, Watches, Automata and other miscellaneous objects of European workmanship, dating from the 18th and early 19th centuries, in the Palace Museum and the Wu Ying Tien, Peking.* Palace Museum, Peiping, 1933.

HARRIS, JOHN (1). *Astronomical Dialogues.* London, 1719.

HARTNER, W. (1). 'The Astronomical Instruments of Cha-Ma-Lu-Ting (Jamāl al-Dīn); their identification and their relations to the Instruments of the Observatory of Marāghah.' *Isis,* 1950, **41**, 184.

HAYASHI, T. (1). 'Brief History of Japanese Mathematics.' *Nieuwe Archief voor Wiskunde*, 1905, **6**, 296, and 1907, **7**, 105.

HIRTH, F. (1). *China and the Roman Orient.* Kelly & Walsh, Shanghai, 1885; Hirth, Leipzig and München, 1885; photographically reproduced in China, 1939.

HO PING-YÜ (1). 'The Astronomical Chapter of the *Chin Shu*; a Translation' (unpub.). Inaug. Diss. Singapore, 1957.

HOWGRAVE-GRAHAM, R. P. (1). 'Some Clocks and Jacks, with Notes on the History of Horology.' *Archaeologia*, 1927, **77**, 257.

HUGHES, A. J. (1). *History of Air Navigation.* Allen & Unwin, London, 1946.

HUMMEL, A. W. (1). *Eminent Chinese of the Ch'ing Period.* 2 vols. Library of Congress, Washington, 1944.

JÄGER, F. (1). 'Das Buch von den wunderbaren Maschinen [the *Ch'i Ch'i T'u Shuo* and the *Chu Ch'i T'u Shuo*]; ein Kapital aus der Geschichte der abendländisch-chinesischen Kulturbeziehungen.' *Asia Major* (n.F.), 1944, **1**, 78.

KARLGREN, B. (1). 'The Book of Documents (*Shu Ching*).' *Bull. Mus. Far Eastern Antiq.* (Stockholm), 1950, **22**, 1.

KHANIKOV, N. (1). 'Analysis and Extracts of the "Book of the Balance of Wisdom" (*al-Kitāb Mīzān al-Ḥikma*), an Arabic work on the Water-Balance, written by al-Khāzinī in the 12th century A.D.' *Journ. Amer. Orient. Soc.* 1860, **6**, 1.

KRACKE, E. A. (1). *Civil Service in Early Sung China (960–1067 A.D.), with particular emphasis on the development of controlled sponsorship to foster administrative responsibility.* Harvard Univ. Press, Cambridge, Mass. 1953 (Harvard–Yenching Institute Monograph series, no. 13).

KRACKE, E. A. (2). 'Sung Society; Change within Tradition.' *Far Eastern Quart.* 1954, **14**, 479.

KRACKE, E. A. (3). 'Family versus Merit in Chinese Civil Service Examinations under the Empire [of the Sung].' *Harvard Journ. Asiat. Stud.* 1947, **10**, 103.

KUBITSCHEK, W. (1). *Grundriss d. antiken Zeitrechnung.* Beck, München, 1928 (Handbuch d. Altertumswissenschaft, ed. W. Otto, Abt. I, Teil 7).

LAI CHIA-TU (1). *Chang Hêng.* (A biography, in Chinese.) Shanghai People's Pub. Co., Shanghai, 1956.

LECOMTE, L. (1). *Memoirs and Observations...made in a late Journey through the Empire of China.* 2nd ed. London, 1698.

LEGGE, J. (1). *The Texts of Confucianism, translated: Pt. I, the 'Shu Ching'....* Oxford, 1879 (Sacred Books of the East, no. 3); repr. in various editions, Com. Press, Shanghai.

LEGGE, J. (2). *The Texts of Confucianism, translated: Pt. II, the 'I Ching'.* Oxford, 1899 (Sacred Books of the East, no. 16).

LEGGE, J. (3). *The Texts of Taoism.* 2 vols. Oxford, 1891 (Sacred Books of the East, nos. 39 and 40).

LEGGE, J. (4). *The Texts of Confucianism translated: Pt. III, the 'Li Chi'.* 2 vols. Oxford, 1885 (Sacred Books of the East, nos. 27 and 28).

LI KUANG-PI & CH'IEN CHÜN-HUA. *Chung-Kuo K'o-Hsüeh Chi-Shu Fa-Ming ho K'o-Hsüeh Chi-Shu Jen Wu Lun Chi* ('Essays on Chinese Discoveries and Inventions in Science and Technology, and on the Men who made them'). San Lien Shu Tien, Peking, 1955.

LI KUANG-PI & LAI CHIA-TU. 'Han Tai ti Wei Ta K'o-Hsüeh Chia; Chang Hêng' ('A Great Scientist of the Han Dynasty; Chang Hêng'); art. in Li Kuang-Pi & Ch'ien Chün-Hua, *q.v.*, pp. 249 ff.

LIU HSIEN-CHOU (1). 'Chung-Kuo tsai Yuan Tung Li fang-mien-ti Fa-Ming' ('Chinese Inventions in Power-Source Engineering'), *Ch'ing-Hua Ta-Hsüeh Chi-Chieh Kung-Ch'êng Hsüeh-Pao* (Ch'ing-Hua University Engineering Journal), 1953 (n.s.), **1** (no. 1), 3. In Chinese.

LIU HSIEN-CHOU (2). 'Chung-Kuo tsai Ch'uan Tung Chi chien fang-mien-ti Fa-Ming' ('Chinese Inventions in Power Transmission'), *Ch'ing-Hua Ta-Hsüeh Chi-Chieh Kung-Ch'êng Hsüeh Pao* (Ch'ing-Hua University Engineering Journal), 1954 (n.s.), **2** (no. 1), 1; with Addendum, 1954, **2** (no. 2), 219. In Chinese.

LIU HSIEN-CHOU (3). 'Chung-Kuo tsai Chi Shih Ch'i fang-mien-ti Fa-Ming' ('Chinese Inventions in Horological Engineering'), *Ch'ing-Hua Ta-Hsüeh Chi-Chieh Kung-Ch'êng Hsüeh Pao* (Ch'ing-Hua University Engineering Journal), 1956 (n.s.), **4**, 1. In Chinese.

LIU HSIEN-CHOU (4). 'Wang Chêng yü Wo Kuo ti-i Pu Chi-Chieh Kung-Ch'êng-Hsüeh' ('Wang Chêng and the First Book on [Modern] Mechanical Engineering in China'). *Chen Li Tsa Chih,* 1943, **1** (no. 2), 215.

LLOYD, H. ALAN (1). *Giovanni de Dondi's Horological Masterpiece of A.D. 1364.* Priv. pr., no place, no date. (London, 1954.)

LLOYD, H. ALAN (2). 'George Graham, Horologist and Astronomer.' *Horol. Journ.* 1951, **93**, 708.

LLOYD, H. ALAN (3) (pref. and ed.). *Tercentenary Exhibition of the Pendulum Clock of Christiaan Huygens; Catalogue.* Pubs. Antiquarian Horological Society, no. 2. At the Science Museum, South Kensington, London, 1956.

LU ERH-K'UEI *et al.* *Tz'u Yuan* (encyclopaedia). Com. Press, Shanghai, 1915; enlarged edition, ed. Fang I, Com. Press, Shanghai, 1939.

McGOWAN, D. J. (1). 'On Chinese Horology, with Suggestions on the Form of Clocks adapted for the Chinese Market.' *Report of the (U.S.) Commissioner of Patents for the Year 1851* (Armstrong, Washington, 1852), 32nd Congress, 1st Session; Pt. 1, Arts and Manufactures, Sect. IX (1), p. 335, with two plates at the back of the volume. Reprinted in *(Silliman's) American Journal of Science and Arts*, 1852 (2nd ser.), **13**, 241; and, as 'American Clocks for China', in *(Chambers') Edinburgh Journal*, 1853. It is interesting that the Commissioner of Patents to whom McGowan's letter was addressed was none other than that Thomas Ewbank whose *Hydraulics and Mechanics* is still a book appreciated and used today.

MASPERO, H. (1). 'Les Instruments Astronomiques des Chinois au Temps des Han.' *Mélanges Chin. et Bouddh.* (Brussels), 1939, **6**, 183.

MASPERO, H. (2). 'L'Astronomie Chinoise avant les Han.' *T'oung Pao*, 1929, **26**, 267.

MATSCHOSS, C. & KUTZBACH, K. (1). *Geschichte des Zahnrades.* VDI-Verlag, Berlin, 1940.

MAYERS, W. F. (1). 'Bibliography of the Chinese Imperial Collections of Literature.' *China Rev.* 1878, **6**, 213, 285.

MICHEL, H. (1). 'Les Jades Astronomiques Chinois.' *Bull. des Mus. Royaux d'Art et d'Histoire* (Brussels), 1947, 31; *Communications de l'Acad. de Marine* (Brussels), 1949, **4**, 111; *Popular Astron.* 1950, **58**, 222; *Oriental Art*, 1950, **2**, 156; *Mélanges Chin. et Bouddh.* 1951, **9**, 153.

MICHEL, H. (2). 'Les Ancêtres du Planétarium.' *Ciel et Terre*, 1955, **71** (nos. 3–4), 1.

MICHEL, H. (3). 'Les Tubes Optiques avant le Télescope.' *Ciel et Terre*, 1954, **70** (nos. 5–6), 3.

MICHEL, H. (4). 'A propos des premières Montres; Over de eerste Horloges.' *Technica (Bull. du Comité Belge de Bijouterie et de l'Horlogerie)*, 1956, March, p. 129.

NEEDHAM, J. (1). *Science and Civilisation in China.* Univ. Press, Cambridge. 7 vols. in course of publication, 1954–.

NEEDHAM, J. (2). 'The Peking Observatory in A.D. 1280, and the Development of the Equatorial Mounting', art. in *Vistas in Astronomy*, vol. 1, the Presentation Volume for Prof. F. J. M. Stratton, F.R.S., Pergamon, London, 1955, p. 67.

NEEDHAM, J., WANG LING & PRICE, D. J. (1). 'Chinese Astronomical Clockwork.' *Nature*, 1956, **177**, 600. Chinese tr. by Hsi Tsê-Tsung in *K'o-Hsüeh T'ung Pao*, 1956 (no. 6), 100.

NEUGEBAUER, O. (1). 'The Early History of the Astrolabe.' *Isis*, 1949, **40**, 240.

ORRERY, THE COUNTESS OF CORK AND. *The Orrery Papers.* 2 vols. Duckworth, London, 1903.

PELLIOT, P. (1). 'L'Horlogerie en Chine' (an essay-review of Chapuis, Loup & de Saussure, *q.v.*). *T'oung Pao*, 1921, **20**, 61.

PELLIOT, P. (2). *Les Débuts de l'Imprimerie en Chine.* Impr. Nat. et Maisonneuve, Paris, 1953 (Oeuvres Posthumes, no. 4).

PFISTER, L. (1). *Notices Biographiques et Bibliographiques sur les Jésuites de l'ancienne Mission de Chine, A.D. 1552–1773.* Mission Press, Shanghai, 1932 (Variétés Sinologiques, no. 59).

PLANCHON, M. (1). *L'Horloge; son Histoire Rétrospective, Pittoresque et Artistique.* Laurens, Paris, 1899; 2nd ed. 1912.

PRICE, D. J. (1). 'Clockwork before the Clock.' *Horol. Journ.* 1955, **97**, 810; 1956, **98**, 31. Correct al-Battanī to al-Bīrūnī in this paper.

PRICE, D. J. (2). 'A Collection of Armillary Spheres and other Antique Scientific Instruments.' *Annals Sci.* 1954, **10**, 172.

PRICE, D. J. (3). 'An International Check-list of Astrolabes.' *Arch. Internat. d'Histoire des Sci.* 1955, **8**, 243 and 363.

PRICE, D. J. (4). Art. in *A History of Technology*, vol. 3, p. 582; ed. C. Singer *et al.* Oxford, 1957.

PRICE, D. J. (5). 'Two Mediaeval Texts on Astronomical Clocks.' *Antiq. Horol.* 1956, **1**, 156.

PRICE, D. J. (6). *The Equatorie of the Planetis* (probably written by Geoffrey Chaucer), with a linguistic analysis by R. M. Wilson. Univ. Press, Cambridge, 1955.

PRICE, D. J. (7). 'The Prehistory of the Clock.' *Discovery*, 1956, **17**, 153. In this paper, al-Battanī should be corrected to al-Bīrūnī.

PULLEYBLANK, E. G. (1). *The Background of the Rebellion of An Lu-Shan.* Oxford, 1954 (London Oriental Series, no. 4).

RAMBAUT, A. (1). 'Note on some Japanese Clocks lately purchased for the Science and Art Museum.' *Scientific Proceedings of the Royal Dublin Society*, 1889, 332.

[RENAUDOT, E. (1).] *Anciennes Relations des Indes et de la Chine de deux Voyageurs mahometans, qui y allèrent dans le neuvième Siècle, traduites d'Arabe, avec des Remarques sur les principaux Endroits de ces Relations.* Coignard, Paris, 1718; Eng. tr. London, 1733. The first translation of the travels of Sulaimān al-Tājir, who was in China in A.D. 851.

RICO Y SINOBAS, M. (1). *'Libros del Saber de Astronomia' de Rey D. Alfonso X de Castilla.* Aguado, Madrid, 1864.

ROBERTSON, J. D. (1). *The Evolution of Clockwork, with a special Section on the Clocks of Japan, and a comprehensive Bibliography of Horology.* Cassell, London, 1931.

DES ROTOURS, R. (1). *Traité des Fonctionnaires et Traité de l'Armée, traduit de la Nouvelle Histoire des T'ang ['Hsin T'ang Shu'], chs. 46–50.* 2 vols. Brill, Leiden, 1948 (Bibl. de l'Inst. des Hautes Etudes Chinoises, Paris, vol. 6).

RUFUS, W. C. (1). 'Astronomy in Korea.' *Trans. Korea Branch Roy. Asiat. Soc.*, 1936, **26**, 1.

SARREIRA, P. R. (1). 'Horas boas e Horas más para a Civilicão Chinesa.' *Broteria* (Lisbon), 1943, **36**, 518.

SARTON, G. (1). *Introduction to the History of Science.* Williams & Wilkins, Baltimore, vol. 1, 1927; vol. 2 (2 pts.), 1931; vol. 3 (2 pts.), 1947 (Carnegie Institution Pubs. no. 376).

SAUNIER, C. (1). *Die Geschichte d. Zeitmesskunst*, tr. from the French by G. Speckhart, 2 vols. Hübner, Bautzen, n.d. (1904).

DE SAUSSURE, L. (1). 'Le Système Astronomique des Chinois.' *Archives des Sci. Phys. et Nat.* (Geneva), 1919, **124**, 186, 561; 1920, **125**, 214, 325.

DE SAUSSURE, L. (2). 'L'Horométrie et le Système Cosmologique des Chinois', introduction to Chapuis, Loup & de Saussure, *q.v.*, also sep. Attinger, Neuchâtel, 1919.

SAUVAGET, J. (1). *'Akhbār al-Ṣīn wa'l-Hind', Relation de la Chine et de l'Inde, rédigée en 851; Texte établi, traduit et commenté....* Belles Lettres, Paris, 1948 (Collection arabe G. Budé). The most recent translation of the travels of Sulaimān al-Tājir, made from the same MS. as that used by Renaudot, Bib. Nat. (Ar.) 2281, written in Syria towards the end of the twelfth century. It is considered that Sulaimān was only one of the author's informants.

SCHMELLER, H. (1). *Beiträge z. Geschichte d. Technik in der Antike und bei den Arabern.* Mencke, Erlangen, 1922 (Abhdl. z. Gesch. d. Naturwiss. u. d. Med. no. 6).

SCHMIDT, M. C. P. (1). *Kulturhistorische Beiträge, II. Die antike Wasseruhr.* Leipzig, 1912.

SCHRAMM, E. (1). *Griechisch-römische Geschütze; Bemerkungen zu der Rekonstruktion.* Scriba, Metz, 1910; also pub. in *Jahrb. d. Gesellsch. f. lothringer Gesch. u. Altertumskunde*, 1904, **16**, 1, 142; 1906, **18**, 276; 1909, **21**, 86.

SHIRATORI, KURAKICHI (1). 'A New Attempt at the Solution of the Fu-Lin Problem.' *Memoirs of the Research Department of the Toyo Bunko*, 1956, **15**, 156; see especially pp. 302 ff. We are indebted to Prof. L. Carrington Goodrich of Columbia University for bringing to our attention the material on clockwork in this publication, which in the main deals with questions concerning the relations of China with Byzantium.

SOOTHILL, W. E. (1). *The Hall of Light; a Study of Early Chinese Kingship.* Butterworth, London, 1951.

STAUNTON, SIR GEORGE T. (1). *An Authentic Account of an Embassy from the King of Great Britain to the Emperor of China...taken chiefly from the Papers of H.E. the Earl of Macartney, K.B. etc.....* 2 vols. Bulmer & Nicol, London, 1797; repr. 1798; abridged ed. 1 vol., Stockdale, London, 1797.

STEVENSON, E. C. (1). *Terrestrial and Celestial Globes; their History and Construction.* 2 vols. Hispanic Soc. of America, Yale Univ. Press, New Haven, Conn., 1921.

SUN WÊN-CH'ING (1). 'Chang Hêng Chu Shu Nien Piao' (Bibliography of the Writings of Chang Hêng), in Chinese. *Chin-Ling Hsüeh Pao* (Nanking), 1932, **2**, 105.

SUN WÊN-CH'ING (2). 'Chang Hêng Nien P'u' (Life of Chang Hêng), in Chinese. *Chin-Ling Hsüeh Pao*, 1933, **3**, 331, sep. pub.; Com. Press, Shanghai, 1935; 2nd ed. Chungking, 1944; 3rd ed. Shanghai, 1956.

SUN ZEN E-TU & DE FRANCIS, J. (1). *Chinese Social History; Translations of Selected Studies.* Amer. Council of Learned Societies, Washington, D.C., 1956 (A.C.L.S. Studies in Chinese and Related Civilisations, no. 7).

SWANN, N. L. (1). *Food and Money in Ancient China; the Earliest Economic History of China, to A.D. 25* (with transl. of *Ch'ien Han Shu*, ch. 24, together with the related texts, *Ch'ien Han Shu*, ch. 91, and *Shih Chi*, ch. 129). Princeton Univ. Press, Princeton, N.J., 1950.

TAKABAYASHI, HYŌE (1). *Tokei Hattatsu Shi* ('The Development of Time Measurement'), in Japanese. Tōyō Shuppansha, Tokyo, 1924.

TAYLOR, E. G. R. (1). 'The South-Pointing Needle.' *Imago Mundi*, 1951, **8**, 1.

TAYLOR, E. W. & WILSON, J. SIMMS. *At the Sign of the Orrery*, for Messrs Cooke, Troughton & Simms, priv. pr., York, no date (1945).

TÊNG SSU-YÜ & BIGGERSTAFF, K. *An Annotated Bibliography of Selected Chinese Reference Works.* Harvard–Yenching Institute, Peking, 1936 (Yenching Journ. Chinese Studies monograph no. 12).

THORNDIKE, L. (1). 'The Invention of the Mechanical Clock about A.D. 1271.' *Speculum*, 1941, **16**, 242.

TOYNBEE, A. J. (1). *A Study of History*. Royal Instit. Internat. Affairs, London, 1935–9. 6 vols.

TREFFLER, CHRISTOPHER (1). 'Die Sich selbst-bewegende Himmels-Kugel; das ist Ein neu-erfunden künstlich Uhr-Werck—vermog dessen Der Himmels Globus mit seinen Sternen und Planeten, so wol nach der ersten als andern Bewegung herum getriben wird, also dass er mit dem Himmels-Firmament gänzlich übereinstimme—Dabey sind auch zu sehen Unterschiedliche Zeiger welche die Jahr, Monaten, Stund und Minuten, wie auch alle Finsternussen auf 20. Jahr zeigen—Ferner Eine andere Kugel, welche Sphaera Ecliptica genennet wird, auf welche alle Finsternussen können gewisen werden. An Tag gegeben durch Christoff Trefflern, Silberdrechseln und Burgern in Augspurg.' Koppmayer, Augsburg, 1679. The fact that Christopher signed himself in the plural, as if a firm, is interpreted by Bedini (1) to indicate that the brother John Philipp Treffler was the one really responsible, though unable to appear in person on the brochure owing to internal strife among the Augsburg guildsmen.

TRIGAULT, N. (1). *De Christiana Expeditione apud Sinas*. Vienna, 1615; Augsburg, 1615.

TSÊNG CHU-SÊN (TJAN TJOE-SOM) (1). *'Po Hu T'ung'; The Comprehensive Discussions in the White Tiger Hall—a Contribution to the History of Classical Studies in the Han Period.* 2 vols. Brill, Leiden, 1949, 1952 (Sinica Leidensia, vol. 6).

UCCELLI, A. (1). *Storia della Tecnica dal Medio Evo ai nostri Giorni*. Hoepli, Milan, 1945.

UCELLI DI NEMI, G. (1) (ed.). *Le Gallerie di Leonardo da Vinci nel Museo Nazionale della Scienza e della Tecnica* (catalogue) Mus. Naz. Sci. Tec., Milano, 1956.

UETA, JOE (1). 'Shih Shen's Catalogue of Stars; the oldest Star Catalogue in the Orient.' *Pubs. Kwasan Observatory* (Kyoto Imp. Univ.), 1930, **1** (no. 2), 17.

USHER, A. P. (1). *A History of Mechanical Inventions*. McGraw-Hill, New York, 1929; 2nd ed. revised, Harvard Univ. Press, Cambridge, Mass., 1954.

VERANZIO, F. (1). *Machinae Novae Fausti Verantii Siceni*. Venice, 1616. First edition perhaps *c.* 1595.

VERBIEST, F. (1). *Astronomia Europaea sub Imperatore Tartaro-Sinico Cám-Hy* [*K'ang-Hsi*] *appellato, ex umbra in lucem revocata à R. P. Ferdinando Verbiest Flandro-Belgico e Societate Jesu, Academiae Astronomicae in Regia PeKinensi Praefecto....* Bencard, Dillingen, 1687.

WANG CHEN-TO (1). 'Ssu-Nan Chih-Nan Chen yü Lo Ching P'an' ('Discovery and Application of Magnetic Phenomena in China, III. Origin and Development of the Chinese Compass Dial'), in Chinese, *Chinese Journ. Archaeol.* 1951, **5** (n.s. **1**), 101.

WANG CHEN-TO (2). 'Chih-Nan-Ch'ê, Chi-Li-Ku-Ch'ê chih K'ao-Ch'êng yü Mo-Chih' ('Investigations and Reproduction in Model Form of the South-Pointing Carriage and the Hodometer, i.e. *li*-measuring drum carriage'). *National Peiping Academy Historical Journal*, 1937, **3**, 1. In Chinese.

WANG LING & NEEDHAM, J. (1). 'Horner's Method in Chinese Mathematics; its Origins in the Root-Extraction Procedures of the Han Dynasty.' *T'oung Pao*, 1955, **43**, 345.

WARD, F. A. B. (1). *Time Measurement*. Pt. I, *Historical Review*. H.M.S.O., London, 2nd ed., 1937; Pt. II, *Descriptive Catalogue* (of the Collections in the Science Museum, London), H.M.S.O., London, 3rd ed., 1955.

WARD, F. A. B. (2). 'Chinese Astronomical Clockwork; a Lecture.' *Antiquarian Horology*, 1956, **1**, 153.

WIEDEMANN, E. (1). 'Ein Instrument das die Bewegung von Sonne und Mond darstellt, nach al-Bīrūnī.' *Der Islam*, 1913, **4**, 5.

WIEDEMANN, E. & HAUSER, F. (1). 'Über die Uhren im Bereich d. islamischen Kultur.' *Nova Acta; Abhdl. d. k. Leop.-Carol. Deutsch. Akad. d. Naturforsch. Halle*, 1915, **100**, no. 5.

WIEDEMANN, E. & HAUSER, F. (2). *Die Uhr des Archimedes und zwei andere Vorrichtungen.* Ehrhardt Karras, Halle, 1918.

WIEGER, L. (1). *Textes Historiques* (Chinese and French in parallel columns), 2 vols. Mission Press, Hsien-hsien, 1929.

WILBUR, C. M. (1). 'Slavery in China during the Former Han Dynasty, B.C. 206–A.D. 25.' *Field Museum of Nat. Hist. (Chicago) Pubs.* (Anthropological Series), 1943, **34**, 1 (pub. no. 525); also in *Journ. Econ. Hist.* 1943, **3**, 56.

WILHELM, R. (1). *'I Ching'; das Buch der Wandlungen.* 2 vols. Diederichs, Jena, 1924. Eng. tr. by C. F. Baynes, 2 vols. Bollingen, Pantheon, New York, 1950.

WILKINSON, L. & BAPU DEVA SASTRI. *The 'Siddhānta Śiromaṇi' of Bhāskara (c. A.D. 1150).* Calcutta, 1861 (Bibliotheca Indica, n.s., nos. 1, 13, 28).

WILLIAMSON, H. (1). *Wang An-Shih.* 2 vols. Probsthain, London, 1935, 1937.

WINTER, H. J. J. (1). 'Muslim Mechanics and Mechanical Appliances.' *Endeavour*, 1956, **15** (no. 57), 25.

WITTFOGEL, K. A., FÊNG CHIA-SHÊNG *et al.* 'History of Chinese Society (the Liao Dynasty)' (A.D. 907–1125). *Trans. Amer. Philos. Soc.* 1948, **36**, 1–650.

WOODCROFT, B. (1). *The 'Pneumatics' of Heron of Alexandria.* Whittingham, London, 1851.

WYLIE, A. (1). *The Mongol Astronomical Instruments in Peking*, Travaux du IIIème Congrès des Orientalistes, 1876; incorporated in *Chinese Researches*, Shanghai, 1897; photographically reproduced, Peiping, 1936.

YABUUCHI, KIYOSHI (1). 'Chūgoku no Tokei' ('Ancient Chinese Timekeepers'), in Japanese. *Japanese Journal of the History of Science*, 1951, no. 19, p. 19.

YAMAGUCHI, RYŪJI (1). *Nihon no Tokei; Tokugawa Jidai no Wadokei no Kenkyu* ('Time Measurement in Japan; Studies on the Clocks of the Tokugawa Period'), in Japanese. Nihon Hyoronsha, Tokyo, 1942. Contains, reproduced in facsimile, the *Karakuri Zui* ('Illustrated Schema of Horological (lit. Mechanical) Ingenuity'), of 1796, by Hosokawa Hanzō Yorinao, with modern transliteration of the text.

YANG LIEN-SHÊNG (1). 'Notes on the Economic History of the Chin Dynasty.' *Harvard Journ. Asiat. Stud.* 1945, **9**, 107.

ZINNER, E. (1). 'Gerbert und das Seerohr.' *Ber. Naturforsch. Ges. Bamberg*, 1952, **33**, 39 (Kl. Veröffentl. d. Remeis Sternwarte, no. 7).

ZINNER, E. (2). 'Aus der Frühzeit der Räderuhr von der Gewichtsuhr zur Federzugsuhr.' Oldenbourg, München, 1954 (*Abhdl. u. Berichte d. Deutsches Museum*, 1954, **22**, no. 3).

TABLES OF CHINESE CHARACTERS

I TECHNICAL TERMS IN THE MAIN TEXT

1	*shui yün i hsiang t'ai*	水運儀象臺	water-powered armillary (sphere) and celestial (globe) tower
2	*hu t'i*	胡梯	balustrade (lit. Persian, or foreign, ladder)
3	*liu ho i*	六合儀	component of the six cardinal points
4	*san ch'en i*	三辰儀	component of the three arrangers of time
5	*ssu yu i*	四游儀	component of the four displacements
6	*yang ching shuang kuei* (or *huan*) (=*t'ien kuei*, no. 7 =*t'ien ching shuang kuei* (or *huan*), nos. 13 and 101)	陽經雙規(環)	split-ring meridian circle
7	*t'ien kuei*	天規	celestial ring (i.e. meridian circle)
8	*yin wei tan kuei* (or *huan*) (=*ti hun huan*, no. 9 =*ti hun tan kuei*, no. 14 =*yin wei huan*, no. 131)	陰緯單規(環)	terrestrial circle (i.e. single-ring horizon circle)
9	*ti hun* (*huan*)	地渾(環)	horizon circle
10	*t'ien yün* (*tan*) *huan*	天運(單)環	diurnal motion gear-ring
11	*hun hsiang*	渾象	celestial globe
12	*ti kuei*	地櫃	casing (of globe)
13	*t'ien ching shuang kuei* (or *huan*)	天經雙規(環)	split-ring meridian circle
14	*ti hun tan kuei* (or *huan*)	地渾單規(環)	single-ring horizon circle
15	*chi lun chu*	機輪軸	time-keeping shaft
16	*chou yeh chi lun*	晝夜機輪	day and night time-keeping wheels
17	*t'ien shu*	天束	upper bearing beam
18	*t'ien lun*	天輪	celestial gear-wheel
19	*ch'ih tao ya*	赤道牙	equatorial gear-ring
20	*chou shih chung ku lun*	晝時鐘鼓輪	wheel for striking (equal double-) hours of the day by bells and drums
21	*k'o*	刻	'quarter' (-hours)
22	*shih k'o chung ku lun*	時刻鐘鼓輪	wheel for striking 'quarters' by bells and drums
23	*shih ch'u chêng ssu-ch'en lun*	時初正司辰輪	wheel for jacks striking beginnings and middles of (double-) hours
24	*pao k'o ssu-ch'en lun*	報刻司辰輪	wheel for jacks reporting 'quarters'
25	*yeh lou chin chêng lun*	夜漏金鉦輪	wheel for striking night-watches on gongs
26	*yeh lou kêng ch'ou ssu-ch'en lun*	夜漏更籌司辰輪	wheel for jacks reporting night-watches and their divisions
27	*yeh lou chien lun*	夜漏箭輪	wheel for float-indicators of night-watches (lit. night clepsydra indicator-rod wheel)
28	*shu lun* (=*shuang shu lun*)	樞輪 雙樞輪	great (driving-) wheel
29	*fu*	輻	spokes
30	*hung*	洪	scoop-holders

31	*wang*	輞	reinforcement rings
32	*hu* (=*shou shui hu*, no. 44)	壺	scoops
33	*ku*	轂	hub
34	*t'ieh shu* (*lun*) *chu* (=*shu chu* =*t'ieh shu chu*)	鐵樞軸	driving-shaft
35	*ti ku*	地轂	earth wheel (lit. earth drum, gear-wheel)
36	*yün po*	運撥	enmeshes (lit. moved by poking or pushing, i.e. engaging peg gearing)
37	*ti lun* (=*hsia lun*)	地輪 下輪	lower wheel
38	*t'ien chu*	天柱	transmission-shaft (N.B. this term is used for the main posts of the framework, as well as for the vertical transmission-shaft; cf. no. 52)
39	*chung lun*	中輪	middle wheel
40	*chi lun*	機輪	time-keeping wheels
41	*shang lun*	上輪	upper wheel
42	*t'ien ch'ih*	天池	upper reservoir
43	*p'ing shui hu*	平水壺	constant-level tank
44	*shou shui hu*	受水壺	water-receiving scoops
45	*t'ui shui hu*	退水壺	sump (lit. water-withdrawing tank)
46	*shêng shui hsia hu*	昇水下壺	lower reservoir (lit. lower water-raising tank)
47	*shêng shui hsia lun*	昇水下輪	lower noria (lit. lower water-raising wheel)
48	*shêng shui shang hu*	昇水上壺	intermediate reservoir (lit. upper water-raising tank)
49	*shêng shui shang lun*	昇水上輪	upper noria (lit. upper water-raising wheel)
50	*ho ch'ê*	河車	manual wheel (lit. stream paddle-wheel)
51	*t'ien ho*	天河	upper flume
52	*t'ien chu*	天柱	main posts (cf. no. 38)
53	*shu liang*	樞梁	main beams
54	*t'ien liang*	天梁	upper beams
55	*t'ien kuan* (=*kuan* =*t'ien hêng kuan*, no. 68)	天關	upper stopping device=upper stop
56	*t'ien chi*	天極	'celestial poles' (framework pieces)
57	*po ya chi lun*	撥牙機輪	time-keeping gear-wheel
58	*ti chi*	地極	lower bearing beam
59	*t'ieh shu chiu* (=*shu chiu*)	鐵樞臼	iron mortar-shaped end-bearing
60	*tsuan* (for no. 60*a*)	纂	pointed cap (of bearing)
60*a*	*tsuan*	鑽	
61	*mu chia*	木架	wooden stand
62	*t'ien hêng*	天衡	upper balancing lever
63	*nao*	腦	head (of lever) (lit. brain) (cf. no. 116)
64	*t'ien ch'üan*	天權	upper weight
65	*wei*	尾	tail (of lever)
66	*t'ien t'iao*	天條	connecting rod or chain of linked rods (lit. celestial rod)

67	*kuan shê* (=*t'ien kuan shê*)	關舌	trip-lever (lit. upper stopping tongue)
68	*t'ien hêng kuan*	天衡關	upper balance stop (cf. no. 55)
69	*shu hêng*	樞衡	lower balancing lever
70	*ko ch'a*	格叉	checking fork
71	*shu ch'üan*	樞權	lower weight (lit. main weight)
72	*t'ien so*	天鎖	upper lock
73	*k'o wu*	渴烏	siphon (lit. thirsty crow)
74	*chü* (verb) (for no. 74*a*) (cf. no. 108)	距	stop
74*a*	*chü*	拒	
75	*t'ieh po* (=*t'ieh po ya*)	鐵撥	iron pin (lit. iron poking tooth)
76	*hou t'ien ku* (=*t'ien ku hou lun*)	後天轂	back gear-wheel
77	*t'ien ku*	天轂	armillary sphere shaft
78	*ch'ien t'ien ku* (=*t'ien ku ch'ien lun*)	前天轂	front gear-wheel
79	*ya chü*	牙距	teeth
80	*tsê chi lun yu hui*	則激輪右回	counter-clockwise
81	*hun hsiang t'ien yün lun* (=*t'ien yün lun*)	渾象天運輪	celestial globe drive-wheel
82	*hsiang chieh*	相接	engage with
83	*hsien*	銜	engage with
84	*t'ien chu* (=*t'ieh t'ien chu*)	天軸	celestial idler (lit. celestial axle)
85	*po ch'i ya chü*	撥其牙距	enmeshes with its teeth
86	*san-pai-liu-shih-wu tu yu chi*	三百六十五度有奇	365 degrees and a fraction
87	*chung wai kuan hsing*	中外官星	stars and constellations north and south of the equator
88	Tzu Wei Yuan (=Tzu Kung)	紫微垣 紫宮	polar region (lit. Purple Palace)
89	*hsiu*	宿	lunar mansions (equatorial belt)
90	*hsiang hsien*	相銜	engages with
91	*po ya*	撥牙	trip-lugs (leaf teeth)
92	*po*	撥	engages
93	*ssu-ch'en*	司辰	jack
94	*chieh*	截	intercepts
95	*chin chêng*	金鉦	gong
96	*hsiang ying*	相應	coincides with
97	*tieh*	疊	combined with (lit. piled up with)
98	*chia ch'ih*	夾持	grasps
99	*yuan hsiang*	圓項	cylindrical necks
100	*t'ieh yang yüeh*	鐵仰月	crescent (-shaped) bearings (lit. iron upward-looking crescents)
101	*t'ien ching* (=*yang ching shuang kuei*, no. 6)	天經	meridian circle
102	*t'ien t'i*	天梯	chain drive (lit. celestial ladder)
103	*t'ien t'o*	天托	gear-box (of celestial ladder)

104	*hsia ku* (=*t'ien t'i hsia ku*)	下轂	lower chain wheel
105	*chun shui chien*	準水箭	water-level marker (lit. arrow, i.e. indicator-rod measuring the water-level)
106	*ch'a shou chu*	杈手柱	crutched post (lit. fork hand post)
107	*ni hsing*	逆行	contrary direction
108	*chü* (noun) (cf. no. 74)	距	peg teeth
109	*shou pa*	手把	handles
110	*hu tou*	戽斗	buckets (of norias)
111	(*t'ieh*) *kuan chu*	(鐵)關軸	(iron) shutting axle
112	*hêng kuang*	橫桄	cross-bar
113	*t'o fêng*	駞峯	camel back
114	*t'ieh hsia*	鐵頰	iron cheeks
115	*t'ieh ho hsi*	鐵鶴膝	iron rods and chain forming parallel linkage (lit. 'crane bird's knee')
116	*shou*	首	head (of lever) (cf. no. 63)
117	*ho t'ai*	合臺	astronomical clock-tower
118	*t'ieh kua lien chou tsa*	鐵括聯周匝	endless circuit of iron links
119	(*t'ien t'i*) *shang ku*	天梯上轂	upper chain wheel (of celestial ladder)
120	*kua*	括	link
121	*shuang ch'a*	雙叉	double fork
122	*shui fu*	水趺	water-level base
123	*chü ch'ih*	曲尺	bent
124	*shang t'ien ku*	上天轂	upper pinion
125	*hsiang chü* (for *chü*, no. 74*a*)	相距	engages with
126	*chung t'ien ku*	中天轂	middle pinion
127	*hsia t'ien ku*	下天轂	lower pinion
128	*kuei piao*	圭表	gnomon device
129	*kuei tso*	圭座	gnomon shadow scale
130	*erh-shih-ssu ch'i*	二十四氣	twenty-four fortnightly periods
131	*yin wei* (*tan*) *huan* (=*ti hun*, no. 9)	陰緯(單)環	single-ring horizon circle
132	*wang t'ung*	望筒	sighting-tube
133	*jih ju*	日入	sunset
134	*hun*	昏	dusk
135	*tai tan*	待旦	waiting for dawn
136	*hsiao*	曉	dawn
137	*jih ch'u*	日出	sunrise
138	*ch'u kêng*	初更	beginnings of night-watches
139	*ch'ou*	籌	subdivisions of night-watches
140	*ti tsu*	地足	stands (at base of framework)
141	*tai*	待	serve (of gear-wheels)
142	*chou t'ien tu fên chih fa*	周天度分之法	expressing the number of mean solar days in the year as an integral factor
143	*shih*	時	double-hour
144	*mu ko*	木閣	wooden pagoda (façade)
145	*ssu ch'en*	司辰	time-reporting jack
146	*t'ien ch'ang huan*	天常環	single-ring equatorial circle (in outer nest of armillary sphere)

147	*san ch'en i shuang huan*	三辰儀雙環	split-ring solstitial colure circle (in middle nest of armillary sphere)
148	*ch'ih tao tan huan*	赤道單環	single-ring equatorial circle (in middle nest of armillary sphere)
149	*huang tao shuang huan*	黃道雙環	split-ring ecliptic circle (in middle nest of armillary sphere)

2 TECHNICAL TERMS AND PHRASES IN THE ANCILLARY TEXTS

150	*hun i*	渾儀
151	*ch'en*	辰
152	*hun hsiang*	渾象
153	*ming i pu chêng*	名亦不正
154	*hun t'ien i*	渾天儀
155	*kuei t'ien chü ti*	規天矩地
156	*hun t'ien*	渾天
157	*shui chuan t'ung hun*	水轉銅渾
158	(*t'ung*) *hou* (*i*)	銅候儀
159	*chi hêng*	璣衡 or 機衡
160	(*t'ung*) *hun* (*i*)	銅渾儀
161	*hun t'ien hsiang*	渾天象
162	*hsüan chi*	璇璣 or 璿璣
163	(*yü*) *hêng*	玉衡
164	*chi*	機
165	*shou ch'uang ch'i shih*	首創其式
166	T'ai-P'ing Hun I	太平渾儀
167	*shêng*	繩
168	*pei* (or *pi*)	髀
169	*ku*	股
170	*piao*	表
171	*li*	里
172	*kou ku*	鈎股 or 句股
173	*ch'ung ch'a*	重差
174	*t'ui*	推
175	*kuei ying chi yu*	晷影極游
176	*t'ien shu*	天數
177	*chü suan shu, an ch'i hsiang*	據算術案器象
178	*mu yang chi lun*	木樣機輪
179	*ch'i fan*	器範
180	*ch'iao ssu*	巧思
181	*mu yang*	木樣
182	*hou t'ien*	候天
183	*chih tu*	製度
184	*hsiao yang*	小樣
185	*ta mu yang*	大木樣
186	*chih yü chi shui yün chi, ch'i yung tsê i*	至于激水運機其用則一
187	*kou chu i chün t'iao*	苟注挹均調
188	*tsê ts'an chiao hsüan chuan chih shih*	則參校旋轉之勢
189	*wu yu ch'a ch'uan*	無有差舛
190	*i lou shui chuan chih*	以漏水轉之
191	*ling t'ai*	靈臺
192	*hsüan chi so chia*	琁璣所加
193	*chi*	璣
194	*yuan t'ien chih hsiang*	圓天之象
195	Shui Yün Hun T'ien Fu Shih T'u	水運渾天俯視圖
196	*erh-shih-pa hsiu*	二十八宿
197	*lun chu kuan chu*	輪軸關柱
198	(*chih*) *shen*	直神
199	*chieh ch'i*	節氣
200	*hou*	候
201	*shu chi lun chu*	樞機輪軸
202	*ling ch'ang tsai chien hsing ch'an tz'u chih nei*	令常在見行躔次之內
203	*fou chien lou*	浮箭漏
204	*ch'êng lou*	稱漏
205	*ch'en chien lou*	沈箭漏
206	*pu hsi lou*	不息漏
207	*pai*	稗
208	*kuan yü i chih yu*	觀玉儀之游
209	*hun ming chu shih, nai ming chung hsing chê yeh*	昏明主時乃命中星者也
210	*chi*	急
211	*shu*	舒
212	*t'iao*	調
213	*hsien t'ien erh t'ien pu wei, hou t'ien erh fêng t'ien shih*	先天而天不違後天而奉天時
214	*tzu ming chung*	自鳴鐘
215	*shih yü chien lüeh*	失於簡略
216	*shih yü nan yung*	失於難用
217	*san ch'en i shê ch'ih yü huan pei*	三辰儀設齒於環背
218	*hêng hsiao*	橫簫
219	*ch'ih*	齒

220	*yuan i*	圓儀
221	*yu i shuang huan*	游儀雙環
222	*huang tao hun i*	黃道渾儀
223	*k'o lou kuei piao*	刻漏圭表
224	*hou t'ai*	候臺
225	*huang tao yu i*	黃道游儀
226	*ju po ch'i tzu*	如博棊子
227	*shih-erh shih chung*	十二時鐘
228	*ying shih tzu ming*	應時自鳴
229	*kan*	干
230	*chih*	支
231	*p'ing chun lun*	平準輪
232	*shui nieh*	水臬
233	*kua*	卦
234	*wei*	維
235	*lung chu*	龍柱
236	*t'ien lun*	天輪
237	*p'ing*	平
238	*ts'ê*	側
239	*ti lun*	地輪
240	*ti chu*	地軸
241	*hêng lun*	横輪
242	*ts'ê lun*	側輪
243	*hsieh lun*	斜輪
244	*ting shen kuan*	定身關
245	*chung kuan*	中關
246	*hsiao kuan*	小關
247	*t'ien ting*	天頂
248	*t'ien ya*	天牙
249	*t'ien chih*	天指
250	*t'ung t'ieh chien sê*	銅鐵漸澁
251	*shih-erh shih p'an*	十二時盤
252	*ts'ê hou*	測候
253 *a*	*chu shui chi lun*	注水激輪
253 *b*	*ling ch'i tzu chuan*	令其自轉
254	*i jih i yeh t'ien chuan i chou*	一日一夜天轉一周
255	*ling tê yün hsing*	令得運行
256	*tzu jan*	自然
257	*lun chu*	輪軸
258	*kou chien chiao ts'o*	鈎鍵交錯
259	*kuan so hsiang ch'ih*	關鎖相持
260	*i shui chi chih, huo i shui yin chuan chih*	以水激之或以水銀轉之
261	*hun i t'u*	渾儀圖
262	*pa fên*	八分
263	*yü lun shang*	於輪上
264	*kung chiang hsing ming*	工匠姓名
265	*p'an hsia*	盤下
266	*hun kuei chih ch'i*	渾規之器
267	*t'ien hsiang*	天象
268	*yang kuan t'ai*	仰觀臺
269	*shih-erh ch'en ch'ê*	十二辰車
270	*hui yuan chêng nan*	廻轅正南
271	*wu*	午
272	*ssu fang hui chuan, pu shuang hao li*	四方廻轉不爽毫釐
273	*pu chia jen li, i shui chuan chih*	不假人力以水轉之
274	*ma shang k'o lou*	馬上刻漏
275	*wo ch'i*	欹器
276	*i chi tung chih*	以機動之
277	*ch'üan*	權
278	*hêng ch'ü*	衡渠
279	*hsing lou (ch'ê)*	行漏車
280	*ma shang pên ch'ih*	馬上奔馳
281	*ch'üan ch'i*	權器
282	*yü hu, yü kuan, liu chu*	玉壺玉管流珠
283	*shui yin*	水銀
284	*shêng*	升
285	*ch'êng chung*	秤重
286	*chin*	斤
287	*ch'êng lou*	稱漏
288	*k'uei jih kuei*	揆日晷
289	*hsia lou k'o*	下漏刻
290	*fou lou*	浮漏
291	*lun lou*	輪漏
292	*ch'üan hêng*	權衡
293	*shui ch'êng*	水秤
294	*i mu wei hêng*	以木爲衡
295	***t'ien ho***	天河
296	*shui chien*	水箭
297	*huan*	鐶
298	*t'ung fu ho*	銅覆荷
299	*t'ung so san t'iao*	銅索三條
300	*t'ung hu*	銅壺
301	*shui kuei*	水櫃
302	*shui hai*	水海
303	*li hsiang hsing*	立象形
304	*yü t'ieh lien fu chung*	於鐵蓮跗中
305	*fang huan*	方鐶
306	*chi kan*	雞竿
307	*i tsu shêng wan ch'üan shang ta t'ung huan*	以組繩挽權上大銅鐶
308	*ch'êng*	秤
309	*ju chung chiu chih chih*	如鐘簴之制
310	*t'ieh hu mên*	鐵胡門
311	*shih p'ai*	時牌

312	*ti tsai t'ien nei*	地在天內
313	*chih li lou k'o, i shui chuan i*	置立漏刻以水轉儀
314	*yü t'ien hsiang ying*	與天相應
315	*shao*	少
316	*i shang ying kuei tu*	以上應晷度
317	*hsing chi*	行極
318	*hun hsiang chih fa*	渾象之法
319	*ti tang tsai t'ien nei*	地當在天內
320	*erh ho yü li*	而合於理
321	*chi tso t'ung hun t'ien i yü mi shih chung*	旣作銅渾天儀於密室中
322	*yü t'ien chieh ho, ju fu ch'i yeh*	與天皆合如符契也
323	*nei wai kuei*	內外規
324	*yü tien shang shih nei*	於殿上室內
325	*hsing chung ch'u mu, yü t'ien hsiang ying*	星中出沒與天相應
326	*yin ch'i kuan li, yu chuan jui lun*	因其關戾又轉瑞輪
327	*kuan li*	關戾
328	*kuan li*	關捩
329	*kuan li*	關棙
330	*kuan*	關
331	*li*	戾
332	*li*	捩
333	*p'i p'a*	琵琶
334	*ming chieh*	蓂莢
335	*ch'ü yi*	屈軼
336	*tuan*	端
337	*hou t'ai t'ung i*	候臺銅儀
338	*hsiao hun*	小渾
339	*shui tui*	水碓
340	Yuan-Yu Hun T'ien I Hsiang	元祐渾天儀象
341	*su shu*	素書
342	San Yuan	三垣
343	*mao*	卯
344	*yu*	酉
345	*ch'ih o*	持扼
346	*shu tou*	樞斗
347	*chu shui chi lun*	注水激輪
348	*chi lun*	機輪
349	T'ai Wei Yuan	太微垣
350	T'ien Shih Yuan	天市垣
351	*kou chien chiao ts'o hsiang ch'ih*	鉤鍵交錯相持
352	Tsao Wu Chê	造物者
353	*ssu ch'en* Shou Hsing	司辰壽星
354	*chu lung*	燭龍
355	*yuan hsiang*	圓象
356	*shui yün chih fa*	水運之法
357	*p'o k'ao shui yün chih tu*	頗考水運制度
358	*t'ung ch'iu*	銅蚪
359	*tien*	點
360	*kêng*	更
361	*mu t'u*	木圖
362	*yung shui chuan chih*	用水轉之
363	*i fa t'ien yün*	以法天運
364	*fên*	分
365	*chih chü*	直距
366	*kuan chu chih lei*	關軸之類
367	*lun*	輪
368	*chü po lun* (=no. 57)	距撥輪
369	*hsüan hsiang*	懸象
370	*chung ku ssu ch'en k'o pao* (i.e. *pao k'o*)	鐘鼓司辰刻報
371	*chien i*	簡儀
372	*yang i*	仰儀
373	*têng lou*	燈漏
374	*yün chu*	雲珠
375	*k'o i shen p'ing shui chih huan chi*	可以審平水之緩急
376	*k'o ch'a chun shui chih chün t'iao*	可察準水之均調
377	*têng ch'iu*	燈球
378	*chin pao*	金寶
379	*huan*	寰
380	Pao Shan Lou	寶山漏
381	*ch'ih tao*	赤道
382	*huang tao*	黃道
383	*pai tao*	白道
384	*chi yün lun ya, yin yü kuei chung*	機運輪牙隱於櫃中
385	*t'ung ch'iu*	銅毬
386	*kung lou*	宮漏
387	*hu*	壺
388	*yün shui shang hsia*	運水上下
389	*hsien*	仙
390	*kuei lou*	晷漏
391	*t'ung hu*	銅壺
392	*hsiang chuan*	香篆
393	*kuei piao*	圭表
394	*kun t'an*	輥彈
395	*i lun wei yung*	以輪爲用
396	*chih shih t'an k'ou i ch'êng shêng*	至時彈扣以成聲
397	*kun tzu*	輥子
398	*hun*	混
399	*hun*	渾

400	*Hsi Yang tzu ming chung*	西洋自鳴鐘
401	*k'o lou*	刻漏
402	Ting Shih T'ai	定時臺
403	*fei hsien t'ai yü*	飛仙臺隅
404	*wei chi chung hsing*	爲擊鐘形
405	*i chi chuan chih*	以機轉之
406	*p'an*	盤
407	*ch'u*	初
408	*chung*	中
409	Tao	道
410	*ch'i chi fa yin yü kuei chung, i shui chi chih*	其機發隱於櫃中以水激之
411	*chêng*	正
412	*mei*	妹
413	*hsi*	兮
414	*chao*	朝
415	*jih*	日
416	*mo*	暮
417	*hun*	昏
418	*ming*	明
419	*ta ts'ai*	大采
420	*ta shih*	大食
421	*chung jih*	中日
422	*chê*	昃
423	*hsiao shih*	小食
424	*hsiao ts'ai*	小采
425	*hsi*	夕
426	*hun ku*	昏鼓
427	*ta ts'ao*	大鼜
428	*yeh pang*	夜半
429	*ch'en chieh*	晨戒
430	*tan ming*	旦明
431	*fa hsü*	發昫
432	*ssu hsiang tan huan*	四象單環
433	*yu kuei shuang huan*	游規雙環
434	*fan ch'ê*	翻車
435	*lung ku ch'ê*	龍骨車
436	*ch'ui mao*	鎚矛
437	*lien chia*	鏈耞
438	*t'ieh lien chia pang*	鐵鏈夾棒
439	*ho hsi fêng*	鶴膝風
440	*shui ching k'o lou*	水晶刻漏
441	*sha lou*	沙漏
442	*tou lun*	斗輪
443	*hêng tien chih chu*	衡奠之軸
444	*tsung tien chih lun*	從奠之輪
445	*ts'ê ching (ying) p'an*	測景(影)盤
446	*ch'uan ya hsiang ju*	犬牙相入
447	*tz'u ti yün i ch'ih*	次第運益遲
448	*ying fu tu*	攍附度
449	*lun hu*	輪壺
449*a*	*ts'ao*	槽
449*b*	*t'ung*	筒
449*c*	*ch'ien tan*	鉛彈

3 NAMES OF PERSONS

450	Su Sung	蘇頌
451	Tzu-Jung	子容
452	Han Ch'i	韓琦
453	Wang An-Shih	王安石
454	Fu Pi	富弼
455	Yeh Mêng-Tê	葉夢得
456	Yuan Wei-Chi	袁惟幾
457	Han Kung-Lien	韓公廉
458	Chang Shih-Lien	張士廉
459	Chi Yün	紀昀
460	Lu Hsi-Hsiung	陸錫熊
461	Yu Mou	尤袤
462	Juan T'ai-Fa	阮泰發
463	Ch'ien Tsêng	錢曾
464	Shih Yuan-Chih	施元之
465	Tê-Ch'u	德初
466	Su Tung-P'o	蘇東坡
467	Ouyang Hsiu	歐陽修
468	Chang Hai-P'êng	張海鵬
469	Ch'ien Hsi-Tso	錢熙祚
470	Ch'en Shu-Pao	陳叔寶
471	Chao Hsü	趙煦
472	Ssuma Kuang	司馬光
473	Han Hsien-Fu	韓顯符
474	Shu I-Chien	舒易簡
475	Shen Kua	沈括
476	P'êng Ch'êng	彭乘
477	I-Hsing	一行
478	Wang Fan	王蕃
479	Chang Ssu-Hsün	張思訓
480	Chang Hêng	張衡
481	Liang Ling-Tsan	梁令瓚
482	Wang Yuan-Chih	王沈之
483	Chou Jih-Yen	周日嚴
484	Yü T'ai-Ku	于太古
485	Chang Chung-Hsüan	張仲宣

No.	Name	Characters
486	Miao Ching-Chang	苗景張
487	Tuan Chieh-Chi	端節級
488	Liu Chung-Ching	劉仲景
489	Hou Yung-Ho	侯永和
490	Yü T'ang-Ch'en	于湯臣
491	Yin Ch'ing	尹清
492	Huang Ch'ing-Ts'ung	黃卿從
493	Shih Shen	石申
494	Kan Tê	甘德
495	Wu Hsien	巫咸
496	Ko Hung	葛洪
497	Chang P'ing-Tzu	張平子
498	Lu Kung-Chi	陸公紀
499	Ch'en Cho	陳卓
500	Lu Chi	陸績
501	Wang Ying-Lin	王應麟
502	Li Shun-Fêng	李淳風
503	Sun Chio	孫觳
504	Ch'en Miao	陳苗
505	Ho Ch'êng-T'ien	何承天
506	Liu Yao	劉曜
507	K'ung T'ing	孔挺
508	Ch'ao Ch'ung	晁崇
509	Hsieh Lan	解蘭
510	Wei Chêng	魏徵
511	Chao Hsü	趙頊
512	Lohsia Hung	落下閎
513	Yao Ch'ung	姚崇
514	Chia K'uei	賈逵
515	Yü Yuan	于淵
516	Chou Ts'ung	周琮
517	Ch'ien Ming-I	錢明逸
518	Mai Yün-Yen	麥允言
519	Chao Chên	趙禎
520	T'ao Hung-Ching	陶宏景
521	Chao Hsi-Ku	趙希鵠
522	Fan Chung-Yen	范仲淹
523	Sun Ch'o	孫綽
524	Chao Ching	趙炅
525	Wu Chao-Su	吳昭素
526	Hui-Yuan	惠遠
527	Chao Hêng	趙恆
528	Liu P'ien	柳玭
529	Yuan	袁
530	(Yuan) Tsan-Shan	袁贊善
531	Yang Chiung	楊烱
532	T'ao Ku	陶穀
533	Li Shih-Min	李世民
534	Yin K'uei	殷夔
535	Li Lung-Chi	李隆基
536	An Lu-Shan	安祿山
537	Yang Kuei Fei	楊貴妃
538	(Yang) Yü-Huan	楊玉環
539	Chang Sui	張遂
540	Chou Mi	周密
541	Wei Shu	韋述
542	Lu Ch'ü-T'ai	陸去泰
543	Huan Chih-Kuei	亘執珪
544	Chang Tsu	張鷟
545	Li Fang	李昉
546	Li Lan	李蘭
547	Wang P'u	王普
548	Wu Chao	武照
549	Kêng Hsün	耿詢
550	Tun-Hsin	敦信
551	Wang Yung	王勇
552	Wang Chih-Chi	王世積
553	Kao Chih-Pao	高智寶
554	Hsiu Wang	秀王
555	Ho Ch'ou	何稠
556	Yang Chien	楊堅
557	Yang Kuang	楊廣
558	Yüwên Hua-Chi	宇文化及
559	Yang Chia	楊甲
560	Yüwên K'ai	宇文愷
561	Hsü Chien	徐堅
562	Shen Yo	沈約
563	Ssuma Tê	司馬德
564	Liu Yü	劉裕
565	Ko Hêng	葛衡
566	Ko Hsüan	葛玄
567	Pao P'u Tzu	抱朴子
568	Sun Shêng	孫盛
569	Liu Chih	劉智
570	Li Shan	李善
571	Fang Hsüan-Ling	房玄齡
572	Liu Chih	劉志
573	Liu Yu	劉祐
574	Liu Pao	劉保
575	Lu Ch'ui	陸倕
576	Hsiao T'ung	蕭統
577	Ma Kuo-Han	馬國翰
578	Yen K'o-Chün	嚴可均
579	Chou Ch'ü-Fei	周去非
580	Fu Shêng	伏勝
581	Wang Ch'ung	王充
582	Kêng Shou-Ch'ang	耿壽昌
583	Yü Hsi	虞喜

584	Fan Yeh	范曄
585	Ts'ai Yung	蔡邕
586	Liu Tao-Hui	劉道會
587	Ssuma Yao	司馬曜
588	Ma Chün	馬鈞
589	Huangfu Yü	皇甫愈
590	Wang An-Li	王安禮
591	Ouyang Fa	歐陽發
592	Hsü Chiang	許將
593	Chu Pien	朱弁
594	Ts'ai Pien	蔡卞
595	Ch'ao Tuan-Yen	晁端彥
596	Lin Tzu-Chung	林子中
597	Chang Tun	章惇
598	Ts'ai Ching	蔡京
599	Wang Fu	王黼
600	Wang	王
601	Liu Chêng	劉拯
602	Wang Tsêng	王曾
603	Chao Chi	趙佶
604	Chao Huan	趙桓
605	Chêng K'ang-Ch'êng	鄭康成
606	Chêng Hsüan	鄭玄
607	Liang Shih-Ch'êng	梁師成
608	Wei Han-Chin	魏漢津
609	Chu Mien	朱勔
610	Wang Lao-Chih	王老志
611	Wang Tzu-Hsi	王仔昔
612	Lin Ling-Su	林靈素
613	Liu Hun-K'ang	劉混康
614	Li Chi-Tsung	李繼宗
615	Ting Shih-Jen	丁師仁
616	Li Kung-Chin	李公謹
617	Chao Kou	趙構
618	Chu Hsi	朱喜
619	Hsieh Chi	謝伋
620	Ch'in Kuei	秦檜
621	Su Hsi	蘇攜
622	Yuan Chêng-Kung	袁正功
623	Shao O	邵諤
624	Huan T'an	桓譚
625	Yang Hsiung	楊雄
626	Tsêng Nan-Chung	曾南仲
627	Min-Chan	民瞻
628	Tsêng Min-Hsing	曾敏行
629	Sun	孫
630	Toktaga (T'o-T'o)	脫脫
631	Ouyang Hsüan	歐陽玄
632	Yang Yün-I	楊雲翼
633	Wanyen Ching	完顏璟
634	Wanyen Hsün	完顏珣
635	Kuo Shou-Ching	郭守敬
636	Sung Lien	宋濂
637	Khubilai (Hu-Pi-Lieh)	忽必烈
638	Huangfu Chung-Ho	皇甫仲和
639	Shang Lu	商輅
640	Togan, or Toghan, Timur (T'o-Huan T'ieh-Mu-Erh)	妥歡貼睦爾
641	Hsiao Hsün	蕭洵
642	Ch'en Jui	陳瑞
643	Wang P'an	王泮
644	Liu Chieh-Chai	劉節齋
645	Ma T'ang	馬堂
646	Tung T'ing-Ch'in	董廷欽
647	Hsü Ch'ao-Chün	徐朝俊
648	Hsü Kuang-Ch'i	徐光啓
649	Hung Ch'u-Yin	洪處尹
650	Ts'ui Yu-Chih	崔攸之
651	Chin Hsing-Shu	金性洙
652	Wang Chün	王鋆
653	Hsüeh Chi-Hsüan	薛季宣
654	Juan Wên-Ta	阮文達
655	Wang Jen-Chün	王仁俊
656	Juan Yuan	阮元
657	Fêng Shih-K'o	馮時可
658	Wang Shih-Chên	王士禎
659	Huang Wei	黃暐
660	Yü Yüeh	俞樾
661	Chu Kao-Chih	朱高熾
662	Sung Ch'i	宋祁
663	Liu Hsü	劉昫
664	Li Ch'un	李春
665	Mao Pang-Han	毛邦翰
666	Ch'ien Lo-Chih	錢樂之
667	Liu I-Lung	劉義隆
668	Ch'ao Mei-Shu	晁美叔
669	Lu Erh-K'uei	陸爾奎
670	Pai Hsing-Chien	白行簡
671	Li Fan	李梵
672	Ts'ui Tzu-Yü	崔子玉
673	Lu Yu	陸游
674	Wang Chih-Ch'un	王之春
675	Wang Kuei	王珪
676	Wang Tan	王旦
677	P'ei Sung-Chih	裴松之
678	Chao Ta	趙達
679	Tao-Chêng	道証
680	Ch'üan Hêng	權衡

681 Nakane Genkei 中根元圭
682 Hsia Ho-Liang 夏賀良
683 Kan Chung-K'o 甘忠可
684 Wang Mang 王莽
685 Tso Ssu 左思
686 Liu Yuan-Lin 劉淵林
687 Chang T'ing-Yü 張延玉
688 Li T'ien-Ching 李天經
689 Chan Hsi-Yuan 詹希元
690 Chou Shu-Hsüeh 周述學
691 Ch'en Jen-Hsi 陳仁錫
692 Chêng Chün-Yung 鄭君永
693 Ssuma Jang-Chü 司馬穰苴
694 Chuang Chia 莊賈
695 Têng Yü-Han 鄧玉函
696 Wang Chêng 王徵
697 Kao Ssu-Tê 高斯得
698 Chang Hsi-Ming 張璽明

4 TITLES OF BOOKS AND OTHER WRITINGS

700 *Shih Lin Yen Yü* 石林燕語
701 *Pên Ts'ao T'u Ching* 本草圖經
702 *Hsin I Hsiang Fa Yao* 新儀象法要
703 *Yung-Lo Ta Tien* 永樂大典
704 *Ssu K'u Ch'üan Shu Tsung Mu T'i Yao* 四庫全書總目提要
705 *Sung Shih* 宋史
706 *Sui Ch'u T'ang Shu Mu* 遂初堂書目
707 *Shao-Shêng I Hsiang Fa Yao* 紹聖儀象法要
708 *Shui Yün Hun T'ien Chi Yao* 水運渾天機要
709 *Yeh Shih Yuan Shu Mu* 也是園書目
710 *Tu Shu Min Ch'iu Chi* 讀書敏求記
711 *Mo Hai Chin Hu* 墨海金壺
712 *Shou Shan Ko Ts'ung Shu* 守山閣叢書
713 *Mo K'o Hui Hsi* 墨客揮犀
714 *Sui Shu* 隋書
715 *Hun T'ien Hsiang Shuo* 渾天象說
716 *Chou Pei* (or *Pi*) *Suan Ching* 周髀算經
717 *Chiu Chang Kou Ku Ts'ê Yen Hun T'ien Shu* 九章鈎股測驗渾天書
718 *Hun I Chu* 渾儀注
719 *K'ai Yuan Chan Ching* 開元占經
720 *Hun T'ien* 渾天
721 *Hun I T'u Chu* 渾儀圖注
722 *Hsing Ching* 星經
723 *Hun T'ien T'u Chu* 渾天圖注
724 *Hun T'ien I Shuo* 渾天儀說
725 *Hun T'ien T'u* 渾天圖
726 *I Ching* 易經
727 *Hsiao Hsüeh Kan Chu* 小學紺珠
728 *Shang Shu Wei K'ao Ling Yao* 尚書緯考靈耀
729 *Yüeh Ling* 月令
730 *Ssu Shih Chung Hsing T'u* 四時中星圖
731 *Lun Hun Hsiang* 論渾象
732 *Hun I Tsung Yao* 渾儀總要
733 *Tung T'ien Ch'ing Lu* 洞天清錄
734 *Hun I Fa Yao* 渾儀法要
735 *Fêng Ch'uang Hsiao Tu* 楓窗小牘
736 *Ch'ing I Lu* 清異錄
737 *Fa Hsiang Chih* 法象志
738 *Mêng Ch'i Pi T'an* 夢溪筆談
739 *Ch'i Tung Yeh Yü* 齊東野語
740 *Yü Hai* 玉海
741 *Chi Hsien* (*Shu Yuan*) *Chu Chi* 集賢書院注記
742 *Ling Hsien* 靈憲
743 *Ch'ao Yeh Ch'ien Tsai* 朝野僉載
744 *T'ai-P'ing Kuang Chi* 太平廣記
745 *Niao Ch'ing Chan* 鳥情占
746 *Kuo Shih Chih* 國史志
747 *Pei Shih* 北史
748 *Hsü Shih Shuo* 續世說
749 *T'ai-P'ing Yü Lan* 太平御覽
750 *Liang Shu* 梁書
751 *Nan Shih* 南史
752 *T'ien I Shuo Yao* 天儀說要
753 *Shih Ching* 詩經
754 *T'ang Yü Lin* 唐語林
755 *Ta Ch'ing Hui Tien* 大清會典
756 *Ch'u Hsüeh Chi* 初學記
757 *Kuan Shu K'o Lou T'u* 官術刻漏圖
758 *Liu Ching T'u* 六經圖
759 *Shih Wu Chi Yuan* 事物紀原
760 *Chiu T'ang Shu* 舊唐書
761 *Hsin T'ang Shu* 新唐書
762 *Yuan Chien Lei Han* 淵鑑類函
763 *Lou K'o Fa* 漏刻法
764 *Chou Li* 周禮
765 *Sung Shu* 宋書
766 *I-Hsi Ch'i Chü Chu* 義熙起居注
767 *Ku Wei Shu* 古微書
768 *Chin Yang Ch'un Ch'iu* 晉陽春秋
769 *Lun T'ien* 論天

770	*Ch'ou Jen Chuan*	疇人傳
771	*Suan Wang Lun*	算罔論
772	*Fei Niao T'u*	飛鳥圖
773	*Wên Hsüan*	文選
774	*Chiu Chang Suan Shu*	九章算術
775	*Chin Shu*	晉書
776	*Hsin Lou K'o Ming*	新漏刻銘(新刻漏銘)
777	*Yü Han Shan Fang Chi I Shu*	玉函山房輯佚書
778	*Ch'üan Shang Ku San Tai Ch'in Han San Kuo Liu Ch'ao Wên*	全上古三代秦漢三國六朝文
779	*Lou Shui Chuan Hun T'ien I Chih*	漏水轉渾天儀制
780	*Ling Wai Tai Ta*	嶺外代答
781	*Hsi Hu Chih*	西湖志
782	*P'ei Wên Yün Fu*	佩文韻府
783	*Shang Shu Ta Chuan*	尚書大傳
784	*Lun Hêng*	論衡
785	*An T'ien Lun*	安天論
786	*Hou Han Shu*	後漢書
787	*Chin Ch'i Chü Chu*	晉起居注
788	*T'ien Wên Shu*	天文書
789	*Hun I*	渾儀
790	*K'o Lou*	刻漏
791	*Kuei Ying Fa Yao*	晷影法要
792	*Ch'ü Wei Chiu Wên*	曲洧舊聞
793	*Hsüan-Ho Po Ku T'u Lu*	宣和博古圖錄
794	*Chiu Yü T'u Chih*	九域圖志
795	*Sung Shih Chi Shih Pên Mo*	宋史紀事本末
796	*Tu Hsing Tsa Chih*	獨醒雜志
797	*Yuan Shih*	元史
798	*Hsü T'ung Chien Kang Mu*	續通鑑綱目
799	*Ku Kung I Lu*	故宮遺錄
800	*Tzu Ming Chung Piao T'u Fa*	自鳴鐘表圖法
801	*Tz'u Yuan*	辭源
802	*Shih-erh Yen Chai Sui Pi*	十二硯齋隨筆
803	*Hung Mao Shih Ch'en Piao*	紅毛時辰表
804	*Ko Chih Ku Wei*	格致古微
805	*P'êng Ch'uang Hsü Lu*	蓬窗續錄
806	*Ch'ih Pei Ou T'an*	池北偶談
807	*P'êng Ch'uang Lei Chi*	蓬窗類紀
808	*Ch'a Hsiang Shih San Ch'ao*	茶香室三鈔
809	*Liao Shih*	遼史
810	*Chin Shih*	金史
811	*Shu Ching*	書經
812	*T'ang Hui Yao*	唐會要
813	*Hsün Tzu*	荀子
814	*T'ien Ti Yin Yang Ta Lo Fu*	天地陰陽大樂賦
815	*Chuang Tzu*	莊子
816	*Ssu Fên Li*	四分曆
817	*Lao Hsüeh An Pi Chi*	老學庵筆記
818	*Kuo Ch'ao Jou Yuan Chi*	國朝柔遠記
819	*Liang Ch'ao Kuo Shih*	兩朝國史
820	*San Kuo Chih*	三國志
821	*Samguk Sagi*	三國史記
822	*Kêng Shen Wai Shih*	庚申外史
823	*Kuo Yü*	國語
824	*Ch'ien Han Shu*	前漢書
825	*Ssuma Fa*	司馬法
826	*Wu Tu Fu*	吳都賦
827	*Nung Shu*	農書
828	*Wu Ching Tsung Yao*	武經總要
829	*Ming Shih*	明史
830	*Ming Wên Ch'i Shang*	明文奇賞
831	*Ku Wên Ch'i Shang*	古文奇賞
832	*Wu Lun Sha Lou Ming Hsü*	五輪沙漏銘序
833	*Huang Ti Nei Ching, Ling Shu*	黃帝內經靈樞
834	*Shih Chi*	史記
835	*Ch'i Ch'i T'u Shuo*	奇器圖說
836	*Chu Ch'i T'u Shuo*	諸器圖說
837	*Shang Yu Chi*	尚友集

5 NAMES OF PLACES

850	Nan-an	南安
851	Ch'uan-chow	泉州
852	Tan-t'u	丹徒
853	Wu-hsing	吳興
854	Shou-chow	壽州
855	Yuan-wu	原武
856	Chêng-chow	鄭州
857	Ch'êng-tu	成都
858	Ch'ang-an	長安
859	Hai-chow	海州
860	Tan-yang	丹陽
861	K'un-Lun Shan	崑崙山
862	K'ai-fêng	開封
863	Yü-Chang	豫章
864	Nan-ch'ang	南昌
865	Chiang-hsiang	江鄉

866	Mu-pei	睦陂
867	Lu-ling	廬陵
868	Pien-ching	汴京
869	Yen	燕
870	Ch'ien Fo Tung	千佛洞
871	Shang-tu	上都
872	Chao-ch'ing	肇慶
873	Hsiang-shan Ao	香山澳
874	Luan-yang	灤陽
875	Hsin-an	新安
876	P'u-yang	浦陽

6 TITLES, INSTITUTIONS, BUILDINGS, EXPRESSIONS, ETC.

900	*tzu*	字
901	Chi Hsien Chiao Li Kuan	集賢校理官
902	Tu Chih P'an Kuan	度支判官
903	Lien-T'ui	廉退
904	*ku chün-tzu*	古君子
905	Yu P'u Shê	右僕射
906	Chung Shu Mên Hsia Shih Lang	中書門下侍郎
907	Kuang Lu Ta Fu	光祿大夫
908	Shang Hu Chün	上護軍
909	Wu-kung Chün K'ai Kuo Hou	武功郡開國侯
910	T'ai Tzu Shao Shih	太子少師
911	*chin-shih*	進士
912	Kuan Chü Sêng	官局生
913	San Ch'ü Tso Hsiao Chai	三衢坐嘯齋
914	Ssu Chien	司諫
915	T'ien Wên Yuan	天文院
916	Han Lin Yuan	翰林院
917	T'ai Shih Chü	太史局
918	Pi Fu	秘府
919	Chung Ch'ang Shih	中常侍
920	Pi Shu Ko Chü	秘書閣局
921	Ch'un Kuan Chêng	春官正
922	Hsia Kuan Chêng	夏官正
923	Ch'iu Kuan Chêng	秋官正
924	Tung Kuan Chêng	冬官正
925	Tu Tso Jen Yuan	都作人員
926	Chi Ying Tien	集英殿
927	Wu Ch'êng Tien	武成殿
928	Ch'ao T'ang	朝堂
929	*shu shih*	術士
930	*shu jen*	術人
931	Chieh Hu	挈壺
932	Li Pu Shang Shu	吏部尙書
933	Shih Tu	侍讀
934	Wên Ming Tien	文明殿
935	Ssu T'ien Chien	司天監
936	T'ai-Ch'u	太初
937	Wên Tê Tien	文德殿
938	Lung T'u Ko	龍圖閣
939	Ssu T'ien Chien Hsüeh-Sêng	司天監學生
940	Ning Hui Ko	凝暉閣
941	Li Chêng Shu Yuan	麗正書院
942	Chi Hsien Yuan	賢集院
943	Fu Chêng Mên	敷正門
944	*chiang*	匠
945	*k'o*	客
946	Jen Tsung	仁宗
947	Chen Tsung	眞宗
948	T'ai Tsung	太宗
949	Hsüan Tsung	玄宗
950	Chê Tsung	哲宗
951	Shen Tsung	神宗
952	Wu Tsê T'ien	武則天
953	Wu Hou	武后
954	Hou Chu	後主
955	Kao Tsu	高祖
956	Yang Ti	煬帝
957	Ch'ien Yang Tien	乾陽殿
958	Chieh Hu Chêng	挈壺正
959	An Ti	安帝
960	Wu Ti	武帝
961	Wên Ti	文帝
962	Kuan Hsiang Tien	觀象殿
963	Huan Ti	桓帝
964	Shun Ti	順帝
965	Ch'u Kung	儲宮
966	Yung An Kung	永安宮
967	Chiao Shu Lang	校書郎
968	Chien I Ta Fu	諫議大夫
969	T'ê Chin Shao Tsai	特進小宰
970	*fang wai chih shih*	方外之士
971	Ying Fêng Ssu	應奉司
972	Hui Tsung	徽宗
973	Ch'in Tsung	欽宗
974	Ming T'ang	明堂
975	Yü Fu	御府
976	Chung Ku Yuan	鐘鼓院
977	*tao kuan*	道觀

978	Ts'ê Yen Hun I K'o Lou So	測驗渾儀刻漏所
979	Kao Tsung	高宗
980	Ssu T'ien T'ai	司天臺
981	Ts'ê Yen So	測驗所
982	Pi Shu Shêng	秘書省
983	Chang Tsung	章宗
984	Hsüan Tsung	宣宗
985	Ta Ming Tien	大明殿
986	Shih Tsu	世祖
987	Pei Chi Ko	北極閣
988	Shun Ti	順帝
989	Hui Tsung	惠宗
990	Kung Pu Lang Chung	工部郎中
991	I Ho Yuan	頤和園
992	Li Ma-Tou P'u-Sa	利瑪竇菩薩
993	Chien An Wang	建安王
994	Hsiao Tsung	孝宗
995	Hou	后
996	Fei	妃
997	P'in	嬪
998	Mei Jen	美人
999	Nü Yü	女御
1000	Nü Shih	女史
1001	Ying Tsung	英宗
1002	T'ai Tsu	太祖
1003	Hsiao Chao Wang	孝昭王
1004	Yuan Ming Yuan	圓明園
1005	Ssu Wu Shih	司寤氏
1006	Hou Chih Shih	候之士

INDEX

by Muriel Moyle

(Figures in **bold type** refer to the Supplement. Figures in parentheses refer to footnotes.)

INDEX

INDEX

INDEX

INDEX

INDEX